METHODS IN MOLECULAR BIOLOGY

Series Editor
John M. Walker
School of Life and Medical Sciences,
University of Hertfordshire, Hatfield,
Hertfordshire, AL10 9AB, UK

For further volumes:
http://www.springer.com/series/7651

RNA Nanostructures

Methods and Protocols

Edited by

Eckart Bindewald

Frederick National Laboratory for Cancer Research
Leidos Biomedical Research
Frederick, MD, USA

Bruce A. Shapiro

RNA Structure and Design Section, RNA Biology Laboratory
National Cancer Institute, National Institutes of Health
Frederick, MD, USA

Humana Press

Editors
Eckart Bindewald
Frederick National Laboratory
for Cancer Research
Leidos Biomedical Research
Frederick, MD, USA

Bruce A. Shapiro
RNA Structure and Design Section, RNA Biology Laboratory
National Cancer Institute, National Institutes of Health
Frederick, MD, USA

ISSN 1064-3745 ISSN 1940-6029 (electronic)
Methods in Molecular Biology
ISBN 978-1-4939-7137-4 ISBN 978-1-4939-7138-1 (eBook)
DOI 10.1007/978-1-4939-7138-1

Library of Congress Control Number: 2017943677

Printed on acid-free paper

This Humana Press imprint is published by Springer Nature
The registered company is Springer Science+Business Media LLC
The registered company address is: 233 Spring Street, New York, NY 10013, U.S.A.

Preface

RNA nanotechnology is a young field with many potential applications. The goal is to utilize designed RNA strands, such that the obtained constructs have specific properties in terms of shape and functionality. RNA has potential functionalities that are comparable to that of proteins, but possesses (compared to proteins) simpler design principles akin to DNA. The promise is that designed RNA complexes may make possible novel types of molecular assemblies with applications in medicine (as therapeutics or diagnostics), material science, imaging, structural biology, and basic research.

Indeed, using this approach, scientists have shown that they can design RNAs that self-assemble into predefined shapes (such as rings, cubes, tetrahedrons, or lattices). Furthermore, designed RNAs can be programmed to impart different functionalities such as gene knockdown via RNA interference, temperature-specific behavior or RNA-based logic or multi-functional assemblies.

These successes, however, are typically only possible due to the use of specialized computational and experimental approaches. Repeating achievements based on regular research papers is frequently challenging if the methods are described only briefly. It is therefore particularly useful that detailed protocols provided by leading experts in the field are compiled as a unit, thus making the current state of the art accessible to scientists entering the field. Here we present in 23 chapters a spectrum of computational and experimental protocols pertaining to the creation, characterization, and utilization of RNA nanostructures.

Content

Computational approaches are an integral part of RNA design. Sun and Chen describe the use of a coarse-grained computational model in order to predict interactions between metal ions and the three-dimensional structure of RNAs (Chapter 1). Next, the use of RNA motifs for the generation of RNA ring structures is presented by Parlea et al. (Chapter 2). This is accomplished by an enumeration of motifs and linker helix lengths. It is shown how to process the obtained ring structures using a web interface. Coarse-grained models can be refined utilizing all-atom simulations in general and molecular dynamics simulation in particular. Kim et al. present a protocol for the simulation of RNAs and RNA lipid interactions using molecular dynamics and the Amber software package (Chapter 3). Jang and Ahn describe the use of circular DNA templates for the self-assembly of multimeric RNAi structures via rolling circle transcription (Chapter 4). The resulting RNAi transcripts can be hybridized with DNA strands that are conjugated with folate, leading to nucleic acid nanoparticles than can target folate-expressing cancer cells. Raghava describes the use of a computational approach for the prediction of the immunogenicity of RNA structures with respect to toll-like receptors TLR7 and TLR8 (Chapter 5). This is particularly interesting for predicting the immunogenicity of RNA nanostructures because an immune response is, depending on the application, to be avoided or desired.

Designed RNAs are frequently generated from DNA using RNA polymerase. Kireeva et al. demonstrate how the conditions for utilizing T7 RNA polymerase can be optimized for the efficient production of RNA nanostructures (Chapter 6). This allows for the incorporation of chemically modified nucleotides, resulting in RNA nanostructures that are resistant to ribonuclease degradation.

An important aspect of RNA nanotechnology is the ability to characterize the generated particles. An important quantity for characterizing the stability of RNA nanostructures is the melting temperature. Benkato et al. describe the use of thermal gradient gel electrophoresis for the determination of the melting temperatures of RNA complexes (Chapter 8). Boerneke and Herrmann demonstrate that X-ray crystallography can be used to determine the structure of RNA complexes at atomic resolution (Chapter 9). Their protocol includes detailed steps for the design of the RNA structures (an RNA square and two types of RNA triangles), experimental preparation of the RNA as well as analysis of X-ray crystallography results. Dabkowska et al. describe the self-assembly of RNA scaffolds on the surface of lipid bilayers (Chapter 7). The authors utilize a variety of technologies (such as atomic force microscopy, ellipsometry, and confocal fluorescence microscopy) in order to characterize the RNA nanostructure in the context of an interacting lipid bilayer. Interestingly, the cationic lipid promotes the adsorption and assembly of the polyanionic RNA.

One potential obstacle with respect to biomedical applications is the efficient delivery of nucleic acid material into cells. A variety of contributions pertain to the delivery of nucleic acid nanostructures into cells. While delivery agents differ widely, they have in common, shielding of the electrophilic nature and the negative charge of RNA during cell entry, and protection from environmental factors such as nucleases.

Lokesh et al. provide a protocol for the generation and validation of X-aptamers. X-aptamers can be viewed as oligonucleotide-inspired biopolymers that offer, compared to RNA and DNA, an extended sequence alphabet. X-aptamers have applications in targeting proteins and in cellular delivery of small molecules (Chapter 10).

Kruspe and Giangrade present aptamer-siRNA chimeras (AsiCs) (Chapter 11). Such aptamer-siRNA chimeras have potential applications against cancer if the aptamer part is chosen to bind to cell surface receptors and to facilitate uptake by tumor cells, and the conjugated siRNA is chosen to downregulate an oncogene.

Gupta et al. demonstrate the use of bolaamphiphiles for cellular delivery of RNA. Bolaamphiphiles are linear molecules that consist of a hydrophobic aliphatic chain and polar hydrophilic head groups on either end (Chapter 12). The cellular delivery of mRNA is particularly interesting and challenging (due to their, compared to small RNAs, dramatically larger size). Li and Dong describe the use of a class of lipid-like (also called lipidoid) nanoparticles that they recently developed for the cellular delivery of mRNA (Chapter 13). They describe the preparation of mRNA-loaded nanoparticles, characterization of their structure (via cryo-TEM), cytotoxicity, and transfection properties. Another moiety for cellular delivery of RNA is based on chitosan. Chitosan is a natural polysaccharide that can form RNA-containing nanoparticles, as described in the protocol contributed by Denizli et al. (Chapter 14). Wang et al. describe the synthesis of poly(lactide-co-glycolide) (PLGA) based polymers, the synthesis of cationic lipid G0-C14 based on 0th generation poly (amidoamine) (PAMAM) dendrimer, and the synthesis of RNA-containing PLGA-PEG/lipid hybrid nanoparticles (Chapter 15). The use and characterization of siRNA-containing oxime ether lipid nanoparticles is described by Puri et al. (Chapter 16). This includes the characterization of the transfection efficiency, stability, and assembly properties of the nanoparticles.

These protocols pertaining to RNA delivery show that there are a variety of options available for accomplishing this goal. Taken together, one can conclude that cellular delivery of RNA and DNA payloads is, while challenging, not a prohibitive obstacle for the development of RNA therapeutics.

A nanopore device can be utilized for the sequence-specific detection of RNA and DNA strands. Tian et al. describe the use of a nanopore in order to selectively identify miRNAs while providing an approach for avoiding "contamination" of the nanopore signal due to the presence of different additional RNAs (Chapter 17). Martins et al. provide a protocol for utilizing RNA/DNA hybrids that can re-associate in target cells into active siRNA and lead to knockdown of target genes. This technology is utilized as potential anti-HIV-1 therapy. They provide detailed protocol steps for the design, purification, and assembly of RNA/DNA hybrids as well as for the inhibition of HIV-1 genes (Chapter 18).

An important extension of the concept of static RNA scaffolds is to utilize dynamic RNA complexes in general and RNA switches in particular. Green describes the design and experimental verification of a prokaryotic RNA switch that operates on the level of translation (Chapter 19). A reporter mRNA is designed such that the translation start codon is hidden from the translation initiation complex via a stable hairpin structure. This hairpin is designed to undergo a structural change upon binding of a biomarker RNA. This then exposes the start codon, facilitating translation of the reporter RNA such as a green fluorescent protein.

Zakrevsky et al. provide the protocol for a two-stranded RNA switch that corresponds to the conditional activation of RNA silencing (Chapter 20). Such a switch is particularly challenging because it involves designing an RNA nanostructure where the rather stable Dicer-substrate helix with more than 21 base pairs is energetically outcompeted via several shorter helices. This was made possible via a novel computational approach that allows the estimation of the free energy of folding of multistranded pseudoknotted RNA complexes.

Tanaka et al. demonstrated that the utilization of motifs and domains in designed RNAs and RNA complexes can be rather large entities: they modified a group I ribozyme so that it undergoes self-dimerization (Chapter 21). They present protocol steps for the rational design and experimental assembly of the RNA complex. They also provide several ribozyme activity assays in order to verify the enzymatic activity of the ribozyme. An exciting possibility is the use of CRISPR: Zalatan presents a protocol that demonstrates the use of CRISPR guide RNA in order to recruit desired proteins and thereby control transcription (Chapter 22). The designed RNA acts in this fashion as a "conductor" for recruiting desired protein factors.

Utilizing RNAs for therapeutic purposes entails challenges not only with respect to cellular delivery but also with respect to biodistribution. Haque et al. demonstrate the different biodistribution properties of a variety of different RNA nanoparticles that are based on the very stable phi29 three-way junction motif (Chapter 23). They not only create a spectrum of different RNA nanostructures but also extensively characterize mouse in vivo properties such as biodistribution. This large number of techniques for the production and characterization of a family of different nanostructures represents the current state of the art of RNA nanotechnology.

In summary, we are excited to present this compilation of protocols contributed by renowned experts in the field of RNA nanotechnology. Sharing details not only of the results but also of how to achieve and reproduce those results is important for sustained progress in this research domain. These are exciting times for the creative application of RNA design because the ground work has been laid and important functionalities are readily available in order to be incorporated into designs. We are confident that this volume and its wide spectrum of contributions will be a valuable aid for the design and production of RNA nanostructures.

Frederick, MD, USA

Eckart Bindewald
Bruce A. Shapiro

Contents

Contributors

KIRILL A. AFONIN • *RNA Biology Laboratory, National Cancer Institute, National Institutes of Health, Frederick, MD, USA; Nanoscale Science Program, Department of Chemistry, University of North Carolina at Charlotte, Charlotte, NC, USA; The Center for Biomedical Engineering and Science, University of North Carolina at Charlotte, Charlotte, NC, USA*

HYUNG JUN AHN • *Center for Theragnosis, Biomedical Research Institute, Korea Institute of Science and Technology, Seongbuk-gu, Seoul, South Korea*

BURCU ASLAN • *Department of Experimental Therapeutics, The University of Texas MD Anderson Cancer Center, Houston, TX, USA; Center for RNA Interference and Non-Coding RNA, The University of Texas MD Anderson Cancer Center, Houston, TX, USA*

KHEIRIA BENKATO • *Department of Chemistry, Ball State University, Muncie, IN, USA*

ECKART BINDEWALD • *Frederick National Laboratory for Cancer Research, Leidos Biomedical Research, Frederick, MD, USA*

MARK A. BOERNEKE • *Department of Chemistry and Biochemistry, University of California-San Diego, La Jolla, CA, USA*

MY N. BUI • *Department of Chemistry, Ball State University, Muncie, IN, USA*

KUMARDEEP CHAUDHARY • *Bioinformatics Centre, Institute of Microbial Technology, Chandigarh, India*

SHI-JIE CHEN • *Department of Physics, Department of Biochemistry, and MU Informatics Institute, University of Missouri, Columbia, MO, USA*

ARKADIUSZ CHWOROS • *Centre of Molecular and Macromolecular Studies, Polish Academy of Sciences, Łódź, Poland*

ALEKSANDRA P. DABKOWSKA • *Division of Physical Chemistry, Department of Chemistry, Lund University, Lund, Sweden*

MERVE DENIZLI • *Department of Experimental Therapeutics, The University of Texas MD Anderson Cancer Center, Houston, TX, USA*

SANDEEP KUMAR DHANDA • *Bioinformatics Centre, Institute of Microbial Technology, Chandigarh, India*

YIZHOU DONG • *Division of Pharmaceutics and Pharmaceutical Chemistry, College of Pharmacy, The Ohio State University, Columbus, OH, USA*

ERIC O. FREED • *HIV Dynamics and Replication Program, NCI, Frederick, MD, USA*

PALOMA H. GIANGRANDE • *Department of Internal Medicine, University of Iowa, Iowa City, IA, USA; Department of Radiation Oncology, University of Iowa, Iowa City, IA, USA; Molecular and Cellular Biology Program, University of Iowa, Iowa City, IA, USA; Medical Scientist Training Program, University of Iowa, Iowa City, IA, USA; Abboud Cardiovascular Research Center, University of Iowa, Iowa City, IA, USA*

DAVID G. GORENSTEIN • *AM Biotechnologies, LLC, Houston, TX, USA*

ALEXANDER A. GREEN • *Biodesign Center for Molecular Design and Biomimetics, The Biodesign Institute and the School of Molecular Sciences, Arizona State University, Tempe, AZ, USA*

BENJAMIN GRIFFEL • *Otto H. York Department of Chemical, Biological and Pharmaceutical Engineering, New Jersey Institute of Technology, Newark, NJ, USA; Bard High School Early College, Newark, NJ, USA*

AMY GU • *Department of Bioengineering, University of Missouri, Columbia, MO, USA; Dalton Cardiovascular Research Center, University of Missouri, Columbia, MO, USA*

LI-QUN GU • *Department of Bioengineering, University of Missouri, Columbia, MO, USA; Dalton Cardiovascular Research Center, University of Missouri, Columbia, MO, USA*

PEIXUAN GUO • *Division of Pharmaceutics and Pharmaceutical Chemistry, College of Pharmacy, The Ohio State University, Columbus, OH, USA; Department of Physiology and Cell Biology, College of Medicine, The Ohio State University, Columbus, OH, USA; Dorothy M. Davis Heart and Lung Research Institute, The Ohio State University, Columbus, OH, USA*

KSHITIJ GUPTA • *RNA Structure and Design Section, RNA Biology Laboratory, National Cancer Institute, National Institutes of Health, MD, USA; Department of Inorganic and Physical Chemistry, Indian Institute of Science, Bangalore, India*

FARZIN HAQUE • *Division of Pharmaceutics and Pharmaceutical Chemistry, College of Pharmacy, The Ohio State University, Columbus, OH, USA; Department of Physiology and Cell Biology, College of Medicine, The Ohio State University, Columbus, OH, USA; Dorothy M. Davis Heart and Lung Research Institute, The Ohio State University, Columbus, OH, USA*

THOMAS HERMANN • *Department of Chemistry and Biochemistry, University of California, San Diego, La Jolla, CA, USA; Center for Drug Discovery Innovation, University of California, San Diego, La Jolla, CA, USA*

YOSHIYA IKAWA • *Graduate School of Science and Engineering, University of Toyama, Toyama, Japan*

LUC JAEGER • *Department of Chemistry and Biochemistry, Biomolecular Science and Engineering Program, University of California, Santa Barbara, CA, USA*

MIHUE JANG • *Center for Theragnosis, Biomedical Research Institute, Korea Institute of Science and Technology, Seongbuk-gu, Seoul, South Korea*

DANIEL L. JASINSKI • *Division of Pharmaceutics and Pharmaceutical Chemistry, College of Pharmacy, The Ohio State University, Columbus, OH, USA; Division Department of Physiology and Cell Biology, College of Medicine, The Ohio State University, Columbus, OH, USA; Dorothy M. Davis Heart and Lung Research Institute, The Ohio State University, Columbus, OH, USA*

VAISHNAVI JAWAHAR • *Department of Biological Sciences, University of North Carolina at Charlotte, Charlotte, NC, USA; Nanoscale Science Program, Department of Chemistry, University of North Carolina at Charlotte, Charlotte, NC, USA*

DAHAI JIANG • *Center for RNA Interference and Non-Coding RNA, The University of Texas MS Anderson Cancer Center, Houston, TX, USA; Department of Gynecologic Oncology and Reproductive Medicine, The University of Texas MD Anderson Cancer Center, Houston, TX, USA*

MIKHAIL KASHLEV • *RNA Biology Laboratory, National Cancer Institute, National Institutes of Health, Frederick, MD, USA*

WOJCIECH K. KASPRZAK • *Basic Science Program, Leidos Biomedical Research, Inc., Frederick National Laboratory for Cancer Research, Frederick, MD, USA*

WEINA KE • *Nanoscale Science Program, Department of Chemistry, University of North Carolina at Charlotte, Charlotte, NC, USA*

EMIL F. KHISAMUTDINOV • *Department of Chemistry, Ball State University, Muncie, IN, USA*

TAEJIN KIM • *Department of Chemistry, New York University, New York, NY, USA*

Maria L. Kireeva • *RNA Biology Laboratory, National Cancer Institute, National Institutes of Health, Frederick, MD, USA*
Sven Kruspe • *Department of Internal Medicine, University of Iowa, Iowa City, IA, USA*
Curtis H. Lam • *AM Biotechnologies, LLC, Houston, TX, USA*
Bin Li • *Division of Pharmaceutics and Pharmaceutical Chemistry, College of Pharmacy, The Ohio State University, Columbus, OH, USA*
Hui Li • *Division of Pharmaceutics and Pharmaceutical Chemistry, College of Pharmacy, The Ohio State University, Columbus, OH, USA; Department of Physiology and Cell Biology, College of Medicine, The Ohio State University, Columbus, OH, USA; Dorothy M. Davis Heart and Lung Research Institute, The Ohio State University, Columbus, OH, USA*
Ganesh L. Lokesh • *Institute of Molecular Medicine for the Prevention of Human Diseases, The University of Texas Health Science Center, Houston, TX, USA; Department of Nanomedicine and Biomedical Engineering, McGovern Medical School, The University of Texas Health Science Center, Houston, TX, USA*
Gabriel Lopez-Berestein • *Department of Experimental Therapeutics, The University of Texas MD Anderson Cancer Center, Houston, TX, USA; Center for RNA Interference and Non-Coding RNA, The University of Texas MD Anderson Cancer Center, Houston, TX, USA; Department of Cancer Biology, The University of Texas MD Anderson Cancer Center, Houston, TX, USA*
Lingegowda S. Mangala • *Center for RNA Interference and Non-Coding RNA, The University of Texas MD Anderson Cancer Center, Houston, TX, USA; Department of Gynecologic Oncology and Reproductive Medicine, The University of Texas MD Anderson Cancer Center, Houston, TX, USA*
Angelica N. Martins • *Department of Biological Sciences, University of North Carolina at Charlotte, Charlotte, NC, USA*
Shigeyoshi Matsumura • *Graduate School of Science and Engineering, University of Toyama, Toyama, Japan*
Agnes Michanek • *Division of Physical Chemistry, Department of Chemistry, Lund University, Lund, Sweden*
Gandharva Nagpal • *Bioinformatics Centre, Institute of Microbial Technology, Chandigarh, India*
Tommy Nylander • *Division of Physical Chemistry, Department of Chemistry, Lund University, Lund, Sweden*
Benjamin O'Brien • *Department of Chemistry, Ball State University, Muncie, IN, USA*
Lorena Parlea • *RNA Structure and Design Section, RNA Biology Laboratory, National Cancer Institute, National Institutes of Health, Frederick, MD, USA*
Anu Puri • *RNA Structure and Design Section, RNA Biology Laboratory, Center for Cancer Research, National Cancer Institute, National Institutes of Health, Frederick, MD, USA*
Michael Pennella • *Department of Bioengineering, University of Missouri, Columbia, MO, USA; Dalton Cardiovascular Research Center, University of Missouri, Columbia, MO, USA*
Gajendra Pal Singh Raghava • *Bioinformatics Centre, Institute of Microbial Technology, Chandigarh, India*
Cristian Rodriguez-Aguayo • *Department of Experimental Therapeutics, The University of Texas MD Anderson Cancer Center, Houston, TX, USA; Center for RNA Interference and*

Non-Coding RNA, The University of Texas MD Anderson Cancer Center, Houston, TX, USA
BRUCE A. SHAPIRO • *RNA Structure and Design Section, RNA Biology Laboratory, National Cancer Institute, National Institutes of Health, Frederick, MD, USA*
RISHABH SHARAN • *RNA Structure and Design Section, RNA Biology Laboratory, National Cancer Institute, National Institutes of Health, Frederick, MD, USA*
RUICHENG SHI • *Department of Bioengineering, University of Missouri, Columbia, MO, USA; Dalton Cardiovascular Research Center, University of Missouri, Columbia, MO, USA*
ANIL K. SOOD • *Center for RNA Interference and Non-Coding RNA, The University of Texas MD Anderson Cancer Center, Houston, TX, USA; Department of Gynecologic Oncology and Reproductive Medicine, The University of Texas MD Anderson Cancer Center, Houston, TX, USA; Program in Cancer Biology and Cancer Metastasis, The University of Texas Graduate School of Biomedical Sciences, Houston, TX, USA; Department of Cancer Biology, The University of Texas MD Anderson Cancer Center, Houston, TX, USA*
EMMA SPARR • *Division of Physical Chemistry, Department of Chemistry, Lund University, Lund, Sweden*
CARYN STRIPLIN • *Nanoscale Science Program, Department of Chemistry, University of North Carolina at Charlotte, Charlotte, NC, USA*
MORRIAH STRIPLIN • *Nanoscale Science Program, Department of Chemistry, University of North Carolina at Charlotte, Charlotte, NC, USA*
LI-ZHEN SUN • *Department of Physics, Department of Biochemistry, and MU Informatics Institute, University of Missouri, Columbia, MO, USA*
TAKAHIRO TANAKA • *Department of Chemistry and Biochemistry, Graduate School of Engineering, Kyushu University, Fukuoka, Japan*
VARATHARASA THIVIYANATHAN • *Institute of Molecular Medicine for the Prevention of Human Diseases, The University of Texas Health Science Center, Houston, TX, USA; Department of Nanomedicine and Biomedical Engineering, McGovern Medical School, The University of Texas Health Science Center, Houston, TX, USA*
KAI TIAN • *Department of Bioengineering, University of Missouri, Columbia, MO, USA; Dalton Cardiovascular Research Center, University of Missouri, Columbia, MO, USA*
MATHIAS VIARD • *RNA Structure and Design Section, RNA Biology Laboratory, Center for Cancer Research, National Cancer Institute, National Institutes of Health, Frederick, MD, USA; Leidos Biomedical Research Inc., Center for Cancer Research, National Cancer Institute, National Institutes of Health, Frederick, MD, USA*
DAVID E. VOLK • *Institute of Molecular Medicine for the Prevention of Human Diseases, The University of Texas Health Science Center, Houston, TX, USA; Department of Nanomedicine and Biomedical Engineering, McGovern Medical School, The University of Texas Health Science Center, Houston, TX, USA*
HONGYU WANG • *Institute of Molecular Medicine for the Prevention of Human Diseases, The University of Texas Health Science Center, Houston, TX, USA; Department of Nanomedicine and Biomedical Engineering, McGovern Medical School, The University of Texas Health Science Center, Houston, TX, USA*
LEI WANG • *Otto H. York Department of Chemical, Biological and Pharmaceutical Engineering, New Jersey Institute of Technology, Newark, NJ, USA; College of Chemistry, Chemical Engineering and Materials Science, Soochow University, Suzhou, Jiangsu, People's Republic of China*

NANCY WARD • *AM Biotechnologies, LLC, Houston, TX, USA*
CONGCONG XU • *Division of Pharmaceutics and Pharmaceutical Chemistry, College of Pharmacy, The Ohio State University, Columbus, OH, USA; Department of Physiology and Cell Biology, College of Medicine, The Ohio State University, Columbus, OH, USA; Dorothy M. Davis Heart and Lung Research Institute, The Ohio State University, Columbus, OH, USA*
XIAOYANG XU • *Otto H. York Department of Chemical, Biological and Pharmaceutical Engineering, New Jersey Institute of Technology, Newark, NJ, USA*
PAUL ZAKREVSKY • *RNA Structure and Design Section, RNA Biology Laboratory, National Cancer Institute, National Institutes of Health, Frederick, MD, USA*
JESSE G. ZALATAN • *Department of Chemistry, University of Washington, Seattle, WA, USA*
SERENA ZAMPINO • *RNA Structure and Design Section, RNA Biology Laboratory, Center for Cancer Research, National Cancer Institute, National Institutes of Health, Frederick, MD, USA*

Chapter 1

A New Method to Predict Ion Effects in RNA Folding

Li-Zhen Sun and Shi-Jie Chen

Abstract

The strong interaction between metal ions in solution and highly charged RNA molecules is critical for RNA structure formation and stabilization. Metal ions binding to RNA can induce RNA structural changes that are important for RNA cellular functions. Therefore, quantitative modeling of the ion effects is essential for RNA structure prediction and RNA-based molecular design. Recently, inspired by the increasing experimental evidence that supports the importance of ion correlation and fluctuation in ion–RNA interactions, we developed a new computational model, Monte Carlo Tightly Bound Ion (MCTBI) model. The validity of the model is shown by the improved accuracy in the predictions for ion binding properties and ion-dependent free energies for RNA structures. In this chapter, using homodimeric tetraloop-receptor docking as an illustrative example, we showcase the MCTBI method for the computational prediction of the ion effects in RNA folding.

Key words Metal ion effects, Ion–RNA interactions, Tightly Bound Ion model, RNA folding

1 Introduction

RNA functions are intrinsically related to RNA three-dimensional (3D) structures and their folding stabilities [1–5]. In the folding process, counterions accumulate around RNA to neutralize and screen the strong intra-chain Coulomb repulsion between the negatively charged RNA backbone [6, 7]. Such an ion-mediated force is essential for the stabilization of RNA functional structures. A change in the ionic condition in the solution can shift the conformational equilibrium and induce RNA structural changes. In general, counterions around an RNA can be classified into three different states of RNA-binding. In one extreme, site-specific bound (or chelated) ions [8] can be fully or partially dehydrated and trapped at specific binding sites (such as three-dimensional pockets in the RNA structure) [9]. The direct interaction between the bound ions and the RNA can often provide a stabilizing force for specific RNA tertiary contacts. In the other extreme, diffuse ions in the bulk solution away from the RNA surface usually remain hydrated and highly mobile [10]. Unlike site-bound ions, diffuse

Eckart Bindewald and Bruce A. Shapiro (eds.), *RNA Nanostructures: Methods and Protocols*, Methods in Molecular Biology, vol. 1632, DOI 10.1007/978-1-4939-7138-1_1, © Springer Science+Business Media LLC 2017

ions are not in direct contact with RNA. However, the large number of diffuse ions can cause significant electrostatic effects and play a dominant role in the electrostatic contribution to RNA folding stability.

Between the site-bound ions and the mobile ions in the bulk solvent are the diffuse ions that are in close proximity to the RNA surface. The strong Coulomb attraction between the negatively charged polyanionic RNA and the positively charged diffuse ions can result in a high local concentration of the ions close to the RNA surface, especially around the electro-negative atoms or groups such as the phosphate groups [11]. The high ion concentration can lead to a strong correlation between the ions through excluded volume and Coulombic interactions [12–14]. The correlation effect is more pronounced for multivalent ions (e.g., Mg^{2+} ions) because they carry higher charges and hence involve stronger interactions. We refer to such strongly correlated diffuse ions as the Tightly Bound (TB) ions and to the weakly correlated diffuse ions as Diffusely Bound (DB)ions. Physically, a TB ion is less mobile than a DB ion [12] due to inter-ion correlation.

Although the importance of ions in RNA folding has been long recognized [15, 16], the problem of accurate and efficient modeling of the ion effects, especially for the multivalent ions such as Mg^{2+} ion, still remains [17–20]. Molecular dynamics (MD) simulations can provide atomic details for the ion binding events [14, 21–23]. However, the need to sample the huge ensemble of ion distributions renders the method to be computationally challenging [24]. Analytical models, such as the classical Counterion Condensation (CC) theory [25] and the nonlinear Poisson–Boltzmann (NLPB) theory [26–29], can predict the ion effects through more efficient computations. However, these methods neglect the correlation and fluctuation effects of ions, resulting in underestimation of the Mg^{2+} ion effects in RNA folding stability [30].

Inspired by the importance of the correlation and fluctuation effects for multivalent ions (such as Mg^{2+} ions) in RNA folding [14, 30, 31], several new theories have been developed [32–34]. For example, the three-dimensional reference interaction site model (3D-RISM) uses Ornstein and Zernike integral equation theory to account for the ion–ion correlation [33]. A generalized counterion condensation theory explicitly treats Mg^{2+} ions and ion–ion correlations [34] with the framework of the counterion condensation theory. A hybrid model, Tightly Bound Ion (TBI) model [12], was proposed to account for the discrete, correlated, many-body distributions for the (strongly correlated) TB ions. Since the model was first reported in 2005, it has been under continuous development in the past decade for better accuracy and efficiency [12, 30, 35–40].

Recently, we developed a new generation of the TBI-like model, Monte Carlo Tightly Bound Ion (MCTBI) model [41]. A major advance in this new model is the new Monte Carlo-based sampling algorithm for the ion distribution. Specifically, we applied a two-step ("insertion–deletion") algorithm to sample the distribution of the TB ions: We first generate the TB ion distributions by randomly inserting the ions one by one, we then remove unstable (high energy) ions randomly. We apply the Metropolis Monte Carlo algorithm for random sampling. This new enhanced sampling method enables us to treat large RNA structures ($>$ 200 nts) that have been unattainable with the previous generations of the TBI model. Furthermore, the MCTBI model, which is based on much higher spatial resolution for the ion distributions than the previous model, offers a useful tool to predict ion binding sites with higher accuracy.

In this article, using the tetraloop-receptor (TL-R) complex as an illustrative example, we introduce the MCTBI-based method to predict ion-mediated electrostatic effects in RNA folding. We select the TL-R complex because of its widespread structural and functional roles [42–47]. Experimental results have shown that high Mg^{2+} concentration (several millimolar) [46, 47] can help stabilizing the docked homodimeric TL-R complex structure. Our computational method can provide quantitative insights into the contributions of Mg^{2+} ions in the homodimeric docking process.

2 RNA Structure

To compute the ion effects for a given RNA structure, we first need to have the PDB file for the RNA structure. For known (experimentally determined) structures, such as the docked structure of the homodimeric TL-R complex (*see* Fig. 1; PDB ID: 2I7Z [47]), the structure files can be downloaded from the Protein Data Bank (PDB) Website (*see* **Note 1**). The homodimeric TL-R complex is comprised of two identical chains. In this chapter, we will compute the [Mg^{2+}]-dependence of the electrostatic free energy of the dimer system as a function of the inter-molecular separation as characterized by the distance R between the two (identical) monomeric TL-R structures. Here R is determined from the distance between the centers of mass of the two monomeric TL-R structures (*see* **Note 2**). We set R equal to 0 for the reference state, the native docked state of the homodimer (*see* Fig. 1; PDB ID: 2I7Z [47]).

For systems whose RNA structures are not available in the PDB database, we can build the all-atom 3D structure using structure prediction models [49–55] such as the Vfold model [56] (*see* **Note 3**): We first input the RNA sequence into the Vfold2D module (*see* Fig. 2a and **Note 4**). For the given sequence, the Vfold2D module predicts the most probable 2D structure (base pairs) of the RNA (*see* Fig. 2b). We then provide the sequence and the predicted 2D structure (base pair) information to the Vfold3D

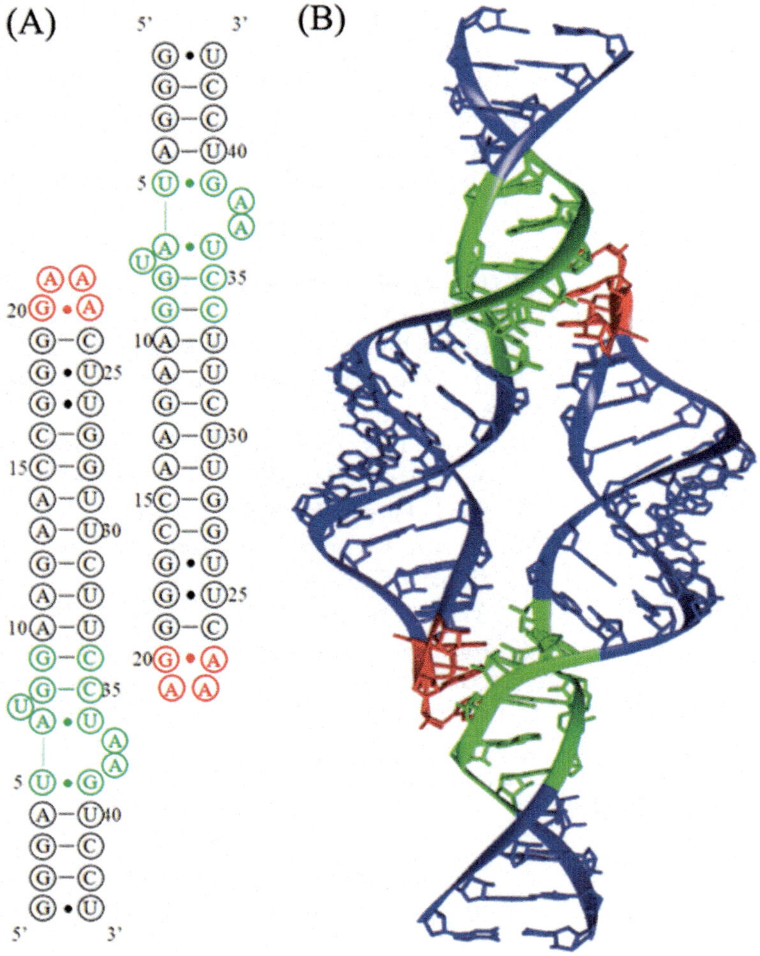

Fig. 1 The secondary structure (**a**) and 3D structure (**b**) of the homodimeric TL-R complex (PDB ID: 2I7Z [47]). The GAAA tetraloop and receptor are highlighted in *red* and *green*, respectively

module (*see* Fig. 2c and **Note 5**). Finally, the Vfold3D module may output the 3D structures for the given RNA sequence (Fig. 2d). The basis of Vfold for 3D structure construction is to use the rigid A-form helix for the helical part and PDB-derived templates for loops/junctions. Alternatively, without using the loop/junction template structures, we can use the coarse-grained RNA folding model to generate the ensemble of loop/junctions conformations and then add all atoms to the coarse-grained conformations [57, 58].

In the current version of the MCTBI model, the coarse-grained charge model is used for the assignment of RNA charges (*see* **Note 6**). Moreover, ions are assumed to remain hydrated (*see* **Note 7**).

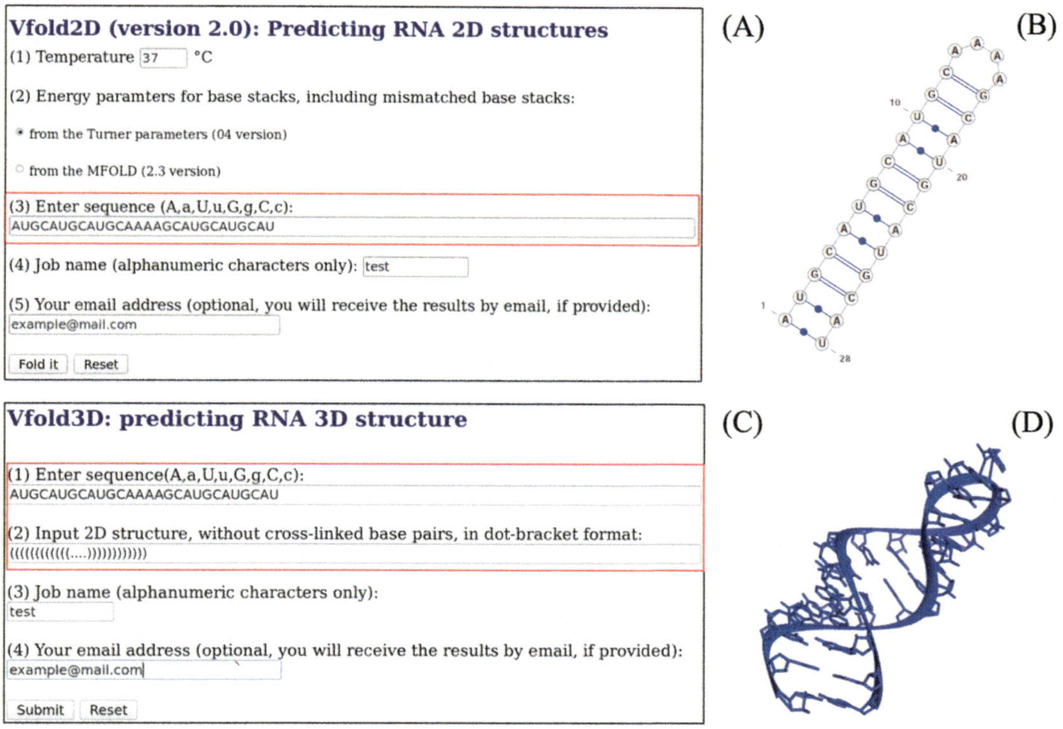

Fig. 2 (**a**) A screenshot of the Vfold2D module. The RNA sequence should be inserted in the place highlighted in *red*. URL: http:/rna.physics.missouri.edu/vfold2D/index.html (**b**) The predicted 2D structure for the sequence. (**c**) The screenshot of the Vfold3D module. The sequence and the basepair information should be input in the places highlighted in *red*. URL: http:/rna.physics.missouri.edu/vfold3D/index.html (**d**) The 3D structures predicted by the Vfold3D module

3 The Monte Carlo Tightly Bound Ion (MCTBI) Model

We compute the electrostatic free energy ΔG of the whole system (RNA and the ionic solution) from the partition function Z: $\Delta G = \text{-ln}\,(Z)$. In this section, we focus on the calculation of the partition function for a given RNA structure in an ionic solution.

3.1 Classification of Tightly Bound (TB) and Diffusely Bound (DB) Regions

In the MCTBI model, the space around RNA is classified into two regions: a Tightly Bound (TB) region for the TB ions (strong correlation) and a Diffusely Bound (DB) region for the DB ions (weak correlations). The correlation for monovalent ions such as K^+ is usually weak because of their relatively weak Coulomb interactions. We treat all the monovalent ions as DB ions. In contrast, multivalent ions such as Mg^{2+} ions accumulated in the close vicinity of the RNA surface usually involve strong Coulomb interactions and are more likely to be strongly correlated. Therefore, as shown in the steps listed below, we classify the TB and DB ions (regions) according to the multivalent ions such as Mg^{2+} ions.

1. To minimize the boundary effect, the RNA is placed at the center of an $L \times L \times L$ box whose size L is larger than six times of Debye length [12].
2. We estimate the spatial distribution of the Mg^{2+} ion concentration in the whole box by solving the non-linear Poisson-Boltzmann (NLPB) equation.
3. According to the Mg^{2+} ion concentration and the corresponding spatial distribution of the ion-correlation strength, we divide the whole box into the TB region around the RNA surface for the TB ions and the DB region for ions away from the RNA surface (*see* **Note 8**). Depending on the RNA structure and the ion concentrations, the TB region can be a thin layer (within ~ 1–5 Å) around the RNA.

The TB region is dependent on the RNA structure and the bulk ion concentrations, as shown in Fig. 3. Compared with an extended structure, a compact structure has a higher RNA (negative) charge density and thus a stronger tendency to induce ion accumulation around the RNA. Therefore, the TB region of a compact structure is usually larger than that of an extended structure, especially for a dilute Mg^{2+} solution (*see* Fig. 3a and c). Moreover, comparisons between Fig. 3a, b and c, d show that a high (bulk) ion concentration leads to a higher affinity for ion binding and thus a larger TB region.

After establishing the demarcation between the TB and DB regions, the partition function of the whole system, including the RNA and the ionic solution, can be calculated from the sum over N_b, the number of the TB ions:

$$Z = \sum_{N_b} Z_b(N_b) \times Z_d(N_d)$$

Here, Z_b and Z_d are the partition functions of the TB and the DB ions, respectively; *See* **Notes 9** and **10** for a detailed description of the partition function calculation. The average interaction between the TB and DB ions is considered through an approximation in Z_d. In the above formula, N_d is the number of the DB ions. The charge neutrality condition gives $N_{2+} = N_b + N_d$, where N_{2+} is the total number of Mg^{2+} ions in the solution. We first compute Z_b with the different TB ion numbers N_b. We then calculate Z_d for the DB ions for each given N_b.

3.2 Partition Function of Tightly Bound (TB) Ions

The partition function is the ensemble average of the Boltzmann weight over all the possible ion distributions. To compute the TB ion partition function Z_b, we first need to sample the TB ion distributions. We use a simple cubic lattice grid to configure the ion coordinates in the TB region. For a given RNA with fixed N phosphates, the number of TB (Mg^{2+}) ions N_b is assumed to vary from 0 to N. For $N_b = N/2$, the RNA backbone is fully

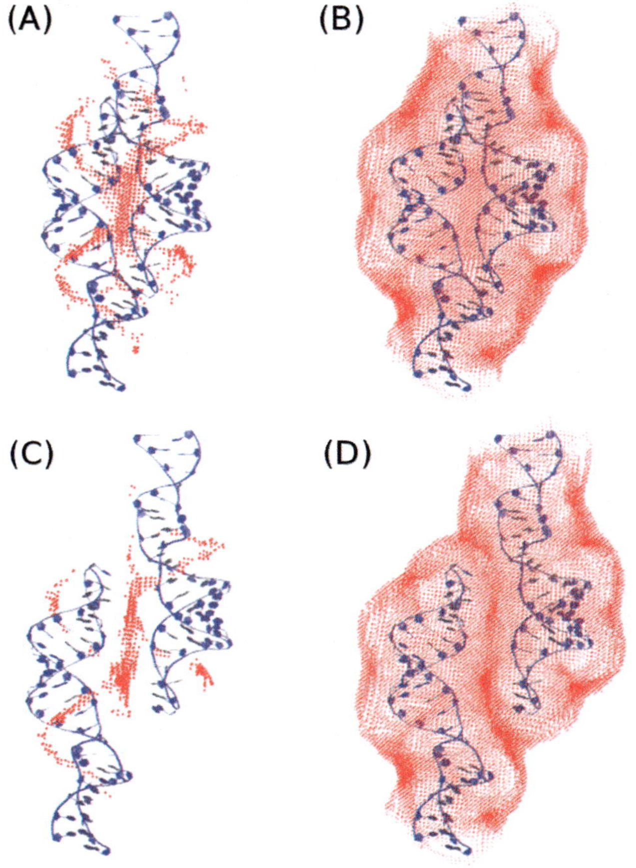

Fig. 3 The TB regions of the homodimeric tetraloop-receptor structure in the docked state at $[Mg^{2+}] = 10\ \mu M$ (**a**) and $[Mg^{2+}] = 10$ mM (**b**), and in the undocked state ($R = 15$ Å) at $[Mg^{2+}] = 10\ \mu M$ (**c**) and $[Mg^{2+}] = 10$ mM (**d**). Here, the *red dots* represent the grid points of the TB region. This figure shows that a compact structure of RNA and a high bulk ion concentration lead to a larger Tightly Bound (TB) region

neutralized by the tightly bound (TB) Mg^{2+} ions. In contrast, for $N_b = N$, on average there is one Mg^{2+} ion bound to a nucleotide, causing the RNA to be overcharged. Overcharging is unlikely to occur under physiological ionic conditions. Therefore, we assume $N_b \leq N$ in our calculation. Each grid site in the TB region can be occupied by at most one ion. Following the Rosenbluth–Rosenbluth (RR) method [59], the partition function Z_b for an N_b-ion system can be calculated through a process with ions added one by one (*see* **Note 11**). However, the original RR method neglects the backtrack effect, namely, the redistribution of the pre-inserted ions in response to the insertion of a new ion. To account for the backtrack effect, we developed the following "insertion-deletion" algorithm for the computation of the partition function for the TB ions [41].

1. For the insertion of the ith ion ($i = 1, 2, \ldots, N$), we enumerate all the available sites $k = 1, 2, 3, \ldots m_i$ in the TB region, where m_i is the number of available sites for the ion. We calculate the total interaction energy $\Delta U_i(k)$ (*see* **Note 12**) between the ith ion (if the ion is placed at site k) and the other charged particles (phosphates and the pre-existing TB ions).
2. From the energy $\Delta U_i(k)$, we compute the probability $p_f(i, k)$ for placing the ith ion at site k (*see* **Note 13**). We then use the Monte Carlo sampling method to randomly select a site for the ith ion according to the probability $p_f(i, k)$. The low energy sites are more likely to be visited by the ions.
3. The TB ions are added to the TB region one by one until all the N ions are inserted ($N_b = N$).
4. We remove the ions one by one from the N-ion distribution generated in the above steps. The sampling is biased toward the removal of higher energy ions (see the probability function p_b in *see* **Note 14**).
5. With the removal of 0, 1, 2, …, $(N - i)$, …, $N - 1$ ions (one by one), low-energy distributions are generated for N, $(N - 1)$, $(N - 2)$, …, i, …, 1-ion systems. For a distribution of the i-ion system, after the ion to be removed is decided, we can calculate the statistical weight for this ion from the sum over all its m_i available sites: $w(i) = \sum_{k=1}^{m_i} e^{-\Delta U_i(k)/k_B T}$
6. We repeat **steps 4–5** for a large number (M_b) of times and compute the average weight $\overline{w}(i)$ ($i \in [1, N_b]$).
7. Repeat all the above steps for a large number (M_f) of times. The average gives the partition function of the N_b TB ions:

$$Z(N_b) = \frac{1}{M_f} \sum \frac{\prod_{i=1}^{N_b} \overline{w}(i)}{N_b!}$$

Through extensive tests, we find that $M_b = N$ and $M_f = N_s$ can provide the optimal balance between efficiency and accuracy [41].

3.3 Partition Function of Diffusely Bound (DB) Ions

The partition function $Z_d(N_d)$ of DB ions is dependent on the distribution of the TB ions in the TB region. However, given the large number of TB ion distributions, solving the NLPB calculation for each TB ion distribution can be computationally demanding. Therefore, we treat the DB ions with an approximation.

1. From the ensemble of TB ion distributions (*see* **steps 1–7** above), for each N_b, we calculate the Boltzmann-averaged mean distribution for the TB ions.
2. Based on the mean distribution of the N_b TB ions, we solve the NLPB equation. From the solution of the NLPB for the

DB ion distribution, we calculate the free energy (*see* **Note 15**): $\Delta G_{\rm d}$ [12].

3. We compute the partition function of the DB ions: $Z_{\rm d}(N_{\rm d})$.

4 Results

4.1 Free Energy Landscape for the Homodimeric TL-R Complex

As a simplified approach to the conformational sampling of the system, we fix the orientations of each monomeric structure and change the separation distance between the two structures. Specifically, we increase the R value from 0 Å to 15 Å. The conformational ensemble corresponds to a specific pathway for the docking of the TL-R complex. For each R, we compute the electrostatic free energy of the homodimeric system. In our calculation, we fix the K^+ concentration at 20 mM and increase the Mg^{2+} concentration from 0 to 20 mM. To investigate the contribution of the Mg^{2+} ions to the docking of the homodimeric TL-R complex, we calculate the electrostatics free energy difference between systems with and without Mg^{2+} ions: $\Delta\Delta G = \Delta G([Mg^{2+}]) - \Delta G(0)$. The result of $\Delta\Delta G$ for the various $[Mg^{2+}]$ and R is shown in Fig. 4. From the free energy landscape, we can also compute ion-dependence of the conformational distribution. The energy landscape shows that a higher $[Mg^{2+}]$ would result in a lower free energy $\Delta\Delta G$ and the

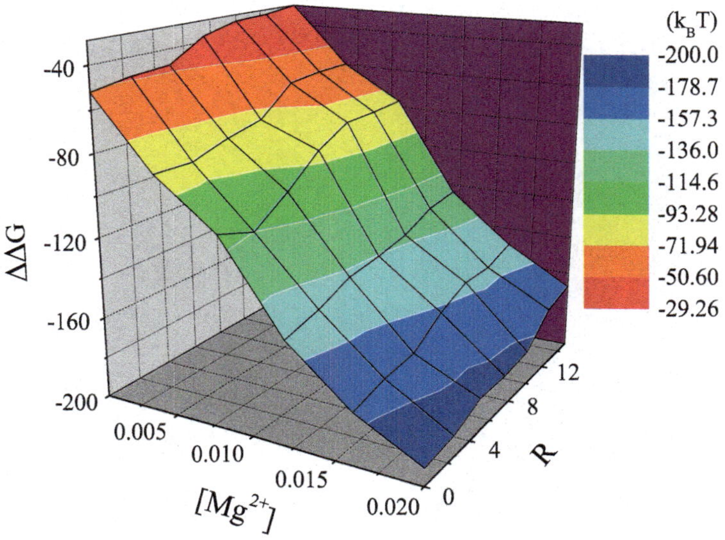

Fig. 4 The electrostatic free energy landscape as a function of Mg^{2+} bulk concentrations and the distances between the two monomeric TL-R structures. The units for the free energy (and the color scale), Mg^{2+} ion concentration, and the inter-molecular distance R are k_BT (at room temperature), M, and Å, respectively

docked state ($R = 0$) is at the bottom of the landscape. The result suggests that Mg^{2+} ions promote the docking of the TL-R complex.

4.2 Distribution of the Bound Ions

From the statistical weight for each TB ion distribution, we can calculate the probability $p(k)$ for a TB ion to be placed at grid site k (*see* **Note 16**). The information about the ion distribution can provide the mechanism for ion-mediated RNA stability and can further offer guidance for the computational design of RNA structure and ionic conditions for stabilizing or destabilizing specific functional motifs. Based on the experimental solutions (20 mM K^+ and 16 mM Mg^{2+}) [47], we predict the distribution of the TB ions; *See* Fig. 5. We find that the ion binding probability decreases toward the outer boundary of the TB region (*see* Fig. 5a), indicating that as the ions move toward the boundary of the TB region, the ion concentration approaches that of DB ions. This result suggests the possible continuity for the transition TB → DB in the interfacial regime. Furthermore, MCTBI-predicted most probable ion binding sites are in overall agreement with the experimentally observed site-specific ion binding sites (*see* Fig. 5b), suggesting that the non-specific electrostatic interactions help the stabilization of site-specific bound ions.

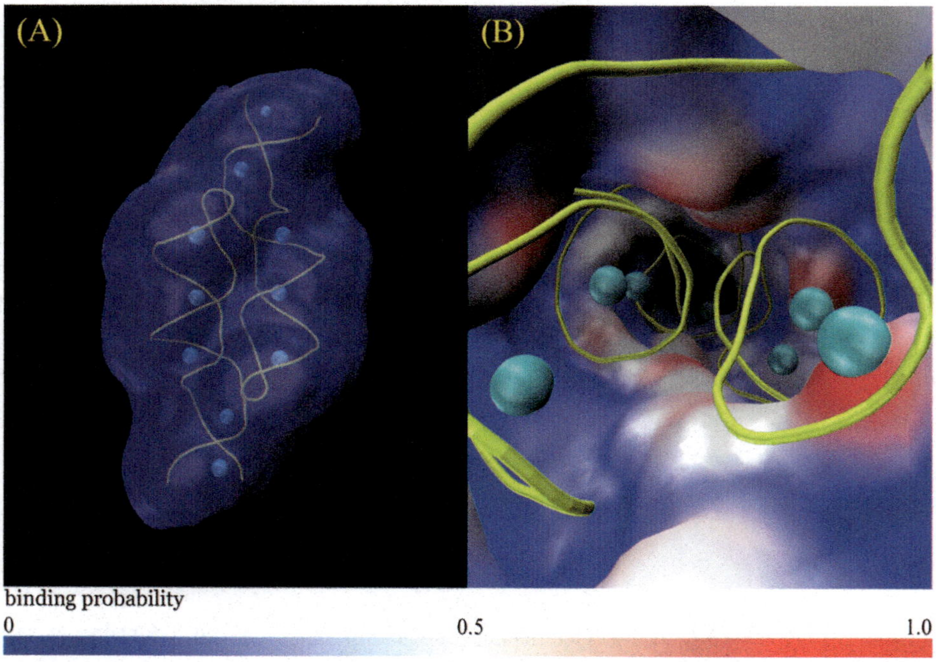

Fig. 5 The predicted TB Mg^{2+} ion distribution in a solution with 16 mM Mg^{2+} and 20 mM K^+. (**a**) and (**b**) show the "far-field" and "near-field" views of the Mg^{2+} ion distribution in the TB region, respectively. The *cyan spheres* and the *yellow backbone* show the site-specific binding ions and the docked TL-R complex structure (PDB ID: 2I7Z)

5 TBI Web Server

Based on the original TBI model, we have developed a web server (*see* **Note 17** and Refs. [12, 39, 60]). The server allows users to predict the most probable and the average ion distributions, binding fractions of the different types of ions, and the electrostatic free energy for a given RNA structure with user-defined ionic solution condition. Here, using the TL-R complex as the example, we illustrate the usage of the TBI web server (*see* Fig. 6).

5.1 Input Parameters

As shown in Fig. 6a, the TBI server has a simple input interface with five input parameters:

- The temperature in the unit of Celsius.
- The concentration of monovalent cation (Na^+ or K^+).
- The concentration of Mg^{2+} ion.
- The 3D structure of the RNA in the PDB format.
- Input the job name with alphanumeric characters.
- The email address of the user (optional). If the email address is provided, the results will be sent to users via email, otherwise, the user can download the results from the server directly.

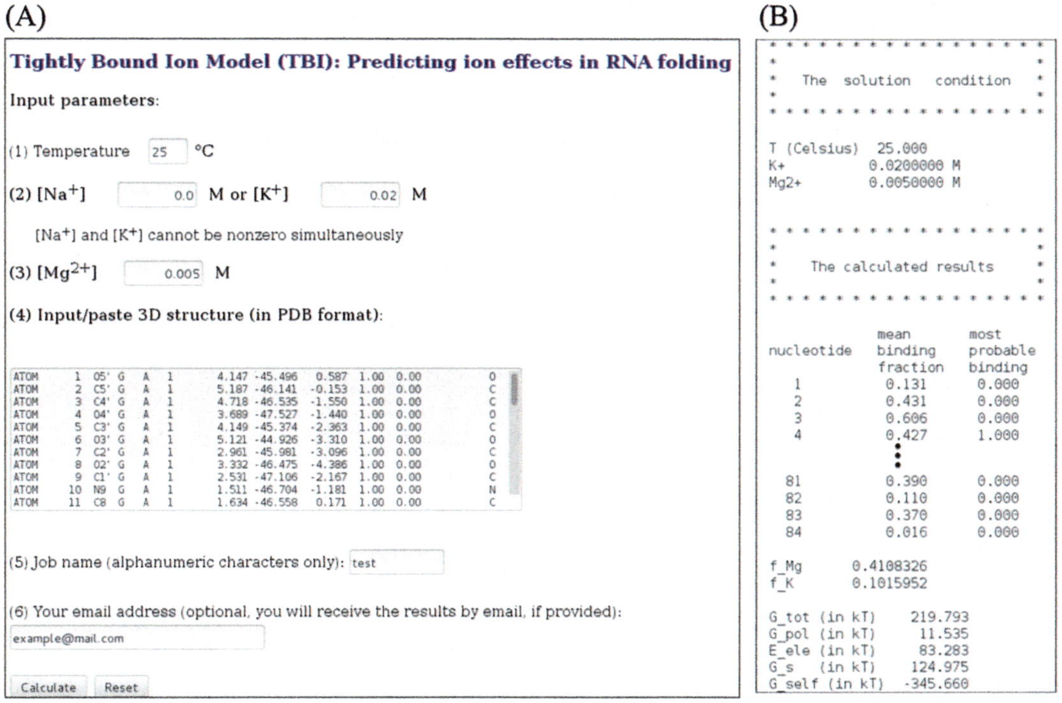

Fig. 6 (**a**) The web interface for parameter input on the TBI server. (**b**) The text file of the output results. The URL of the web server is http:/rna.physics.missouri.edu/tbi_index.html

5.2 Output Results

The results of the server consist of three output files: a text file (with filename extension ".dat") as shown in Fig. 6a and two figure files. The text file shows the temperature and ion concentrations (the first three input parameters above) and the results calculated from the TBI model:

- The basic information about ion binding. The first column is the index number of the nucleotides. The second column gives the Boltzmann averaged ion-binding fraction for each nucleotide. The third column shows the most probable (lowest free energy) ion binding mode.
- The numerical results of the predicted fractional number of bound ions.
- The electrostatic free energies of the metal ion-RNA solution. The results include both the total free energy and the different free energy components (*see* **Note 18**). All the free energies are given in the unit of $k_B T$, where k_B is the Boltzmann constant.

6 Notes

1. URL: http://www.rcsb.org/pdb/home/home.do
2. The two chains are restricted to keep parallel configurations without rotation. The distance R is defined as $R = R_{\text{undocked}} - R_{\text{docked}}$, where R_{undocked} and R_{docked} are the distances between the centers of mass of the two monomeric structures for the undocked and the docked states, respectively.
3. URL: http://rna.physics.missouri.edu/index.html
4. The sequence is 5′AUGCAUGCAUGCAAAAGCAUGCAUG CAU3′.
5. For the predicted 2D structure, the basepair information is ((((((((((((....)))))))))))), where a symmetric pair of "(" and ")" indicates a base pair between the corresponding nucleotides. Unpaired nucleotides are represented by dots.
6. As a coarse-grained RNA charge model, each phosphorus atom is assumed to carry an electronic charge ($-e$) and the remaining atoms are assumed to be neutral.
7. The radii of the three ion species are $r_{Mg^{2+}} = 4.5\,\text{Å}$, $r_{K^+} = 4.0\,\text{Å}$, and $r_{Cl^-} = 4.0\,\text{Å}$, respectively.
8. The ion-correlation strength $\Gamma(x)$ at position $\mathbf{x}$ is determined by the local ion concentration $c_{2+}(x)$:

$$\Gamma(\mathrm{x}) = \frac{(z\,e)^2}{\varepsilon_{\mathrm{w}} a_{\mathrm{ws}}(\mathrm{x})}, \quad \frac{4\pi}{3}[a_{\mathrm{ws}}(\mathrm{x})]^3\left[c_{2+}(\mathrm{x}) - c_{2+}^0\right] = 1$$

Here, $z\,e$ and ε_{w} (=78) are the charge of an ion ($z = 2$ for Mg^{2+}) and the dielectric constant of water. Position $\mathbf{x}$ belongs to the TB region if $\Gamma(\mathrm{x}) > 2.6$ and DB region otherwise [12].

9. For a given number of the tightly bound (TB) ions N_{b}, the partition function for the TB ions $Z_{\mathrm{b}}(N_{\mathrm{b}})$ is calculated from the ensemble average $\langle \ldots \rangle_M$ over all the possible N_{b}-ion binding modes (distributions) M: $Z_{\mathrm{b}}(N_{\mathrm{b}}) = \langle e^{-\Delta G_{\mathrm{b}}^{(M)}/k_{\mathrm{B}}T} \rangle_M$, where $\Delta G_{\mathrm{b}}^{(M)}$ is the interaction energy for the TB ions (and the RNA charges) for each given mode M:

$$\Delta G_{\mathrm{b}}^{(M)} = \Delta U_{\mathrm{ele}}^{(M)} + \Delta U_{\mathrm{pol}}^{(M)} + \Delta U_{\mathrm{LJ}}^{(M)}.$$

Here $\Delta G_{\mathrm{b}}^{(M)}$ includes three components:

- Coulombic interaction energy: $\Delta U_{\mathrm{ele}}^{(M)} = \frac{1}{2}\sum_{ij} \frac{Z_i Z_j e^2}{\varepsilon_{\mathrm{R}} r_{ij}}$, where $\sum_{ij}$ denotes summation over all the charged particles in the TB region (TB ions plus RNA phosphates), $Z_{i(\mathrm{or}j)}e$ is the charge of particle $i(\mathrm{or}j)$, ε_{R} is the dielectric constant of RNA, and r_{ij} is the distance between particles i and j.
- The polarization energy of the charged particles is computed from the Generalized Born (GB) model:

$$\Delta U_{\mathrm{pol}}^{(M)} = \left(\frac{1}{\varepsilon_{\mathrm{W}}} - \frac{1}{\varepsilon_{\mathrm{R}}}\right)\left[\frac{1}{2}\sum_{ij} \frac{Z_i Z_j e^2}{\sqrt{r_{ij}^2 + B_i B_j \exp\left(-\frac{r_{ij}^2}{4B_i B_j}\right)}} + \frac{1}{2}\sum_i \left(\frac{1}{B_i} - \frac{1}{B_i^0}\right) Z_i^2 e^2\right],$$

  where $\sum_i$ and $\sum_{ij}$ denote the summations over all the charged particles in the TB region (TB ions plus RNA phosphates), ε_{W} is the dielectric constant of water, $B_{i(\mathrm{or}j)}$ denotes the Born radius of particle $i(\mathrm{or}j)$, and B_i^0 is the Born radius of the particle in the bulk solvent.
- Lennard-Jones (LJ) potential energy, which is the sum over all the TB ions and the RNA (phosphate) charges:

$$\Delta U_{\mathrm{LJ}}^{(M)} = \frac{1}{2}\sum_{ij} u_{\mathrm{o}}\left[\left(\frac{\sigma_{ij}}{r_{ij}}\right)^{12} - \left(\frac{\sigma_{ij}}{r_{ij}}\right)^{6}\right],$$

  where u_{o} is the LJ constant and σ_{ij} is the equilibrium distance between particles i and j.

10. For a given number of the tightly bound (TB) ions N_b and their statistical average of distribution, the partition function $Z_d(N_d) = e^{-\Delta G_d(N_d)/k_B T}$ for the diffusely bound (DB) ions. The interaction energy of DB ions ΔG_d could be classified into an electrical term ΔG_d^E and an entropic term ΔG_d^S, as $\Delta G_d = \Delta G_d^E + \Delta G_d^S$. The two terms are calculated from the Poisson-Boltzmann equation (NLPB):

$$\Delta G_d^E = \frac{1}{2}\int \sum_\alpha c_\alpha(r) z_\alpha e[\psi(r) + \psi'(r)] d^3 r;$$

$$\Delta G_d^S = k_B T \int \sum_\alpha \left[c_\alpha(r) \ln \frac{c_\alpha(r)}{c_\alpha^0} - c_\alpha(r) + c_\alpha^0 \right] d^3 r.$$

Here α is the ion species. $z_\alpha e$ is the charge of ion species α, $\psi(r)$ and $\psi'(r)$ are the electric potentials with and without the DB ions in the solution, and $c_\alpha(r)$ and c_α^0 represent the local and bulk concentrations, respectively.

11. Assuming the TB region has totally N_s lattice grid sites, the sequentially added 1st, 2nd, ..., N_b-th TB ion would have $m_1 = N_s$, $m_2 = N_s - 1$, ..., $m_{N_b} = (N_s\text{-}N_b)$ available sites. In an ideal situation, if all the particles in the TB region are electrically neutral, the partition function of the N_b-ion system would be equal to

$$Z_b(N_b) = \frac{\prod_{i=1}^{N_b} w(i)}{N_b!} = \frac{\prod_{i=1}^{N_b} m_i}{N_b!} = 1$$

Here for non-interacting "ions," the statistical weight for the i-th inserted ion is $w(i) = i$.

12. The total interaction energy includes Coulombic energy, Lennard–Jones (LJ) potential, and polarization energy.

13. $p_f(i,k) = e\text{-}\Delta U_i(k)/k_B T / \sum_k e\text{-}\Delta U_i(k)/k_B T$ for the ith ion located at position k.

14. $p_b(j,k) = e^{\Delta U_j(k)/k_B T} / \sum_k e^{\Delta U_j(k)/k_B T}$ for the ion j at position k.

15. ΔG_d of the DB ions is the sum of the entropic free energy of the diffuse ions (the electrostatic energy), including the (average) electrostatics interaction energy between TB ions and DB ions.

16. The ion occupancy probability is calculated from the following formula:

$$p(k) = \sum_{N_b=0}^{N} \frac{n(N_b,k)}{M_f \times M_b} \times \frac{Z(N_b,N_d)}{Z}.$$

Here $n(N_b, k)$ is the number of N_b-ion distributions (out of the totally $M_f M_b$ sampled distributions) with site k occupied by a TB ion.

17. URL: http://rna.physics.missouri.edu/tbi_index.html.
18. Free energy components:
    - Polarization free energy: $\Delta G_{\text{pol}} = \langle \Delta U_{\text{pol}} \rangle$;
    - Coulomb interaction energy: $\Delta E_{\text{ele}} = \langle \Delta U_{\text{ele}} + \Delta G_{\text{d}}^{E} \rangle$;
    - Entropic free energy $\Delta G_{\text{s}} = \Delta G_{\text{tot}}$-$\Delta G_{\text{pol}}$-$\Delta E_{\text{ele}}$.

    In the above formulas, $\langle \ldots \rangle$ denotes the Boltzmann weighted average over all the ion binding modes. The server also provides the self-polarization energy of the RNA charges:

$$\Delta G_{\text{self}} = \frac{1}{2}\left(\frac{1}{\varepsilon_{\text{W}}} - \frac{1}{\varepsilon_{\text{R}}}\right)\sum_{\text{P}} \frac{Z_{\text{P}}^2 e^2}{B_{\text{P}}^{(0)}},$$

where the subscript P denotes a phosphate charge in the RNA.

Acknowledgements

This research was supported by NIH grant GM063732.

References

1. Doudna JA, Cech TR (2002) The chemical repertoire of natural ribozymes. Nature 418:222–228
2. Bachellerie JP, Cavaille J, Huttenhofer A (2002) The expanding snoRNA world. Biochimie 84:774–790
3. Gong C, Maquat LE (2011) lncRNAs transactivate STAU1-mediated mRNA decay by duplexing with 3′ UTRs via Alu elements. Nature 470:284–288
4. Bartel DP (2009) MicroRNAs: target recognition and regulatory functions. Cell 136:215–233
5. Kertesz M, Iovino N, Unnerstall U, Gaul U, Segal E (2007) The role of site accessibility in microRNA target recognition. Nat Genet 39:1278–1284
6. Brion P, Westhof E (1997) Hierarchy and dynamics of RNA folding. Annu Rev Biophys Biomol Struct 26:113–137
7. Tinoco I Jr, Bustamante C (1999) How RNA folds. J Mol Biol 293:271–281
8. Cate JH, Doudna JA (1996) Metal-binding sites in the major groove of a large ribozyme domain. Sturcture 15:1221–1229
9. Misra VK, Draper DE (2001) A thermodynamic framework for Mg2+ binding to RNA. Proc Natl Acad Sci USA 98:12456–12461
10. Draper DE, Grilley D, Soto AM (2005) Ions and RNA folding. Annu Rev Biophys Biomol Struct 34:221–243
11. Hayes RL, Noel JK, Whitford PC, Mohanty U, Sanbonmatsu KY, Onuchic JN (2014) Reduced model captures Mg2+-RNA interaction free energy of riboswitches. Biophys J 106:1508–1519
12. Tan ZJ, Chen SJ (2005) Electrostatic correlations and fluctuations for ion binding to a finite length polyelectrolyte. J Chem Phys 122:44903
13. Kolculi E, Lee NK, Thirumalai D, Woodson SA (2004) Folding of the Tetrahymena ribozyme by polyamines: importance of counterion valence and size. J Mol Biol 341:27–36
14. Denesyuk NA, Thirumalai D (2015) How do metal ions direct ribozyme folding. Nat Chem 7:793–801
15. Stein A, Crothers DM (1976) Equilibrium binding of magnesium(II) by Escherichia coli tRNAfMet. Biochemistry 15:157–160

16. Stein A, Crothers DM (1976) Conformational changes of transfer RNA. The role of magnesium(II). Biochemistry 15:160–168
17. Sosnick TR, Pan T (2003) RNA folding: models and perspectives. Curr Opin Struct Biol 13:309–316
18. Woodson SA (2005) Metal ions and RNA folding: a highly charged topic with a dynamic future. Curr Opin Chem Biol 9:104–109
19. Draper DE (2008) RNA folding: thermodynamic and molecular descriptions of the roles of ions. Biophys J 95:5489–5495
20. Chen SJ (2008) RNA folding: conformational statistics, folding kinetics, and ion electrostatics. Annu Rev Biophys 37:197–214
21. Auffinger P, Bielecki L, Westhof E (2003) The Mg2+ binding sites of the 5S rRNA loop E motif as investigated by molecular dynamics simulations. Chem Biol 10:551–561
22. Chen AA, Draper DE, Pappu RV (2009) Molecular simulation studies of monovalent counterion-mediated interactions in a model RNA kissing loop. J Mol Biol 390:805819
23. Hayes RL, Noel JK, Mohanty U, Whitford PC, Hennelly SP, Onuchic J, Sanbonmatsu KY. (2012) Magnesium fluctuations modulate RNA dynamics in the SAM-I riboswitch. J Am Chem Soc 134:12043–C12053
24. Dong F, Olsen B, Baker NA (2008) Computational methods for biomolecular electrostatics. Methods Cell Biol 84:843–870
25. Manning GS (1978) The molecular theory of polyelectrolyte solutions with applications to the electrostatic properties of polynucleotides. Q Rev Biophys 11:179–249
26. Zhou HX (1994) Macromolecular electrostatic energy within the nonlinear Poisson-Boltzmann equation. J Chem Phys 100:3152–3162
27. Misra V, Draper DE (1999) The interpretation of Mg2+ binding isotherms for nucleic acids using Poisson-Boltzmann theory. J Mol Biol 17:1135–1147
28. Baker NA, Sept D, McCammon JA (2001) Electrostatics of nanosystems: application to microtubules and the ribosome. Proc Natl Acad Sci USA 98:10037–10041
29. Tjong H, Zhou HX (2006) The dependence of electrostatic solvation energy on dielectric constants in Poisson-Boltzmann calculations. J Chem Phys 125:206101
30. Tan ZJ, Chen SJ (2010) Predicting ion binding properties for RNA tertiary structures. Biophys J 99:1565–1576
31. Bai Y, Greenfeld M, Herschlag D (2007) Quantitative and comprehensive decomposition of the ion atmosphere around nucleic acids. J Am Chem Soc 129:14981–14988
32. Mak CH, Henke PS (2013) Ions and RNAs: free energies of counterion-mediated RNA fold stabilities. J Chem Theory Comput 9:621–639
33. Giambasu GM, Luchko T, Herschlag D, York DM, Case DA (2014) Ion counting from explicit-solvent simulations and 3D-RISM. Biophys J 106:883–894
34. Hayes RL, Noel JK, Mandic A, Whitford PC, Sanbonmatsu KY, Mohanty U, Onuchic JN (2015) Generalized manning condensation model captures the RNA ion atmosphere. Phys Rev Lett 114:258105
35. Tan ZJ, Chen SJ (2006) Ion-mediated nucleic acid helix-helix interactions. Biophys J 91:518–536
36. Tan ZJ, Chen SJ (2006) Electrostatic free energy landscape for nucleic acid helix assembly. Nucleic Acids Res 34:6629–6639
37. Tan ZJ, Chen SJ (2007) RNA helix stability in mixed Na+/Mg2+ solution. Biophys J 92:3615–3632
38. Tan ZJ, Chen SJ (2008) Salt dependence of nucleic acid hairpin stability. Biophys J 95:738–752
39. He Z, Chen SJ (2012) Predicting ion-nucleic acid interactions by energy landscape-guided sampling. J Chem Theory Comput 8:2095–2102
40. Zhu Y, Chen SJ (2014) Many-body effect in ion binding to RNA. J Chem Phys 141:055101
41. Sun LZ, Chen SJ. (2016) Monte Carlo Tightly Bound Ion model: predicting ion binding properties of RNA with ion correlations and fluctuations. J Chem Theory Comput 12:3370–3381
42. Jaeger L, Michel F, Westhof E (1994) Involvement of a GNRA tetraloop in long-range RNA tertiary interactions. J Mol Biol 236:1271–1276
43. Murphy FL, Cech TR (1994) GAAA tetraloop and conserved bulge stabilize tertiary structure of a group I intron domain. J Mol Biol 236:49–63
44. Costa M, Deme E, Jacquier A, Michel F (1997) Multiple tertiary interactions involving domain II of group II self-splicing introns. J Mol Biol 267:520–536
45. Khvorova A, Lescoute A, Westhof E, Jayasena SD (2003) Sequence elements outside the hammerhead ribozyme catalytic core enable intracellular activity. Nat Struct Biol 10:708–712
46. Davis JH, Tonelli M, Scott LG, Jaeger L, Williamson JR, Butcher SE (2005) RNA helical packing in solution: NMR structure of a 30 kDa GAAA tetraloop-receptor complex. J Mol Biol 351:371–382

47. Jared HD, Trenton RF, Marco T, Samuel EB (2007) Role of metal ions in the tetraloop-receptor complex as analyzed by NMR. RNA 13:79–86
48. Chen SW, Honig B (1997) Monovalent and divalent salt effects on electrostatic free energies defined by the nonlinear Poisson-Boltzmann equation: application to DNA binding reactions. J Phys Chem B 101:9113–9118
49. Magnus M, Boniecki MJ, Dawson W, Bujnicki JM (2016) SimRNAweb: a web server for RNA 3D structure modeling with optional restraints. Nucleic Acids Res 44:W315–W319
50. Cheng CY, Chou F, Das R (2015) Modeling complex RNA tertiary folds with Rosetta. Methods Enzymol 553:35–64
51. Krokhotin A, Dokholyan NV (2015) Computational methods toward accurate RNA structure prediction using coarse-grained and all-atom models. Methods Enzymol 553:65–89
52. Purzycka KJ, Popenda M, Szachniuk M, Antczak M, Lukasiak P, Blazewicz J, Adamiak RW (2015) Automated 3D RNA structure prediction using the RNAComposer method for riboswitches. Methods Enzymol 553:3–34
53. Bellaousov S, Reuter JS, Seetin MG, Mathews DH (2013) RNAstructure: web servers for RNA secondary structure prediction and analysis. Nucleic Acids Res 41:W471–W474
54. Gruber AR, Bernhart SH, Lorenz R (2015) The ViennaRNA web services. In: RNA bioinformatics. Methods in molecular biology, vol 1269. Springer, New York, pp 307–326
55. Bindewald E, Kluth T, Shapiro BA (2010) CyloFold: secondary structure prediction including pseudoknots. Nucleic Acids Res 38: W368–W372
56. Xu XJ, Zhao PN, Chen SJ (2014) Vfold: a web server for RNA structure and folding thermodynamics prediction. PLoS One 9: e107504
57. Cao S, Chen SJ (2005) Predicting RNA folding thermodynamics with a reduced chain representation model. RNA 11:1884–1897
58. Cao S, Chen SJ (2011) Physics-based de novo prediction of RNA 3D structures. J Phys Chem B 115:4216–4226
59. Rosenbluth MN, Rosenbluth AW (1955) Monte Carlo calculation of the average extension of molecular chains. J Chem Phys 23:365–369
60. Zhu Y, He Z, Chen SJ (2015) TBI server: a web server for predicting ion effects in RNA folding. PLos One 10:e0119705

Chapter 2

Computational Generation of RNA Nanorings

Rishabh Sharan, Eckart Bindewald, Wojciech K. Kasprzak, and Bruce A. Shapiro

Abstract

A variety of designed RNA ring structures (ranging from triangles to hexagonal rings) have been reported in the scientific literature. Designing self-assembling RNA ring structures from structural motifs is, however, a nontrivial problem as there are many combinations of motifs and linking helices. Moreover, most combinations of motifs and linker helices will not lead to ring closure. A solution to this problem was recently published using a "design-by-catalog" approach where motif combinations that lead to rings are precomputed and tabulated. Here we present a web-browser based workflow for creating RNA rings using Galaxy, a web-based platform that can be used for workflow management. An example of how these RNA rings are generated and processed to create a 3D model of the ring is discussed.

Key words RNA, Ring, Closure, Enumeration, Scaffold

1 Introduction

1.1 RNA Nanotechnology

RNA can be designed to self-assemble into a large variety of geometries (such as RNA tiles, hexameric rings, cubes, bundles, and arrays) [1–8] due to its ability to form a variety of base-pairs with at least two hydrogen bonds, in addition to the standard Watson–Crick base pairs [9]. RNA is also able to self-assemble in cellular conditions [10] through interstrand and intrastrand interactions [11]. RNA is not only able to form many shapes, but is also able to perform enzyme-like functions [12–15]. Also, a polyvalent RNA nanostructure can have different functional attachments, such as siRNAs, aptamers, peptides, or fluorescent tags [8, 10, 16]. Due to existing computational prediction capabilities, RNA is relatively conducive to rational design. In terms of therapeutics and medicine, RNA is biocompatible and biodegradable. Due to these properties, RNA is an attractive material to be used in nanotechnology and biomedical applications.

Eckart Bindewald and Bruce A. Shapiro (eds.), *RNA Nanostructures: Methods and Protocols*, Methods in Molecular Biology, vol. 1632, DOI 10.1007/978-1-4939-7138-1_2, © Springer Science+Business Media LLC 2017

1.2 Traditional Computational Design Approaches

Through the use of computational tools [17–19] the process of designing nanostructures can be simplified and accelerated. We highlight two strategies for the computational design of RNA nanostructures: building a scaffold de novo or starting with existing RNA motifs [4, 16, 20, 21]. Here we discuss both of these approaches as well as a new catalog-based approach.

1.2.1 De Novo Design of RNA Scaffolds

There are some geometries and properties for which there are no naturally occurring RNA structural motifs. An example of this situation is the in silico designed RNA nanocube [8, 22]. For the case of the RNA cube, we could not identify any three-way junctions in the RNAJunction database that had substantial similarity to a cube corner [23]. Several approaches are possible for the creation of 3D models of such RNA nanostructures. One approach is to design the RNA nanocube with a de novo three-way junction [8]. Another approach is to focus on optimizing the placement of helices and then placing bridging RNA strands de novo [22]. Both methods resulted in 3D models of nanocubes [22].

1.2.2 Building Block Approach

Analysis of available RNA tertiary structures has led to the identification of RNA building blocks. These motifs have been compiled into various databases such as the k-turn database [24], SCOR [25], FRABASE [26], Motif Atlas [27], RNA CoSSMos [28], and RNA STRAND [29]. RNAJunction is a database of kissing loops, bulges, and *n*-way junctions [23]. These databases can be used to identify motifs with geometries of interest and serve as a good starting place for the designer. Note that these compilations of RNA motifs often utilize different conventions of how to define and group structural motifs. These motifs can be linked together with helices or single-stranded regions allowing for a building-block based approach. Given the building blocks, the designer is able to create a variety of shapes.

1.2.3 Catalog-Based Design

A new approach is the catalog-based design in the form of the Ring Catalog [30]. Rings are defined as repeated concatenations of one or two motifs, such that the motifs together with their connecting linker helices form a circular RNA complex. As described in reference [23], motifs are RNA structural elements such that each end of all involved RNA strands forms a base pair with another RNA strand end. In this fashion, the structural element possesses terminal base pairs that can be extended with helices. Two or more motifs can be connected with linker helices that extend the terminal base pairs. Most nonbranching concatenations of motifs and linker helices will not correspond to closed ring structures but rather to rods or spirals. The combination of which motifs and linker helices to use in order to obtain ring-closure is most likely not obvious to even the expert designer. A solution to this problem is the complete enumeration of motif combinations and linker helix lengths while cataloging the subset of structures that are generated in this fashion that are forming closed ring structures. One issue is that iterating over many linker helix lengths and motif combinations may lead to a "combinatorial explosion," making searching for rings computationally demanding. A subset of motifs was chosen by creating clusters and using representatives of those clusters in the ring search. A cluster was defined as a set of motifs with the same sequence and same NC-IUBMB nomenclature notation [31]. This ensures there are no redundant motifs. Also, the types of motifs chosen were three-way junctions, internal loops/bulges, and kissing loops. By avoiding the "combinatorial explosion," a trade-off was made in terms of comprehensiveness. NanoTiler, a Java-based multifunctional software that is able to create different RNA-based nanostructures [17], was used to perform the search using its "growscan" command. This used all pairs of motifs and iteratively changed the length of the linking helices. Rings with a ring-closure score below a threshold were stored and then classified based on topology. Ring structure coordinate data is available online at https://rnajunction.ncifcrf.gov/ringdb. This catalog-based approach allows the user to sift through ring structures based on topologies of interest, thus decreasing the effort needed to design ring structures from scratch.

Here, we demonstrate how to generate ring structures, with the square ring shown in the Ring Catalog publication as the example [30], and generate sequences that can be purchased to assemble the ring experimentally.

1.3 Galaxy

Galaxy is a web-based platform for workflow management, data integration, and data analysis [32]. While originally developed for genomic data, Galaxy has become a general workflow management system for bioinformatics and other areas that require numerical processing of data sets. Galaxy functions as a web server and the user interacts with Galaxy via a web browser. The system allows the user to upload data sets, to manipulate them with computational tools and to store the results in a history. This is achieved by encapsulating each of the programs or algorithms into web-based tools where the user typically only specifies input parameters via clickable menu items. This simplifies the access to these computational tools compared to a command line interface. Particularly useful for computational analysis is the history feature of Galaxy. Whenever a user launches a computational tool, the corresponding results are stored in a history. This helps with repeating workflows and reproducing results at a later time because the user is able to see exactly what was run as well as what the inputs and outputs were. Galaxy also has an extensive tool library called Tool Shed where users can install tools created by other users onto their local Galaxy instance [33]. Galaxy also handles any dependencies those tools may require, which simplifies the installation process. Here we use Galaxy as a web-based front-end for 3D structure manipulation in general and for the design of RNA rings in particular.

2 Materials

Readers interested in how rings are generated in the Ring Database should follow all of this section and download the junction used to generate the square. Otherwise, Subheadings 2.5 and 3.1 may be skipped, as the downloaded structure from Subheading 2.4 will be the same as the structure generated from Subheading 3.1.

2.1 Virtual Operating System for Hosting a Unix-Like Operating System

The work described in this chapter is based on the VirtualBox system. The VirtualBox virtual machine environment allows one to run a variety of Linux distributions on non-Linux host machines such as Apple OS X or Microsoft Windows.

1. Download a version of VirtualBox that is appropriate for the operating system of your host computer from https://www.virtualbox.org/wiki/Downloads
2. Follow the installation instructions available in the User Manual available on the VirtualBox website.

2.2 Download and Start Custom Linux Distribution

The Galaxy web server requires a Unix-like operating system that can host the other needed software components. In this chapter, we describe a VirtualBox virtual machine-based Linux distribution that is based on Ubuntu-14 and has been customized by us in order to facilitate RNA 3D structural bioinformatics and computational biology. Note that a Linux distribution that runs within VirtualBox is called an appliance.

1. Download the virtual machine image from https://binkley2.ncifcrf.gov/users/bshapiro//software.html.
2. Start the VirtualBox software installed in the previous step.
3. Within VirtualBox, click **File**, then **Import appliance**.
4. Click the folder icon and navigate to the virtual machine image downloaded in the first step.
5. Click **Next**, **Next**, then **Import**.
6. A pop-up will appear indicating the virtual machine is being imported.
7. Once the importing is complete, double click on the new virtual machine.
8. Once the login screen appears log-in with username and password provided with the download instructions.

2.3 Starting and Accessing Galaxy

1. After starting the hosted Linux virtual machine, and a screen prompting for username and password appears, log in with the username rnalab (the password is provided as part of the download instructions). The custom Linux distribution has been prepared by us such that the local Galaxy web server starts automatically upon logging into the prepared user account named rnalab. No additional steps are needed to start Galaxy. If, however, a different user account is utilized or if the Galaxy web server has terminated, Galaxy may need to be started manually. For more details *see* **Note 1**.
2. In order to access the running Galaxy web server, open a web browser (for example by double-clicking the **Firefox** icon on the desktop or by choosing the web browser in the Applications pull-down menu).
3. In the address bar of the web server type http://localhost:8080 or equivalently http://127.0.0.1:8080. This is the web address of the local Galaxy instance.

2.4 Download Junction Structure

1. Download the three-way junction used by opening a web browser and navigating to https://rnajunction.ncifcrf.gov/index.php and typing **12316** in for **Entry ID**.
2. Click **Download** next to **Extracted Structure** to download the PDB-formatted coordinate file of the ring. This webpage also has the sequence of the motif which may be useful later during sequence optimization.

2.5 Download Ring

1. Download draft ring by opening web browser and navigating to https://rnajunction.ncifcrf.gov/ringdb.
2. Click item **Three-way Junctions**, then scroll down and click on the ring image that is square-shaped. (At the time of writing this, it is the fourth image).
3. The ring of interest has the following characteristics:
   - `Motif 1: __2OGM.rnaview.pdb_j3_0-A2073_0-C2179_0-A2204.pdb__`
   - `Motif 2: NA`
   - `Descriptor: 1_1_1_3_5`
4. Once that ring has been found, click **PDB** under **PDB** to download the pdb file of the ring.

3 Methods

If interested in generating the ring (Subheading 3.2), follow Subheading 3.1 and upload the junction from Subheading 2.4. If not, follow Subheading 3.1 and upload the ring from Subheading 2.5, then skip to Subheading 3.3.

3.1 Upload File

1. Start a web browser. Preinstalled with the virtual-machine distribution is the FireFox web browser. This browser can be started by clicking on the FireFox logo or by selecting the pull-down menu Applications → Network → FireFox web browser at the top left of the window of the virtual machine. View within the web browser the status of the Galaxy server by choosing the website http://127.0.0.1:8080. This link is also provided in the bookmarks toolbar of the preinstalled FireFox browser. This web address is referred to as the local Galaxy homepage.
2. It is optional but recommended that you log into the Galaxy web server. Without logging in, some functionality of the Galaxy web server (such as the 3D visualization) will not be available. You can use a prepared guest account: At the top right of the web browser window, choose User → Login. At the prompt, enter as e-mail address guest@rna.lab and the password provided with the download instructions. Once logged in, you can change the password under User → Preferences → Change your password. Alternatively, you can register a new local Galaxy account under User → Register. At the end of this protocol step, you should be logged into the local Galaxy web server instance.
3. On the left side of the local Galaxy homepage, under **Tools**, upload the downloaded pdb file by clicking **Get Data**, then **Upload File**.
4. In the new window that appears, click **Choose local file** and navigate to the location of the downloaded pdb.
5. Click the drop-down menu under **Type** and choose data type **pdb**.
6. Click **Start** and a history element corresponding to a new compute job will appear on the right side of the local Galaxy homepage. Once the background color of this history element is green, the compute job is completed (*see* **Note 2**).

3.2 Generate Draft Ring Structure

3.2.1 Scan for Rings

1. Under **Tools** on the left side, click **NanoTiler**, then **Grow Scan** (*see* Fig. 1-h).
2. For the **Junction** menu of the **Grow Scan** tool, make sure the junction uploaded (as described in Subheading 3.1) is selected. All other settings can be left at default values. Click **Execute** to begin the compute job (*see* **Note 3**).
3. Once the compute job is completed, refresh the history by clicking the icon with two arrows in a circle on the top right of the browser page next to the gear.
4. A few more computational results should appear—they correspond to additional ring structures that the program identified. The data set named **Grow Scan on data<some number>**

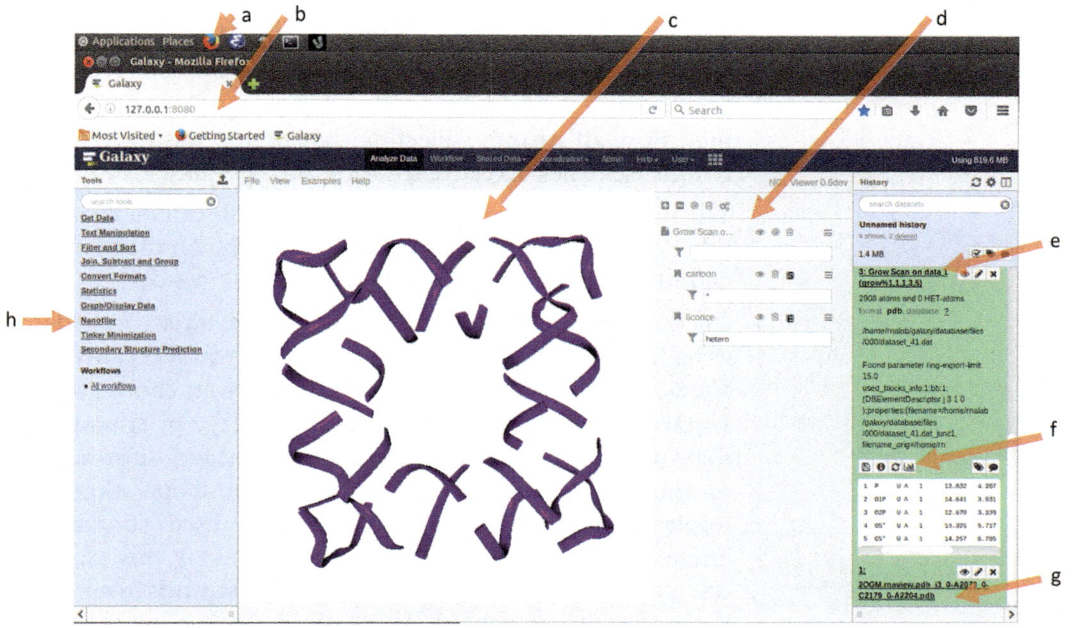

Fig. 1 Screenshot of local Galaxy instance after an initial ring structure has been found. (*a*) icon of the FireFox web browser; (*b*) text area for specifying the website of the browser; (*c*) main window of the Galaxy web server site. It can be used to configure tools to be launched, to inspect data sets or (as shown for this case) display the 3D structure of an RNA via the NGL viewer software; (*d*) parameter settings for the molecular viewer; (*e*) clickable title of the generated RNA ring data set. (*f*) icon that launches the NGL molecular viewer; (*g*) clickable title of the uploaded structure of the three-way junction; (*h*) link and header corresponding to a variety of NanoTiler tools

(grow%1,1,1,3,5) corresponds to the found ring structure (*see* Fig. 1-e and **Note 4**). Additional ring structures as well as versions of generated structures that do not contain linker helices are also available in the history pane.

5. Visualize the generated ring structure by selecting the visualization icon of the data set (the icon has the appearance of a histogram, *see* Fig. 1-f).

3.3 Postprocessing of Ring Structure

3.3.1 Optimize Ring Structure

1. On the left side of the local Galaxy homepage, click **NanoTiler**, then **Improve Ring Closure**, select the file named **Grow Scan on data<some number> (grow%1,1,1,3,5)** (*see* Fig. 1-h).
2. Press **Execute** to launch the program that optimizes ring closure.

3.3.2 Fuse Strands

In the presented example, there are 20 RNA strands. These can be fused resulting in a total of five strands using **Ring fuse**, **Fuse all strands**, and **Fuse strands** (*see* **Note 5**).

1. On the left side of the local Galaxy homepage, click **NanoTiler** (Fig. 1-h), then **Ring fuse**, selecting the file that was the

output of **Improve Ring Closure**. Click **Execute** to begin the compute job.

2. On the left side of the local Galaxy homepage, click **NanoTiler**, then **Fuse all strands**, selecting the file that was the output of **Ring fuse**. Click **Execute** to begin the compute job.
3. The number of chains in the obtained result can be verified on the level of the raw coordinate file by clicking on the "eye" icon within the history pane of Galaxy.
4. Click on the graph icon of the resulting data set in order to visually inspect the result structure using the **NGL Molecule Viewer** (*see* Fig. 1-f). In the molecular viewer, choose a representation that colors the 3D molecular image by strand. Note that structural gaps will be closed at a later stage—instead focus on strand color as indication of the strand fuse operation. If the resulting structure consists of more unfused strands than intended (if there are more than five strands in this example), one can attempt to rerun the **Fuse all strands** tool with a higher distance threshold. Alternatively one can use the molecular viewer in order to determine which strands should be fused explicitly.
5. To fuse specific strands perform the following steps:
   - Within the left tool pane of the local Galaxy homepage, click **NanoTiler**, then **Fuse strands**.
   - Choose the **Fuse all strands on data** as Junction, then enter the letters of the two strands in **Strand that is elongated** and **Strand to append** (*see* **Note 6**). Click **Execute** to begin the compute job.

3.3.3 Extend Helices

1. On the left side of the local Galaxy homepage, click **NanoTiler**, then **Extend Helix** (*see* **Note 7**).
2. Choose the result of the fusing from the last step for the **Structure**, **10** for **Number of basepairs**. For **Sequence and basepair of where to extend** enter the strand and base number separated by a colon (for example, A:1 will extend a helix off of the first base of strand A). Click **Execute** to begin the compute job.
3. Repeat for each applicable strand each time using the newly created file with the extended helices for structure.
4. To fuse the newly created helices to the square structure, use **Fuse all strands** and **Fuse strands**. Use **NGL Molecule Viewer** to see which strands need to be fused. Once each of the helices are fused, there will be five strands again (*see* Fig. 2a and **Note 8**).

3.3.4 Move Nick

1. Sometimes, the RNA strand breaks (i.e., nicks) that are utilized as a result of the previous steps are not at the desired positions.

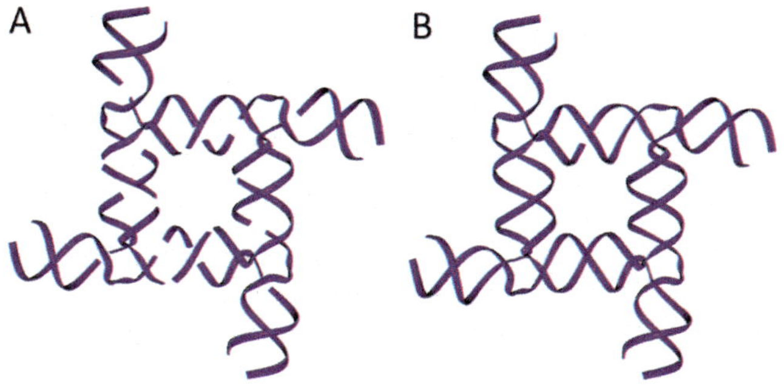

Fig. 2 Structures at different processing stages. (**a**) RNA square structure after helixes has been added. (**b**) The RNA structure after energy minimization

In the case of the RNA square utilized in this example, use the **NGL Molecule Viewer** to find the 5′ end of the inner strand (this strand corresponds to strand E in Fig. 5 in reference [30]). Estimate the number of nucleotides needed to be split and then rejoined for the 5′ end of the strand to move toward the middle of a side of the RNA square. Note down the position of the base **after** the split (*see* **Note 9**).

2. On the left side of the local Galaxy homepage, click **NanoTiler**, then **Split strands**. Click **Execute** to launch the compute job.
3. Type the letter identifier of the strand representing the inner circular strand, for *Strand*.
4. Then use position noted down from **step 1**. The strand is cut before this position (*see* **Note 10**).
5. On the left side of the local Galaxy homepage, click **NanoTiler**, then **Fuse strands**. We will fuse the new strand to the old one. For **Strand that is elongated** type the name of **new strand**, and then the name of **old strand** for **Strand to append**. Click **Execute** to launch the compute job (*see* **Note 11**).
6. Use the NGL viewer and verify the nick is moved near the center.

3.3.5 Optimize Sequences

Initially, all generated linker helices correspond to consecutive all GC base pairings. Therefore, one needs to optimize individual strand sequences next.

1. Sequence optimization is necessary if the ring structure contains generated helical segments consisting of unoptimized placeholder sequences (in this example corresponding to GC base pairs).
2. On the left side of the local Galaxy homepage, click **NanoTiler**, then **Secondary Structure**.

3. Select the file of the structure with the adjusted nick moved in the drop-down menu. For format pick **Server**. Click **Execute** to begin the compute job.
4. Once it is completed, open the file by clicking the eye icon and copy the contents.
5. Point a web browser to the NanoFolder homepage at http://matchfold.abcc.ncifcrf.gov/. For a background of this computational tool for multistrand structure prediction and sequence optimization please consult the original publication [34].
6. Paste the contents from the results of the secondary structure tool (called "Secondary Structure").
7. NanoFolder optimizes bases that are capitalized. Lowercase nucleotides are left unchanged by the optimization algorithm. The aim is to optimize bases that are part of the helices but not part of the motif. This can be achieved by visually inspecting the RNA draft structure and classifying strand regions as either motif regions or as linker helix regions. Helices not part of the motif will initially be only G or only C nucleotides. There are stretches of G's in the motif so it may be helpful to also have the sequence of the motif available as well (*see* **Notes 12–14**).

3.3.6 Mutate 3D Model to Reflect Optimized Sequence

1. On the left side of the local Galaxy homepage, click **NanoTiler**, then **Mutate 3D structure**.
2. Select the file from the output of Subheading 3.3.5.
3. Then click the **Insert Mutations** button five times to bring up five input areas. Each one corresponds to the chain of the same number (except the chains are alphabetically named, so the max is 26), the first is for chain A, 2 is B, 3 is C, etc.
4. Copy the result of optimization into each corresponding area and then click **Execute** to begin the compute job.
5. Verify the mutations using the **Secondary Structure** tool.

3.3.7 Refine 3D Model

1. Molecular dynamics software suites such as Tinker can be used to refine the RNA 3D structure using molecular mechanics energy minimization [35]. Energy minimization will close gaps and reduce structural problems.
2. On the right side of the local Galaxy homepage, click **Tinker Minimization**, then **Prepare pdb**.
3. Set input to output file of **step 4** and click **Execute** to begin the compute job.
4. Once the compute job is completed, on the right side of the local Galaxy homepage, click **Tinker Minimization**, then **Tinker**.

5. Set the input to the output of the previous step and set the minimization termination parameter called "RMSG" to 1 for coarse refinement, and .1 for fine refinement (*see* **Note 15**).
6. The minimization can take several minutes or even an hour to complete, depending on the size of the RNA structure. Visualize or download the resulting structure (*see* Fig. 2b).

4 Notes

1. The installation directory of the Galaxy server is called "galaxy." Modifying the behavior of the local Galaxy web server (such as adding computational tools to it) is entirely possible, but is beyond the scope of this chapter. Please consult the Galaxy documentation for questions related to the administration of the Galaxy server (https://wiki.galaxyproject.org/Learn). The prepared user account is set up in such a way that the Galaxy server starts automatically. In principle, it is possible to start the Galaxy server using two additional ways: (a): Select from the pull-down menu (top left) Applications → RNA Lab Applications → Galaxy. (b): Open a Unix shell terminal window by clicking on the Terminal icon on the desktop. In the command line of the obtained window type cd galaxy; sh run.sh. This starts up an instance of the Galaxy web server at port 8080. Note that the web server startup script only succeeds if no other web server is active at the same port number.
2. A gray background of the element of the Galaxy history pane indicates that the corresponding compute job is waiting to start, while a yellow background indicates that the compute job is currently running. A green background indicates that the compute job is completed.
3. This step may take some time to complete.
4. There will be other files with names like: **Grow Scan on data<some number> (grow%1,1,1,3,5_nohelices)**. These are the same structure with and without included linking helices.
5. The strand "fuse" family of commands merely change strand names, while atom coordinates are not changed.
6. The order does not matter unless the strands need to be appended in a specific order (can be fused on both ends such as in a ring).
7. Adding helix-forming complementary regions to a nucleic acid nanostructure may improve its thermal stability and assembly properties.

8. Because the chain letter labels change after each fusing operation, it is recommended to use a molecular viewer (such as the integrated **NGL Molecule Viewer**) to identify RNA strands that should be fused.
9. Checking the structure in the viewer will show that the inner strand starts and ends close to a junction. This means that during assembly, one end of the helix will only have a few base pairs and the other will have a lot more. Balancing it stabilizes the shorter helix. This can be achieved by splitting off nucleotides at the end of the strand corresponding to the shorter helix and adding them to the strand corresponding to the helix in need of stabilization.
10. To verify that the strand was split, click the "eye" icon in the history pane element corresponding to the data set and inspect the resulting pdb text file. At the end of the file, there will be a chain with the same number of bases as was split off.
11. The old strand is the strand that is deleted, because the old chain is appended to the new strand. If we were to reverse it, we would end back up where we were before this.
12. Any additional sequence regions that would need to be added (such as sequence regions at the 5′ end related to a utilized RNA polymerase (8)), should be added before the sequence optimization step—either in form of an expanded 3D model or in form of an expanded secondary structure text input for the utilized sequence optimization program.
13. Ascertain the quality of the optimized sequence by inspecting the output of the optimization program and by utilizing RNA secondary structure prediction programs.
14. This step may take some time to complete.
15. The smaller the minimization tolerance parameter ("RMSG") is set, the better the structure will end up, but it will take longer (e.g., a run with RMSG set to 0.1 can frequently take more 30 min for the presented example).

Acknowledgment

This work has been funded in whole or in part with Federal funds from the Frederick National Laboratory for Cancer Research, National Institutes of Health, under Contract No. HHSN261200800001E. This research was supported [in part] by the Intramural Research Program of the NIH, National Cancer Institute, Center for Cancer Research. The content of this publication does not necessarily reflect the views or policies of the Department of Health and Human Services, nor does the mention of trade names, commercial products, or organizations imply endorsement by the US Government.

References

1. Guo P (2010) The emerging field of RNA nanotechnology. Nat Nanotechnol 5:833–842
2. Dibrov SM, McLean J, Parsons J, Hermann T (2011) Self-assembling RNA square. Proc Natl Acad Sci 108:6405–6408
3. Geary C, Rothemund PWK, Andersen ES (2014) A single-stranded architecture for cotranscriptional folding of RNA nanostructures. Science 345:799–804
4. Grabow WW, Zakrevsky P, Afonin KA, Chworos A, Shapiro BA, Jaeger L (2011) Self-assembling rNA nanorings based on RNAI/II inverse kissing complexes. Nano Lett 11:878–887
5. Yingling YG, Shapiro BA (2007) Computational design of an RNA hexagonal nanoring and an RNA nanotube. Nano Lett 7:2328–2334
6. Jaeger L, Westhof E, Leontis NB (2001) TectoRNA: modular assembly units for the construction of RNA nano-objects. Nucleic Acids Res 29:455–463
7. Nasalean L, Stombaugh J, Zirbel CL, Leontis NB (2009) RNA 3D structural motifs: definition, identification, annotation, and database searching. In: Walter NG, Woodson SA, Batey RT (eds) Non-protein coding RNAs. Springer, Berlin, pp 1–26
8. Afonin KA, Bindewald E, Yaghoubian AJ, Voss N, Jacovetty E, Shapiro BA, Jaeger L (2010) In vitro assembly of cubic RNA-based scaffolds designed in silico. Nat Nanotechnol 5:676–682
9. Leontis NB, Westhof E (2014) Self-assembled RNA nanostructures. Science 345:732–733
10. Afonin KA, Kireeva M, Grabow WW, Kashlev M, Jaeger L, Shapiro BA (2012) Co-transcriptional assembly of chemically modified RNA nanoparticles functionalized with siRNAs. Nano Lett 12:5192–5195
11. Grabow WW, Jaeger L (2014) RNA self-assembly and RNA nanotechnology. Acc Chem Res 47:1871–1880
12. Cech TR, Zaug AJ, Grabowski PJ (1981) In vitro splicing of the ribosomal RNA precursor of Tetrahymena: involvement of a guanosine nucleotide in the excision of the intervening sequence. Cell 27:487–496
13. Guerrier-Takada C, Gardiner K, Marsh T, Pace N, Altman S (1983) The RNA moiety of ribonuclease P is the catalytic subunit of the enzyme. Cell 35:849–857
14. Kruger K, Grabowski PJ, Zaug AJ, Sands J, Gottschling DE, Cech TR (1982) Self-splicing RNA: autoexcision and autocyclization of the ribosomal RNA intervening sequence of Tetrahymena. Cell 31:147–157
15. Ramakrishnan V (2014) The ribosome emerges from a black box. Cell 159:979–984
16. Khaled A, Guo S, Li F, Guo P (2005) Controllable self-assembly of nanoparticles for specific delivery of multiple therapeutic molecules to cancer cells using RNA nanotechnology. Nano Lett 5:1797–1808
17. Bindewald E, Grunewald C, Boyle B, O'Connor M, Shapiro BA (2008) Computational strategies for the automated design of RNA nanoscale structures from building blocks using nanoTiler. J Mol Graph Model 27:299–308
18. Martinez HM, Maizel JVJ, Shapiro BA (2008) RNA2D3D: a program for generating, viewing, and comparing 3-dimensional models of RNA. J Biomol Struct Dyn 25:669–683
19. Jossinet F, Ludwig TE, Westhof E (2010) Assemble: an interactive graphical tool to analyze and build RNA architectures at the 2D and 3D levels. Bioinformatics 26:2057–2059
20. Chworos A, Severcan I, Koyfman AY, Weinkam P, Oroudjev E, Hansma HG, Jaeger L (2004) Building programmable jigsaw puzzles with RNA. Science 306:2068–2072
21. Bida J, Das R (2012) Squaring theory with practice in RNA design. Curr Opin Struct Biol 22:457–466
22. Afonin KA, Kasprzak W, Bindewald E, Puppala PS, Diehl AR, Hall KT, Kim TJ, Zimmermann MT, Jernigan RL, Jaeger L, Shapiro BA (2014) Computational and experimental characterization of RNA cubic nanoscaffolds. Methods 67:256–265
23. Bindewald E, Hayes R, Yingling YG, Kasprzak W, Shapiro BA (2008) RNAJunction: a database of RNA junctions and kissing loops for three-dimensional structural analysis and nanodesign. Nucleic Acids Res 36:D392–D397
24. Schroeder KT, McPhee SA, Ouellet J, Lilley DM (2010) A structural database for k-turn motifs in RNA. RNA 16:1463–1468
25. Tamura M, Hendrix DK, Klosterman PS, Schimmelman NRB, Brenner SE, Holbrook SR (2004) SCOR: structural classification of RNA, version 2.0. Nucleic Acids Res 32: D182–D184
26. Popenda M, Szachniuk M, Blazewicz M, Wasik S, Burke EK, Blazewicz J, Adamiak RW (2010) RNA FRABASE 2.0: an advanced web-accessible database with the capacity to search the three-dimensional fragments within RNA structures. BMC Bioinformatics 11:1–12

27. Petrov AI, Zirbel CL, Leontis NB (2013) Automated classification of RNA 3D motifs and the RNA 3D motif atlas. RNA. doi:10.1261/rna.039438.113
28. Vanegas PL, Hudson GA, Davis AR, Kelly SC, Kirkpatrick CC, Znosko BM (2011) RNA CoSSMos: characterization of secondary structure motifs—a searchable database of secondary structure motifs in RNA three-dimensional structures. Nucleic Acids Res 40:D439. doi:10.1093/nar/gkr943
29. Andronescu M, Bereg V, Hoos HH, Condon A (2008) RNA STRAND: the RNA secondary structure and statistical analysis database. BMC Bioinformatics 9:1–10
30. Parlea L, Bindewald E, Sharan R, Bartlett N, Moriarty D, Oliver J, Afonin KA, Shapiro BA (2016) Ring catalog: a resource for designing self-assembling RNA nanostructures. Methods 103:128. doi:10.1016/j.ymeth.2016.04.016
31. Lilley DM (2000) Structures of helical junctions in nucleic acids. Q Rev Biophys 33:109–159
32. Afgan E, Baker D, den BM v, Blankenberg D, Bouvier D, Čech M, Chilton J, Clements D, Coraor N, Eberhard C et al (2016) The Galaxy platform for accessible, reproducible and collaborative biomedical analyses: 2016 update. Nucleic Acids Res 44:W3
33. Blankenberg D, Von Kuster G, Bouvier E, Baker D, Afgan E, Stoler N, Taylor J, Nekrutenko A (2014) Dissemination of scientific software with Galaxy ToolShed. Genome Biol 15:1
34. Bindewald E, Afonin K, Jaeger L, Shapiro BA (2011) Multistrand RNA secondary structure prediction and nanostructure design including pseudoknots. ACS Nano 5:9542–9551
35. Ponder JW, Richards FM (1987) An efficient newton-like method for molecular mechanics energy minimization of large molecules. J Comput Chem 8:1016–1024

Chapter 3

Protocols for Molecular Dynamics Simulations of RNA Nanostructures

Taejin Kim, Wojciech K. Kasprzak, and Bruce A. Shapiro

Abstract

Molecular dynamics (MD) simulations have been used as one of the main research tools to study a wide range of biological systems and bridge the gap between X-ray crystallography or NMR structures and biological mechanism. In the field of RNA nanostructures, MD simulations have been used to fix steric clashes in computationally designed RNA nanostructures, characterize the dynamics, and investigate the interaction between RNA and other biomolecules such as delivery agents and membranes.

In this chapter we present examples of computational protocols for molecular dynamics simulations in explicit and implicit solvent using the Amber Molecular Dynamics Package. We also show examples of post-simulation analysis steps and briefly mention selected tools beyond the Amber package. Limitations of the methods, tools, and protocols are also discussed. Most of the examples are illustrated for a small RNA duplex (helix), but the protocols are applicable to any nucleic acid structure, subject only to the computational speed and memory limitations of the hardware available to the user.

Key words RNA, Molecular dynamics simulations, Force field, Antechamber, MM-PB(GB)SA, Generalized born, Elastic network model

1 Introduction

Various RNA nanostructures have been designed experimentally and computationally [1–3] and some of them have been studied to develop therapeutic applications [3–6]. RNA nanostructures can be designed using specialized computational tools such as RNA2D3D and NanoTiler [7, 8]. Once RNA nanostructures are computationally designed, MD simulations can be used for further structural refinement and analysis. Energy minimization can be used to fix the steric clashes in the designed RNA nanostructure models. MD simulations have been used to investigate the importance of structure flexibility in the process of nanostructure self-assembly [9], the stability and dynamics of designed RNA nanostructures [10, 11], as well as interactions between RNA nanostructures and other biomolecules, such as drug delivery agents [12, 13].

Eckart Bindewald and Bruce A. Shapiro (eds.), *RNA Nanostructures: Methods and Protocols*, Methods in Molecular Biology, vol. 1632, DOI 10.1007/978-1-4939-7138-1_3,

Several molecular dynamics packages such as Amber, NAMD, GROMACS, and CHARMM have been widely used for the study of nucleic acids, proteins, membrane, peptides, and biopolymers (*see* **Note 1**) [14–19]. In this chapter, we introduce the MD simulation protocols for the Amber molecular dynamics package. However, the fundamental concepts of MD simulation protocols and computational approaches discussed in this chapter can be easily applied to other molecular dynamics packages with proper modifications.

MD simulations are performed using the atomic potential energy functions:

$$V_{\text{Amber}} = \sum_{\text{bonds}} k_r (r - r_{\text{eq}})^2 + \sum_{\text{angles}} k_\theta (\theta - \theta_{\text{eq}})^2 + \sum_{\text{dihedrals}} \frac{V_n}{2}[1 + \cos(n\phi - \gamma)] + \sum_{i<j} \left[\frac{A_{ij}}{R_{ij}^{12}} - \frac{B_{ij}}{R_{ij}^{6}}\right] + \sum_{i<j} \frac{q_i q_j}{\varepsilon R_{ij}}$$

The first three terms are interactions between two atoms (bonded interaction), three atoms (angular interaction), and four atoms (dihedral interaction), respectively. The last two terms are nonbonded electrostatic and van der Waals interactions, respectively. The modeling of a biological system is determined by the above atomic potential energy with specific force files (FF) and topological information. FF is a group of numerical parameter sets for the potential energy function. The topology file contains atomic connectivity, atom names, atom types, and atomic charges. Each molecular dynamics package has slightly different potential energy functions, FF, and topology format. For more information on the limitations of Force Fields, *see* **Note 2**.

To keep our examples simple, we use a relatively small 25 base pair-long small interfering RNA (siRNA) with two base overhang at the 3′ end (Subheading 3.1 and Fig. 1a). An siRNA regulates the expression of specific genes so that it can be used for cancer therapy. Based on this unique property, the therapeutic potential of the hexagonal nanostructure which is assembled with multiple siRNAs is proposed based on biophysical and biochemical studies [4–6]. The general protocol for MD simulations of siRNA, which is introduced below can be applied to not only RNA nanostructures but also natural RNAs, including hairpin and pseudoknot conformations.

When using RNA nanostructures for therapeutic applications, delivery agents are often required to protect the RNA nanostructures, to overcome the electrostatic repulsion between the cell membrane and the RNA backbone, and to protect RNA nanostructures against degradation (nuclease digestion) in the blood stream. We have demonstrated that bolaamphiphiles can be used for that purpose, utilizing computational and experimental methods to characterize them and their interactions with RNA [12, 13].

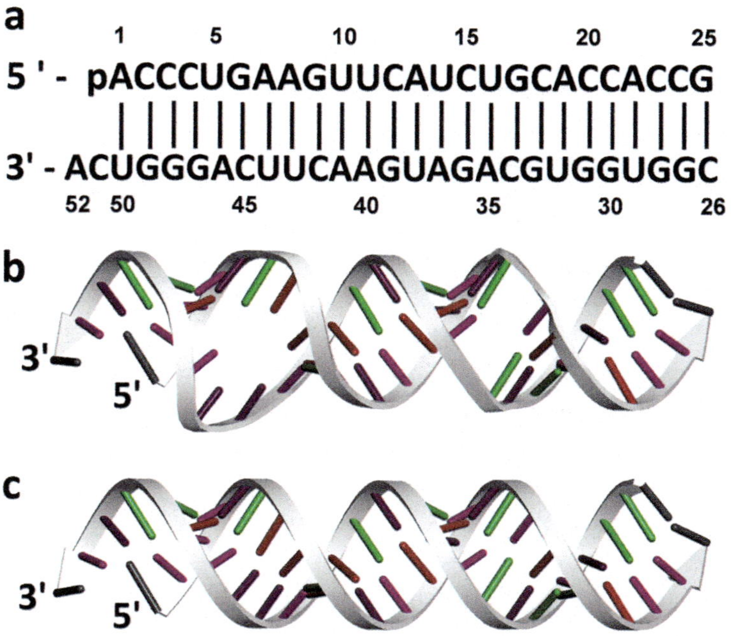

Fig. 1 (**a**) The sequence of an siRNA and (**b**) the initial structure of the siRNA. To illustrate the effect of energy minimization, two parts of the siRNA helix are manually distorted using DSV. (**c**) The siRNA structure after energy minimization. The energy minimization reduces the amount of structural distortions that are apparent in (**b**)

Bolaamphiphiles are cationic lipid-like molecules which have two to four positively charged head groups that are connected by a hydrophobic chain. However, the proper FF and topological information for artificially synthesized small molecules such as bolaamphiphiles are often not available. Therefore, we present a protocol for preparing a force field and topology file for this type of a small molecule.

This chapter gives an introductory outline of a molecular dynamics simulation and analysis in Amber. It starts with the preparation of the necessary input files (topology and coordinates) and preliminary structure refinement via the energy minimization protocol outlined in Subheading 3.1, **steps 1–5**. It continues in Subheading 3.1, **steps 6** and **7** with an outline of the steps needed to create a simulation system with the solute (our RNA nanoparticle) immersed in explicit solvent (water and ions) and presents a protocol used to equilibrate this system. The parameters and commands needed to run a "production" MD simulation, i.e., a dynamics simulation of an equilibrated system are outlined in Subheading 3.1, **step 8**. Subheading 3.1, **steps 9–11** introduce sample post-MD visualization and analysis steps. Subheading 3.2 focuses on calculating parameters necessary for an MD simulation of a small molecule containing atom groups that are not included in the standard Force Field parameters distributed with Amber. This involves applying antechamber, a

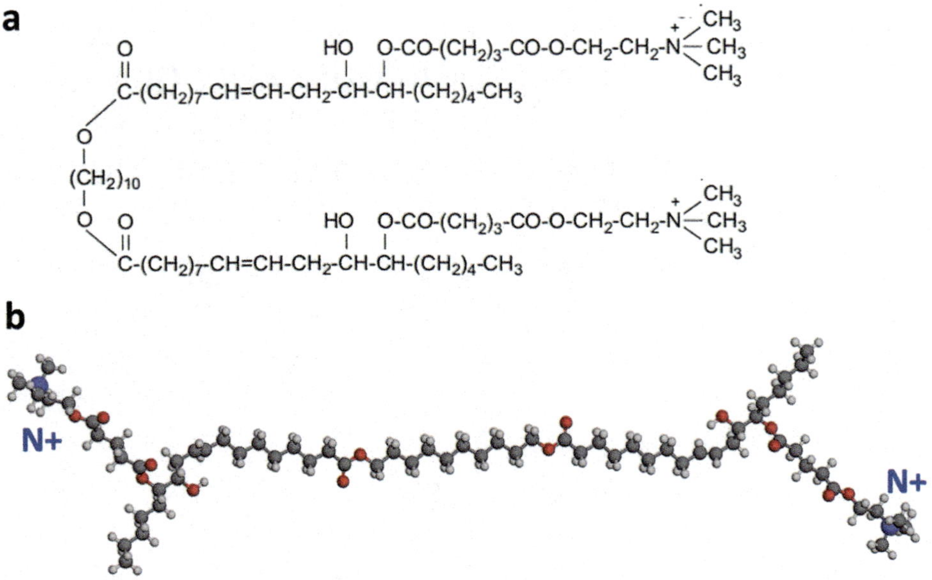

Fig. 2 (**a**) The molecular structure of GLH19 and (**b**) 3D model of GLH19. Positively charged head groups (N+) are indicated with *blue spheres*

program that is part of the Amber package, to the bolamphiphile GLH19 (Subheading 3.2 and Fig. 2a). We also describe the Molecular Mechanics–Poisson-Boltzmann/Generalized Born Surface Area (MM-PB(GB)SA) method for calculating the binding free energy between a negatively charged siRNA and a positively charged GLH19 micelle (Subheading 3.3). Subheading 3.4 presents an alternative approach to MD using the implicit solvent Generalized Born (GB) model in which RNA is still represented in explicit atomic detail, while solvent is modeled as a dielectric continuum, and the solvation free energy is approximated as a sum of polar and nonpolar terms. Current limitations of molecular dynamics simulations are discussed in **Note 3**. Finally, in **Note 4**, an Anisotropic Network Model, a form of an Elastic Network is discussed as a potential coarse-grained alternative to molecular dynamics, depending on the questions one is trying to answer.

2 Materials

In this section, we briefly introduce several software packages such as the Amber molecular dynamics package, Gaussian, Discovery Studio Visualizer (DSV), and Visual Molecular Dynamics (VMD). The detailed information for the installation of these programs is beyond the scope of this chapter and can be found from each corresponding websites. Typically, most molecular dynamics and computational packages including Amber and Gaussian are installed on a high performance computing systems of the institute or individual research laboratory, while DSV and VMD can be installed on a desktop computer running Windows, Linux, or OS X.

2.1 Amber

Amber is one of the most commonly used molecular dynamics packages for biomolecular research [14]. Amber provides not only an MD simulation platform with its own FF and topological library, but also a variety of computational and analytical tools. We introduce essential components of the Amber package for energy minimization and MD simulations (Subheading 3.1), basic analytical tools (ptraj/cpptraj command, Subheading 3.1), antechamber (Subheading 3.2), and MM-PB(GB)SA methods (Subheading 3.3). More information and many tutorials can be found on the Amber website (http://ambermd.org/).

2.2 Gaussian

Gaussian (http://www.gaussian.com/index.htm) is designed for electronic structure modeling. We use Gaussian for partial charge computations for a small molecule, the bolaamphiphile GLH19 (Subheading 3.2).

2.3 Discovery Studio Visualizer (DSV)

DSV can be downloaded from http://accelrys.com/ and installed on Windows and Linux systems. DSV can be used to build, edit, relocate, and visualize molecules. A simple molecule such as GLH19 can be easily built using the DSV menu item *Tools → Small Molecules → Sketch Molecules*. The geometry of generated molecules can be optimized by *Structure → Clean Geometry*. The final structure can be saved with various formats including PDB format. In addition, canonical nucleic acid structures can be easily built using *Tools → Macromolecules → Build and Edit Nucleic Acid*. The siRNA which is used in this chapter is built by DSV. In addition, the positions of a single atom, group of atoms, or the entire molecule can be easily changed by drag and drop. More information can be found in the DSV website and user manual.

2.4 VMD

VMD can be downloaded from http://www.ks.uiuc.edu/Research/vmd/ and installed on Windows, Linux, and OS X. VMD can load various atomic coordinate formats including PDB as well as the MD simulation trajectories of Amber, GROMACS and CHARMM with different coloring and drawing methods. In addition to visualization, users can analyze the trajectory by measuring, for example, distance, angle, and dihedral angle of selected atoms, root mean square deviation (RMSD) between two structures, and salt bridges. VMD can be also used to prepare the initial structure and topology files for MD simulations (Subheadings 3.1, **step 9** and *see* **Note 1**).

2.5 Initial Structures

The natural biomolecular structures captured by X-ray crystallography and NMR spectroscopy are commonly used as the initial structures for MD simulations. Since these structures do not have all the solvent (water) and ions (but may include some important ion placement information, for example), users need to solvate them and add ions using the leap (tleap or xleap variants) module in Amber or VMD, as described below.

Unique RNA nanostructures can be designed with the aid of RNA2D3D and NanoTiler [7, 8] that may include fragments of natural biomolecules. The structural refinement and dynamics simulation protocols described below are equally applicable to natural and designed RNA nanostructures.

3 Methods

3.1 Protocols for Energy Minimization and MD Simulations of RNA

Computationally designed or manually assembled RNA nanostructures can have steric clashes. Since these clashes can overlap the atomic van der Waals radii or change atomic geometry, they can cause potential energy errors and prevent proper MD simulation performance. Energy minimization can fix the steric clashes and optimize the atomic geometry according to the potential energy function. To illustrate the effect of energy minimization, the two parts of an siRNA helix are manually distorted using DSV (Fig. 1b). The energy minimization fixes the distorted backbone and recovers the proper helix as shown in Fig. 1c. The detailed protocol for the energy minimization is presented below:

1. Using DSV, build siRNA (*Tools → Macromolecules → Build and Edit Nucleic Acid*). Users can easily build an RNA duplex by specifying its sequence in *Enter Nucleotide Sequence* window. Check the structure and save siRNA in PDB format.
2. Open the PDB file with a text editor and check atom and residue names. The conventions used for nucleic acid residue and atom names frequently differ between different molecular dynamics packages. Change atom and residue names as needed depending on which version of Amber you are using (refer to http://www.wwpdb.org/documentation/file-format for details).
3. LEaP is one of the main Amber programs for building a new biomolecular system or for modifying an existing system. The plain text version of LEaP can be invoked using the tleap command, while a window interface version will open with the xleap command. Open xleap with the following command:

   ```
   xleap -s -f learc.ff10
   ```

   The *-s* flag indicates that LEaP should ignore any leaprc file it might find, and the *-f* flag indicates the opening force field file, leaprc.ff10. The ff10 parameter set contains the force field parameters for protein and nucleic acids, including ff99bsc0 and chi.OL3 updates for nucleic acids [20]. Please refer to the Amber manual for naming conventions in the particular Amber version. For example, in Amber 14, leaprc.ff14SB loads the same nucleic acid parameters (as well as protein, water, and ion parameters), while in the latest Amber 16 multiple calls to the source leaprc.X files relevant to the system you are simulating are necessary. In other words, instead of the above xleap call, one may want to invoke xleap or tleap and then issue a series of calls:

```
source leaprc.ffSB14 - in Amber 14
```

or two calls for RNA in solvent (water TIP3P) simulation in Amber 16

```
source leaprc.RNA.OL3
source leaprc.water.tip3p
```

In the xleap window interface, type the following commands:

```
mol=loadpdb siRNA.pdb
saveamberparm mol siRNA.prmtop siRNA.inpcrd
quit
```

xleap will generate the initial coordinates for the siRNA (siRNA.inpcrd) and store the topological information and FF parameters in the topology file, siRNA.prmtop. File names, of course, are the user's choice. Detailed information containing the transcript of the leap session with information about the FF, interpretation of the input PDB file and generation of initial coordinates and topology can be found in the leap.log file. Typically, severe errors resulting in no output files being written out can be tracked via the warnings and error messages written out to the leap.log file.

4. Energy minimization can be performed with the following Amber command:

```
sander -O -i min.in -o min.out -p siRNA.prmtop -c siRNA.
inpcrd -r siRNA-Emin.rst
```

Sander (Simulated Annealing with NMR-Derived Energy Restraints), one of the main Amber modules, carries out energy minimization, MD simulations, and NMR refinements. The program utilizes an input file with name min.in and an output file with name min.out. siRNA-Emin.rst is the result coordinate file after energy minimization. The parameters for the input file, min.in are listed in Table 1.

5. Once energy minimization is completed, make sure the resulting structure does not contain steric clashes or structural entanglements. Keep in mind that it is possible to run a minimization

Table 1
The list of input files for energy minimization, heating, and MD simulations

min.in (Subheading 3.1, step 4)	**minN.in (*N* = 1–11, Subheading 3.1, step 7)**
minimization &cntrl imin=1, ntx=1, drms=0.01, irest=0, ntxo=1, cut=15.0, ntpr=100, ntwx=100, ntwe=100, nsnb=20, maxcyc=30000, ncyc=1000, igb=1,saltcon=1.0, ntt=0,offset=0.13, gbsa=1, ntb=0, &end	minimization of water/ions holding RNA &cntrl imin=1, ntx=1, drms=0.01, ntmin=1, dielc=1.0, ntrx=1, ntxo=1, nscm=100, ntwprt=0, ntb=2, ntp=2, irest=0, cut=9.0, ntpr=100, ntwx=100, ntwe=100, nsnb=10, maxcyc=10000, ncyc=10000, ntt=0, ntr=1, scee=1.2, vlimit=20.0, &end Hold RNA fixed K* RES 1 52 END END *K=200, 25, 20, 15, 10, 8, 6, 4, 3, 2, 10

heat.in (Subheading 3.1, step 7)	**mdpmeN.in (*N* = 1 and 2, Subheading 3.1, step 7)**
heat/dynamics holding RNA fixed &cntrl imin=0, ntxo=1, ntrx=1, cut=9.0, tempi=100.0, ntwprt=1665, ntpr=500, ntwx=500, ntwe=500, nstlim=20000, temp0=300.0, dt=0.002, nscm=100,nsnb=10, ntc=2, ntf=2, tol=0.00001, ntx=1, irest=0, ntb=2, ntp=1, pres0=1.0, taup=0.2, scee=1.2, ntt=1, ntr=1, nmropt=1, &end &wt type='TEMP0', istep1=0, istep2=1000, value1=100.0, value2=300.0, &end &wt type='TEMP0', istep1=1000, istep2=20000, value1=300.0, value2=300.0, &end &wt type='END', &end GROUP FOR CONSTRAINTS 200.0 RES 1 52 END END	dynamics holding RNA fixed &cntrl imin=0, ntxo=1, ntrx=1, cut=9.0, tempi=100.0, ntwprt=1665, ntpr=100, ntwx=500, ntwe=500, nstlim=10000, temp0=300.0, nscm=0,nsnb=10, ntc=2, ntf=2, tol=0.00001, tautp=0.2, t=0.0, vlimit=20.0, comp=44.6, ntx=1, irest=0, dielc=1.0, ibelly=0, ntb=2, ntp=1, pres0=1.0, taup=0.2, scee=1.2, dt=0.002, ntt=1, ntr=1, &end GROUP FOR CONSTRAINTS K* RES 1 52 END END K*=200, 250

(continued)

Table 1
(continued)

heat2.in (Subheading 3.1, step 7)

```
heat/dynamics solvent and RNA
&cntrl
imin=0,
ntxo=1, ntrx=1,
cut=9.0, tempi=100.0, ntwprt=1665,
ntpr=500, ntwx=500, ntwe=500,
nstlim=20000, temp0=300.0,
dt=0.002, nscm=100,nsnb=10,
ntc=2, ntf=2, tol=0.00001,
ntx=1, irest=0, heat=0.0,
ntb=2, ntp=1, pres0=1.0, taup=0.2,
scee=1.2, dt=0.002,
ntt=1, vlimit=20.0,comp=44.6,
ig=71277, nmropt=1, t=0.0,
&end
&wt
type='TEMP0', istep1=0, istep2=2000,
value1=100.0, value2=300.0,
&end
&wt
type='TEMP0', istep1=2000, istep2=20000,
value1=300.0, value2=300.0,
&end
&wt
type='END',
&end
&rst
iat=0,
&end
```

min12.in (Subheading 3.1, step 7)

```
minimization of water/ions and RNA
&cntrl
imin=1, ntx=1, drms=0.0001, ntmin=1,
ntrx=1, ntxo=1, nscm=100,
ntb=1, ntf=1,
irest=0, cut=9.0,
ntpr=100, ntwx=100, ntwe=100,
nsnb=10, tol=0.0005,ntc=1,
maxcyc=1000, ncyc=50000,
dx0=0.1, dxm=0.5,
&end
```

md.in (Subheading 3.1, step 8)

```
MD-RUN
&cntrl
imin=0,
ntxo=1, ntrx=1,
cut=9.0, tempi=300.0, ntwprt=1665,
ntpr=500, ntwx=500, ntwe=500,
ntwr=500, nstlim=5000000,
temp0=300.0, dt=0.002,
nscm=100,nsnb=10,dielc=1.0,
ntc=2, ntf=2, tol=0.00001,
ntx=5, irest=1, ntb=2, ntp=1,
pres0=1.0, taup=0.5,
scee=1.2, dt=0.002,
ntt=1, vlimit=20.0,comp=44.6,
&end0
```

For minN.in file, *N* runs from 1 to 11 and for each *N*, *K* is defined as 200, 25, 20, 15, 10, 8, 6, 4, 3, 2, and 1, respectively. For mdpmeN.in file, *N* runs 1–2 and for each *N*, *K* is defined as 200 and 25, respectively. The detailed explanations for input parameters can be found in the recent Amber manual

to a successful completion while still leaving intersecting aromatic ring structures in the resulting file. At least a visual inspection is recommended before proceeding further. Users can easily convert the file siRNA-Emin.rst (Amber coordinate file) to PDB format with the following Unix command line:

```
ambpdb -p siRNA.prmtop < siRNA-Emin.rst > siRNA-Emin.pdb
```

The resulting file siRNA-Emin.pdb can be loaded into DSV or VMD for further verification and processing. If the structure still contains steric clashes, users may consider correcting them manually using DSV and then run another energy minimization.

6. Once minimization is completed, it is necessary to solvate the siRNA in water with physiological salt concentration (0.15 M

NaCl) to run production MD simulations. xleap can be used for solvation and uniform ion distribution. Open xleap with the same command as before (Subheading 3.1, **step 3**). In the xleap window interface, type following commands:

```
mol=loadpdb siRNA-Emin.pdb
check mol
solvatebox mol TIP3PBOX 15.0 0.832
addions mol Na+ 0
addions mol Na+ NumOfIons
addions mol Cl- NumOfIons
saveamberparm mol siRNA-sol.prmtop siRNA-sol.inpcrd
savepdb mol siRNA-sol.pdb
```

The *solvatebox* command solvates the siRNA with the most commonly used water model TIP3P that is also most appropriate for RNA (a simple, planar, and rigid water model described by Coulombic and Lennard–Jones terms). In the above example the minimum distance between solute and the edge of the water box is 15.0 Å, and the minimum distance between solute and nearest water molecule is 0.832 Å. The distance between the siRNA surface and water box planes must be greater than the selected nonbonded interaction cutoff (usually 9–12 Å; 9 Å is used in the examples in Table 1—see parameter *cut*), otherwise the siRNA in the primary water box will interact with counterparts via periodic boundary conditions. The water box volume (*Vol*) information is displayed in the dialog window after solvation is completed. When *NumOfIons* is set to 0, the *addions* command neutralizes the system by placing counterions (Na+) near negatively charged phosphate groups in siRNA. Physiological salt concentration (0.15 M) is achieved by adding extra Na+ and Cl− ions. The required number of extra ions can be calculated based on the water box volume information provided by LEaP and the desired salt concentration, using the following equation:

$$\text{Num of Ions} = \text{Vol} \times 0.15 \times 6.022 \times 10^{-4}$$

Once the *addions* command completes its work, the topology and initial coordinates of the solvated system can be saved with the *saveamberparm* command. The resulting system can be also saved in PDB format with the *savepdb* command.

7. The initial atom coordinate configurations of TIP3P water and ions are not properly equilibrated at this stage. To equilibrate water and ions, multiple (we show 16 steps (*see* Table 2), but this is open to tuning by the user, based on literature searches) relatively short stages of energy minimization, heating to the target temperature, and two steps of short MD simulations are

Table 2
Energy minimization, heat up, and MD simulations for equilibration

Equilibration procedure (Subheading 3.1, step 7)
SANDER -O -i min1.in -o min1.out -p siRNA-sol.prmtop -c siRNA-sol.inpcrd -ref siRNA-sol.inpcrd -r min1.rst
SANDER -O -i heat.in -o heat.out -p siRNA-sol.prmtop -c min1.rst -ref min1.rst -r heat.rst -x heat.x
SANDER -O -i mdpme.in -o mdpme.out -p siRNA-sol.prmtop -c heat.rst -ref heat.rst -r mdpme.rst -x mdpme.x
SANDER -O -i min2.in -o min2.out -p siRNA-sol.prmtop -c mdpme.rst -ref mdpme.rst -r min2.rst -x min2.x
SANDER -O -i mdpme2.in -o mdpme2.out -p siRNA-sol.prmtop -c min2.rst -ref min2.rst -r mdpme2.rst -x mdpme2.x
SANDER -O -i min3.in -o min3.out -p siRNA-sol.prmtop -c mdpme2.rst -ref mdpme2.rst -r min3.rst -x min3.x
SANDER -O -i minN.in -o minN.out -p siRNA-sol.prmtop -c minN-1.rst -ref minN-1.rst -r minN.rst -x minN.x
(N runs 4 to 11 and N-1 runs 3 to 10)
SANDER -O -i min12.in -o min12.out -p siRNA-sol.prmtop -c min11.rst -r min12.rst -x min12.x
SANDER -O -i heat2.in -o heat2.out -p siRNA-sol.prmtop -c min12.rst -r heat2.rst -x heat2.x0

performed, while the siRNA is held by gradually decreasing constraints. After the constraint on the siRNA is completely released, the entire system is minimized before the final heating phase. In this example, the explicit solvent calculations employ the Particle Mesh Ewad algorithm (PME) and utilize the Berendsen thermostat to maintain the temperature and the Berendsen algorithm to maintain the pressure [21–23]. The detailed procedures, commands and input files for the equilibration are listed in Tables 1 and 2.

8. Once the system (siRNA, ion, and water) is fully equilibrated, production MD simulations can be executed with following command.

```
SANDER -O -i md.in -o md_1.out -p siRNA-sol.prmtop -c
heat2.rst -r siRNA-md1.rst -x siRNA-md1.x
```

The parameters for *md.in* are listed in Table 1. MD simulation will produce 10 ns long trajectory (5,000,000 (number of steps, *nstlim*) × 0.002 ps (time step, *dt*)). Additional 10 ns long MD simulation can be performed using siRNA-md1.rst as the file containing the restart coordinates. Note that the parameter *ntwprt* in the equilibration and production MD input files determines the number of atoms written out to the trajectory files, and it can be used to output only the solute (RNA) atoms, RNA and ions, or all the atoms from the solvent box (when not specified or explicitly set to its default value of 0). In the examples shown here (Table 1), only the 1665 RNA atoms are output. This is relevant to the trajectory visualization and analysis steps (paragraphs 9–11), which must be performed with a topology file (.prmtop) corresponding to the output in the trajectory. Thus we can use the RNA-only topology file *siRNA.prmtop* from the leap session preceding the minimization or create a custom topology file in a separate leap session for any customized output format, for example, RNA and ions, but not waters.

9. The MD simulation trajectory can be visualized in VMD. In the VMD menu, load the topology and trajectory files: *File → New Molecule → Browse (select topology file, siRNA.prmtop) → Determine file type (select AMBER7 Parm) → Load → Browse (select trajectory file, siRNA-md1.x) → Determine file type (select AMBER Coordinate) → Load*. VMD provides several convenient hot keys for visualization and simple geometry measurements. Hot keys, 'r', 't', and 's', allow one to rotate, translate, and scale the structure, respectively. Numerical hot keys, '1', '2', '3', and '4' will toggle on/off labels of selected atoms, measure a distance between two selected atoms, measure an angle formed by three selected atoms, and display a dihedral angle formed by four selected atoms, respectively. Detailed information of measurements can be found in menu, *Graphics → Label → Atoms, Bonds, Angles,* and *Dihedrals.*

10. The Amber package also provides various tools for data analysis. Below are examples of the cpptraj commands for the measurements of distance, angle, dihedral angle, and RMSD.

```
cpptraj siRNA.prmtop <Analysis.in> Analysis.out

An example input file with the name Analysis.in is shown
below:

trajin siRNA-md1.x
rms first out RMSD.dat :1-52
distance dist :2@P :2@O5' out Dist.dat
angle ang :2@P :2@O5' :2@C5' out Angle.dat
dihedral dihe :2@P :2@O5' :2@C5' :2@C4' out Dihedral.dat
average siRNA-ave.pdb :1-52 pdb
```

The *rms* command will calculate the RMSD of the siRNA with respect to the first frame of the trajectory. The numerical index 1-52 indicates all 52 residues of the sample siRNA. Geometry measurements are performed by the last three commands shown above. For example, the command, *distance* measures the distance between the P and O5′ atoms in the second residue of the siRNA. The results will be stored in Dist.dat. An average structure over all the coordinate sets in the trajectory is calculated in the last step and output into a PDB format file.

11. Principal component (PC) analysis information that can be extracted from an MD trajectory is very useful at assessing large scale (collective), low frequency (first few components) motions of the molecule subjected to molecular dynamics simulations. Given our Cartesian space MD trajectories, PCs capture variance in the coordinate space. The first PC captures the largest variance in the input data, the second PC captures the second largest variance, etc. Motions along individual PCs can be visualized using VMD or Chimera, but one has to keep in mind that the actual motions of the simulated systems are a combination of the PCs. In other words, a transition from one conformation (state) to another that may appear to be a smooth path when visualized for one PC is in reality more complex. An example protocol is shown below for the sample molecule from the previous paragraphs, although one should keep in mind that more topologically complicated structures, such as multiway junctions could provide more interesting information about the largest, low frequency motions. The first step in the process is generation of the coordinates covariance matrix from the MD trajectory (which is a time series of positional coordinates). To focus only on the internal dynamics of the system, we filter out the full system translations and rotations by using an average structure for the entire trajectory, which in our example is assumed to have been precomputed (as shown in paragraph 10) and optionally minimized, as a reference structure in our input. Note that we are dealing with an RNA-only trajectory here, as discussed in paragraph 10, written out from an explicit solvent MD simulation.

   A call to cpptraj *siRNA.prmtop* <*PC_analysis.in*> *PC_analysis.out*, with the input file containing the commands shown below will calculate the first five PCs, using only the P backbone atoms:

```
trajin siRNA-md1.x
reference siRNA-ave.pdb
center :1-52
image familiar
# Realign the trajectory frames to the average structure to
remove rotations and translations
rms reference out siRNA-FULL-RMS-vs-MDave.dat :1-52@P


# Using the side-effects of RMS re-alignment of all the MD
frames
```

```
# compute the coordinate covariance matrix only for the
backbone P atoms
matrix covar name covmatrix :1-52@P

# Diagonalize the above matrix, followed by Principal
Component analysis. Write out:
# average coordinates, eigenvalues (PC) and eigenvectors,
i.e. eigenmode data (top 5 here)
analyze matrix covmatrix out siRNA-eigenvectors-P.dat vecs
5

# Project snapshots onto modes (to be used by visualizers).
Use the same atom mask as in the matrix !
projection modes siRNA-eigenvectors-P.dat out siRNA-PC-
proj5.ppj beg 1 end 5 :1-52@P
```

To create to a pseudo-trajectory file that can be used to visualize PCs in VMD requires modifying the topology file to match the atom mask used in the PC calculations shown above and creating a pseudo-trajectory for each PC from the siRNA-eigenvectors-P.dat file. The full procedure is presented in Amber cpptraj Tutorial C3 available from the Amber website (https://www.ambermd.org).

3.2 Protocols for the Preparation of Force Field and Topology for a Small Molecule

In this section, we introduce Antechamber, one of the programs of the Amber package, to show how to assign proper FF and topology for a small molecule such as GLH19. Antechamber can determine proper FF parameters, atom type, bond type, and atom connectivity based on the geometry of a molecule. Antechamber is also able to calculate the atomic charges using several different charge methods. In this section, we introduce high level *ab initio* calculations using the Gaussian package to determine precise atomic charges. The detailed procedure for applying antechamber is shown below.

1. The initial structure of GLH19 can be easily built by DSV: *Tools* → *Small Molecules* → *Sketch Molecules*. "*Sketch Molecules*" creates only carbon atoms. However, carbon atoms can be changed to different atom types by *Chemistry* → *Element* → *Select desired atom*. Hydrogen atoms can be added by Tools tap under '*Editing*' menu. The built GLH19 structure can be saved in PDB format. Figure 2b shows the generated 3D model of GLH19. A small molecule model can also be built with the aid of other programs such as Corina (https://www.mn-am.com/online_demos/corina_demo_interactive). However, the energy minimization is still required to clean the geometry of the built structure.
2. Execute the following command to generate ".ac" file containing the chemical formula, atomic coordinates, atom type, and connectivity:

```
antechamber -fi pdb -fo ac -i GLH19.pdb -o GLH19.ac -rn G19
-at gaff
```

–fi is the input file format, *-fo* is the output file format, *-i* is the input molecule file, *-o* is the output file, *-rn* is the residue name, and *–at* is the atom type. The residue name should be three characters and unique. For GLH19, we use the general Amber force field (GAFF), which is designed for small organic molecules. The generated file with the connectivity information has in this example the name GLH19.ac.

3. Execute the following command to generate a "prep" file containing the molecular configuration, atom type, coordinates, and atomic charges:

```
prepgen -i GLH19.ac -o GLH19.prep -f prep -rn G19 -at gaff
```

prepgen command generates a "prep" file, GLH19.prep, based on the input file GLH19.ac. The atomic charges are set to zero, because no charge computation method has been applied yet.

4. Atomic charges can be obtained using *ab initio* calculations as performed by the program Gaussian. The following antechamber command converts the initial PDB file, GLH19.pdb to an input file for the program Gaussian with the name G19.gau:

```
antechamber -fi pdb -fo gcrt -i GLH19.pdb -o GLH19.gau -m 1
-nc 2
```

Here, *-fi* is the input file format, *-fo* is the output file format, -*gcrt* is the Gaussian input file format, *-i* is the input PDB file, *-o* is the output file, which will be used for Gaussian computation. *-m* is the spin multiplicity ($m = 2S + 1$, where S is the total spin of molecule), and *-nc* is the net charge of the molecule. Since GLH19 has no unpaired electrons, the total spin is zero so that the spin multiplicity is 1. The total charge of GLH19 is +2 due to two positively charged head groups.

5. Run Gaussian with the following command:

```
g03 < GLH19.gau > GLH19.out
```

After the computation performed by the program Gaussian is completed, run the following antechamber command:

```
antechamber -fi gout -fo ac -i GLH19.out -o GLH19-charge.ac
-c resp
```

This command will compute atomic charges using the Restrained Electrostatic Potential (RESP) method based on the result of the quantum chemistry computation provided in the file GLH19.out. In the GLH19-charge.ac file, the atomic charges are listed in the ninth column. Copy the atomic charges from the generated ".ac" file (in this example GLH19-charge.

ac) and paste them into the 11th column of the prep file (in this example called GLH19.prep).

6. Execute the parmchk command:

```
parmchk -f prepi -i GLH19.prep -o GLH19.frcmod -rn G19 -p /
your_directory/gaff.dat
```

The *parmchk* command reads the prep file and the gaff file (gaff.dat) to write out an frcmod file for the missing parameters. If there is any missing parameter, it will be listed in the frcmod file and users may need to provide the proper one.

7. Now, all necessary components, GLH19.pdb, GLH19.prep, and GLH19.frcmod, are available for a molecular dynamics simulation. To generate topology (prmtop) and coordinate (inpcrd) files, start the program xleap and execute the following commands within the xleap terminal:

```
source leaprc.gaff
loadamberprep GLH19.prep
loadamberparams GLH19.frcmod
mol=loadpdb GLH19.pdb
saveamberparm GLH19.prmtop GLH19.inpcrd
```

8. Energy minimization described in the Subheading 3.1, **step 4** will optimize the geometry of GLH19. The solvation and MD simulation can be performed using the protocol described in Subheading 3.1, **steps 6–8**.

3.3 Protocols for MM-PB(GB)SA Method

In this section, we introduce how to compute the binding free energy between the siRNA and the GLH19 micelle. During MD simulations, the siRNA and GLH19 micelle which are initially separated by 10 Å bind together to form a complex due to the electrostatic interactions between the negatively charged siRNA backbone and the positively charged hydrophilic head groups in the GLH19 micelle (Fig. 3). The protocols for MD simulations of the GLH19 micelle are described in reference [12].

In general, the binding free energy can be obtained from the total free energy difference between before and after binding. However, it is not easy to calculate the free energy difference since the energy fluctuation due to water is much greater than binding free energy. This difficulty can be overcome using MM-PB(GB)SA methods [24]. In the MM-PBSA method, we use molecular mechanics to calculate the mechanical energy difference ($\Delta G_{\text{Complex, Vacuum}}$) between before and after complex formation. The solvation energy of each system is calculated using the Poisson-Boltzmann (PB) or generalized Born (GB) Surface Area methods ($\Delta G_{\text{Complex,Solv}}$, $\Delta G_{\text{GLH19, Solv}}$, and $\Delta G_{\text{siRNA, Solv}}$). The entropic penalty ($T\Delta S$) due to the binding can be calculated by normal

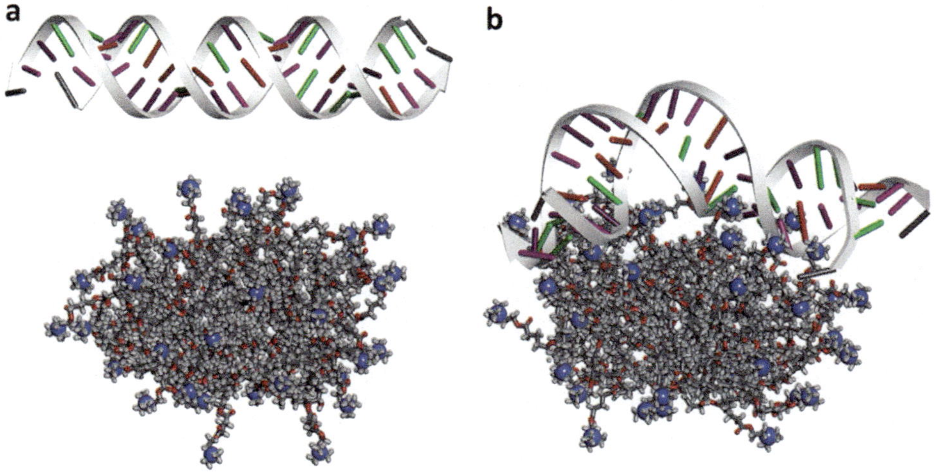

Fig. 3 (**a**) The initial configuration of an siRNA and a GLH19 micelle and (**b**) the complex formation of the siRNA with the GLH19 micelle after 100 ns long MD simulation. Positively charged head groups in the GLH19 micelle are indicated with *blue spheres*

mode analysis. From these computations, the binding free energy (ΔG_{Bind}) can be calculated as below:

$$\Delta G_{\text{Bind}} = \Delta G_{\text{Complex, Vacuum}} + \Delta G_{\text{Complex, Solv}} - \Delta G_{\text{GLH19, Solv}} - \Delta G_{\text{siRNA, Solv}} - T\Delta S$$

1. The initial configuration for the siRNA and the GLH19 micelle can be prepared by DSV. Open both siRNA and GLH19 micelle PDB coordinate files in separate DSV windows. Select and copy the siRNA coordinates and paste it into the GLH19 window. Because it is important to consider the electrostatic interactions between the siRNA and the GLH19 micelle, place the siRNA surface close to that of the GLH19 micelle within the nonbonded cutoff distance of 12 Å. Save the system in PDB format as siRNAG19.pdb (Fig. 3a).
2. Solvate the system using xleap. Since the system contains the siRNA as well as the GLH19 micelle, load ff10 for siRNA and GAFF for GLH19 with the following commands:

```
xleap -f leaprc.ff10
source leaprc.gaff
loadamberprep GLH19.prep
loadamberparams GLH19.frcmod
mol=loadpdb siRNAG19.pdb
```

   Perform subsequent steps of solvation, adding ions, preparation of prmtop and inpcrd files as described in Subheading 3.1, **step 6**. Save the topology and initial coordinate files before solvation as siRNAG19.prmtop and siRNAG19.inpcrd, respectively (they will be used for MM-PBSA computation). Save the solvated system as siRNAG19sol.prmtop and siRNAG19sol. inpcrd. Run an MD simulation for 100 ns with the MD simulation protocol described in Subheading 3.1. The siRNA and GLH19 micelle will bind together (Fig. 3b). The last 20 ns of the 100 ns-long MD trajectory is selected for the MM-PBSA computation described below.

3. To compute the binding free energy, the MM-PBSA method requires topology files and MD trajectories of the siRNA, the G19 micelle, and the siRNA-GLH19 complex. The list of required trajectories and topology files are below.

   Complex: siRNAG19.x (MD trajectory), siRNAG19.prmtop, and siRNAG19-sol.prmtop.
   siRNA: siRNA.x (MD trajectory), siRNA.prmtop, and siRNA-sol.prmtop.
   GLH19: GLH19micelle.x (MD trajectory), GLH19micelle. prmtop, and GLH19micelle-sol.prmtop
   Place all these files in the same folder and execute the following command:

```
MMPBSA.MPI -O -i mmpbsa.in -o MMPBSA_siRNAG19.dat -sp
siRNAG19-sol.prmtop -cp siRNAG19.prmtop -rp G19micelle.
prmtop -lp siRNA.prmtop -srp GLH19micelle-sol.prmtop -slp
siRNA-sol.prmtop -y siRNAG19.x -yr GLH19micelle.x -yl
siRNA.x > progress.log
```

   The option *-i* is used for specifying the input file and *-o* is used for specifying the output file. The options *-sp*, *-srp*, and *–slp* indicate topology files for the solvated systems, siRNAG19, G19micelle, and siRNA, respectively. The flags, *-cp*, *-rp*, and *–lp* indicate the topology files for the unsolvated systems, siRNAG19, G19micelle, and siRNA, respectively. The options *–y*, *-yr*, and *–yl* specify MD trajectories for the complexes, GLH19 micelle, and siRNA, respectively. The parameters of ***mmpbsa.in*** are listed in Table 3.

4. The contribution of entropy on the binding affinity can be estimated by calculating the normal modes of the complex with the following command:

```
MMPBSA.MPI -O -i mmpbsa-nmode.in -o Entropy_siRNAG19.dat -
sp siRNAG19-sol.prmtop
-cp siRNAG19.prmtop -rp G19micelle.prmtop -lp siRNA.prmtop
-srp GLH19micelle-sol.prmtop
-slp siRNA-sol.prmtop -y siRNAG19.x -yr GLH19micelle.x -yl
siRNA.x > progress.log
```

   The parameters of ***mmpbsa-nmode.in*** are listed in Table 3.

Table 3
Input parameters for MM-PBSA and entropy computations

mmpbsa.in (Subheading 3.3, step 3)	mmpbsa-nmode.in (Subheading 3.3, step 4)

&general
  endframe=2000, verbose=1,
# entropy=1, 	
/
&gb
  igb=2, saltcon=0.15
/
&pb
  istrng=0.15,
/
	&general
	  endframe=2000, keep_files=2,
	/
	&nmode
	  nmstartframe=1,
	  nmendframe=2000,
	  nminterval=80, nmode_igb=1,
	  nmode_istrng=0.15,
	/0

The detailed explanations for input parameters can be found in the recent Amber manual

5. The binding free energy can be obtained from the results of MM-PBSA and the entropy calculations with the following equation:

   Binding Free Energy(ΔG_{Bind}) = binding free energy from MM - PBSA – Entropy($T\Delta S$).

3.4 Protocols for Implicit Solvent MD Simulations of Nucleic Acids

In this section we introduce two variants of an implicit solvent molecular dynamics protocol suitable for simulations of nucleic acids. The motivation for considering this option is that simulation systems combining nanostructures such as nanocubes [3, 5, 25] with water and ions in an explicit solvent box, which has to be sufficiently large to be greater than nonbonded interaction cutoff used in the MD simulation (*see* Subheading 3.1), can quickly exceed 250,000 atoms, most of which comprise water molecules. Such large systems tend to slow down simulations even on the fast graphics processing units (GPU). In an implicit solvent simulation, such as the Generalized Born model (GB), solutes, such as nucleic acid-based particles are treated explicitly, while the solvation free energy is approximated as a sum of polar and nonpolar terms. Even though the implicit solvent simulations have gained acceptance for proteins [26], for strongly negative nucleic acids the parametrization of the polar component has been a major obstacle to achieving structure stability in simulations. DNA structural parameters have been shown to be reproducible by the GB-HCT model when tested on a relatively small double-stranded DNA fragment [27–29]. However, problems with maintaining the stability of more topologically complicated structures, such as tRNA have been reported [30]. Strongly paired and stable by design, nanostructures may be more, but not absolutely more resistant to such problems. Recently, a GB-neck2 protocol, initially developed to improve the accuracy of

GB-based protein simulations has been extended to include parameters for nucleic acids and nucleic acid–protein complexes [31]. The parameters have been derived from a large training set of RNA and DNA duplexes, as well as DNA quadruplexes, nucleic acid complexes with proteins and simulations of RNA and DNA hairpin folding. The authors reported major improvements in the accuracy of effective Born radii parameters, leading to more accurate energy calculations and yielding stable nucleic acid structures in simulations up to microsecond time scales. Please refer to **Note 3** for the discussion of our evaluation of the GB protocols applied to RNA and DNA nanocubes. Equilibration parameters that worked well in our tests with nanocubes are illustrated in Tables 4 and 5 and used in a sample GB-neck2 protocol outlined below. Simulation speed gains in our test runs of RNA and DNA nanocubes showed a 2× speed-up over explicit solvent runs on the same single NVIDIA Tesla K40 GPU, and approximately 10× faster execution than simulations on a 2.7 GHz 32 core Intel Xeon Processor. For this reason, it is worth following the reports on GB parameter updates, as some of the known remaining issues in nucleic acid implicit solvent simulations (*see* **Note 3**) are planned to be addressed by the Amber development team next [31].

To run the GB-HCT simulation protocol, use LEaP to generate the initial coordinates (RNA.inpcrd) and topology file (RNA.prmtop—topological information and FF parameters).

```
tleap
source leaprc.ff14SB
set default PBRadii mbondi
mol=loadpdb RNA.pdb
saveAmberParm mol RNA-GB-HCT-nosolv.prmtop RNA-GB-HCT-nosolv.
inpcrd
quit
```

The Amber14 force field file, leaprc.ff14SB also includes the ff99bsc0 and chi.OL3 updates for nucleic acids, as we mentioned earlier. The parameter PBRadii can be set to several sets of atomic radii for generalized Born (GB) or Poisson-Boltzmann (PB) calculations. For nucleic acids, the choice of mbondi must be used with the igb = 1 parameter in the GB-HCT simulations, while the recently revised mbondi3 is compatible with igb = 8 choice in the GB-neck2 protocol. Thus the tleap session parameter files for an implicit solvent GB-neck2 protocol run would look as follows:

```
tleap
source leaprc.ff14SB
set default PBRadii mbondi3
mol=loadpdb RNA.pdb
```

Table 4
Implicit solvent GB-neck2 equilibration and production MD sample protocol script

GB Equilibration and Production MD procedure (Subheading 3.4)
After system specific initialization of Amber environment (for example: module load amber)
the path to your amber installation directory is set and stored in the environment variable AMBERHOME
set ABIN = "$AMBERHOME/bin/pmemd.cuda" # OR pmemd.MPI for non-GPU/CUDA binary
set PRMTOP = "RNA-GBn2-nosolv.prmtop"
set INPCRD = "RNA-GBn2-nosolv.inpcrd"
set PREFIX = "RNA-GBn2"
#————————————————————————————————
echo " Starting Amber Run with GB-neck2 (igb=8 for mbondi3) Equilibration Protocol"
echo " (6 steps for 2.0ns: dt=2fs, ntt=3 with gamma_ln=5.0..10.0): "
echo " Step 1 of 6: Energy minimization"
$ABIN -O -i GBn2-min.in -o min.out -p $PRMTOP -c $INPCRD -ref $INPCRD -r min.rst
echo " Step 2 of 6: Heating to 300K and gamma_ln=15.0 (0.25ns)"
$ABIN -O -i GBn2-heat.in -c min.rst -p $PRMTOP -ref min.rst -o heat.out -r heat.rst -x heat.x
echo " Step 3 of 6: GB MD with harmonic restraints on RNA = 10.0 and gamma_ln=5.0 (0.25ns)"
$ABIN -O -i GBn2-cst10.in -c heat.rst -p $PRMTOP -ref heat.rst -o cst10.out -r cst10.rst -x cst10.x
echo " Step 4 of 6: GB MD with harmonic restraints on RNA = 1.0 and gamma_ln=5.0 (0.5ns)"
$ABIN -O -i GBn2-cst1.in -c cst10.rst -p $PRMTOP -ref cst10.rst -o cst1.out -r cst1.rst -x cst1.x
echo " Step 5 of 6: GB MD with harmonic restraints on RNA = 0.1 and gamma_ln=10.0 (0.5ns)"
$ABIN -O -i GBn2-cst01.in -c cst1.rst -p $PRMTOP -ref cst1.rst -o cst01.out -r cst01.rst -x cst01.x
echo " Step 6 of 6: GB MD with harmonic restraints on RNA = 0.01 and gamma_ln=10.0 (0.5ns)"
$ABIN -O -i GBn2-cst001.in -c cst01.rst -p $PRMTOP -ref cst01.rst -o cst001.out -r cst001.rst -x cst001.x
Produce PDB files for the MINMIZED and EQILIBRATED output
$AMBERHOME/bin/ambpdb -p $PRMTOP < min.rst > $PREFIX-MIN.pdb
$AMBERHOME/bin/ambpdb -p $PRMTOP < cst001.rst > $PREFIX-EQL-02ns.pdb
#————————————————————————————————
A sample production run with two 10ns-long phases
echo " Production GB MD with Langevin thermostat (igb=8 for mbondi3; ntt=3 with gamma_ln=1.0)"
$ABIN -O -i GBn2-MD-10ns.in

(continued)

Table 4
(continued)

GB Equilibration and Production MD procedure (Subheading 3.4)
-p $PRMTOP -c cst001.rst \
-o $PREFIX-1.out -r $PREFIX-1.rst -x $PREFIX-1.x -inf mdinfo-$PREFIX-1.out
$AMBERHOME/bin/ambpdb -p $PRMTOP < $PREFIX-1.rst > $PREFIX-10ns.pdb
$ABIN -O -i GBn2-MD-10ns.in \
-p $PRMTOP -c $PREFIX-1.rst \
-o $PREFIX-2.out -r $PREFIX-2.rst -x $PREFIX-2.x -inf mdinfo-$PREFIX-2.out
$AMBERHOME/bin/ambpdb -p $PRMTOP < $PREFIX-2.rst > $PREFIX-20ns.pdb
echo " Amber Run with GB-neck2 Protocol has finished execution. "0

Table 5
The list of input files for energy minimization, heating, and equilibration phase restrained MD and the production MD simulations shown in Table 4

Equilibration: GBn2-min.in (Subheading 3.4)

```
GBSA minimization (igb=8 for PBRadii = mbondi3)
&cntrl
  imin=1, ntx=1,
  irest=0, ntxo=1,
  drms=0.001,
  cut=999,
  ntpr=100, ntwx=100,
  maxcyc=20000, ncyc=1000,
  igb=8, saltcon=1.0, ntt=0,
  gbsa=1, ntb=0,
&end
```

Equilibration: GBn2-heat.in (Subheading 3.4)

```
Heating from 10 to 300K in 0.1ns, total time of 0.25ns (Langevin ntt=3)
&cntrl
  imin=0, ntx=1,
  irest=0, ntxo=1,
  cut=999,
  dt=0.002, nstlim=125000,
  nscm=100,
  ntpr=1000, ntwx=1000,
  tempi=10.0, temp0=300.0,
  ntf=2, ntc=2,
  gbsa=0, ntb=0,
  igb=8, ntt=3, gamma_ln=5.0, saltcon=1.0
  nmropt=1, ntr=1,
```

(continued)

Table 5
(continued)

Equilibration: GBn2-min.in (Subheading 3.4)

```
&end
&wt
type='TEMP0', istep1=0, istep2=50000, value1=10.0, value2=300.0,
&end
&wt
type='TEMP0', istep1=50001, istep2=125000, value1=300.0,
   value2=300.0,
&end
&wt
type='END',
&end
GROUP FOR RESTRAINTS
15.0
RES 1 312 # 312 nts for the RNA cube example
END
END
```

Equilibration: GBn2-cst001.in (Subheading 3.4)

```
GB-neck2 with Langevin thermostat and restraints (ntr=1) of 0.01 for 0.5
  ns
&cntrl
 imin=0, ntx=5,
 irest=1, ntxo=1,
 cut=999,
 dt=0.002, nstlim=250000, # 125000 step for 0.25ns phases
 ntpr=1000, ntwx=1000,
 temp0=300.0,
 ntf=2, ntc=2,
 gbsa=0, ntb=0,
 igb=8, ntt=3, gamma_ln=10.0, saltcon=1.0, # gamma_ln varying from
  5.0 to 10.0
 ntr=1,
&end
GROUP FOR RESTRAINTS
0.01 # variable restraints R: 10.0, 1.0, 0.1, 0.01 ib GBn2-cstR.in files
RES 1 312 # 312 nts for the RNA cube example
END
```

Production MD: GBn2-MD-10ns.in (Subheading 3.4)

```
MD GB-neck2 run (igb=8 for mbondi3) with Langevin thermostat (ntt=3)
  for 10.0 ns
Write output sparsely every 1000 steps = 2ps (more typical: 500 steps at
  dt=0.002 = 1ps)
&cntrl
 imin=0, ntx=5,
 irest=1, ntxo=1,
 cut=999,
 dt=0.002, nstlim=5000000,
```

(continued)

Table 5
(continued)

Equilibration: GBn2-min.in (Subheading 3.4)

nscm=1000,
ntpr=1000, ntwx=1000,
ntwprt=0,
temp0=300.0,
ntf=2, ntc=2,
igb=8, ntt=3, gamma_ln=1.0, saltcon=1.0, # low gamma_ln for faster conformation exploration
gbsa=0,
ntb=0,
ig=-1,
&end0

Please refer to the Amber manual for explanations for input parameters not discussed in this chapter

```
saveAmberParm mol RNA-GBn2-nosolv.prmtop RNA-GBn2-nosolv.
inpcrd
quit
```

The general outline of the two protocols is identical to the steps shown in a form of and PBS queue script in Table 4. Instead of Amber's SANDER module calls illustrated for the explicit solvent protocol, we are showing it for the PMEMD.cuda module, a performance-optimized subset of SANDER functionality that can be run on GPUs (*see* **Note 5**).

4 Notes

1. Other molecular dynamics packages and visualization software

   Besides Amber, NAMD (http://www.ks.uiuc.edu/Research/namd/), CHARMM (https://www.charmm.org/charmm/), and GROMACS (http://www.gromacs.org/) are commonly used molecular dynamics packages. Like Amber, CHARMM and GROMACS have their own FF and topology libraries, while NAMD is an MD simulation platform, which accepts CHARMM, Amber, and GROMACS FFs. For example, users can employ Amber FF on an NAMD platform by defining Amber topology and coordinate files in an NAMD input file:

```
# Input
amber on
parmfile Amber.prmtop
```

```
ambercoor Amber.inpcrd
```

Although the basic concept and MD protocols of these packages are similar to those of Amber, users will need to familiarize themselves with the FF and unique features of each molecular dynamics package before use.

Besides data analysis and visualization, VMD can be used to prepare the initial MD simulation systems using CHARMM FF. VMD can load CHARMM FF (http://mackerell.umaryland.edu/charmm_ff.shtml) to generate a protein structure file (psf, http://www.ks.uiuc.edu/Training/Tutorials/namd/namd-tutorial-html/node6.html), which corresponds to prmtop topology file in Amber. Once both PDB (coordinate) and topology are loaded to VMD, users can solvate a system using commands under *Extensions → Modeling → Add Solvation Box*. The size and type of solvent can be selected. After solvation, users can neutralize and add extra ions to set the desired salt concentrations using menu: *Extensions → Modeling → Add Ions*. The generated coordinate and topology files can be saved and used for MD simulations on an NAMD platform.

Besides DSV and VMD, pyMOL (https://www.pymol.org/) and Chimera (https://www.cgl.ucsf.edu/chimera/) are also widely used in biomolecular research.

2. Limitations of the force fields.

As the limitations of FF have been revealed by MD simulations over longer time ranges and noncanonical structures, proper modifications and updates have followed. For example, the original Amber force field, ff94, was updated as ff99 with a modified charge set. Later, the α and γ backbone torsion angles of DNA were updated in ff99bsc0. However, for longer MD simulations of RNA, it was found that incorrect loop geometries, locking RNA sugar pucker in the north conformation, and an opening of RNA duplex into a ladder-like configuration are due to strong *anti* conformation of the χ angle. To fix these issues, the force field for RNA has recently been updated to ff99OL3 in which χ angle parameters were tuned for RNA [20]. However, these modifications were causing some issues in DNA quadruplex structures so that a separate FF, OL15 was prepared for DNA.

Besides these issues, most FFs have fundamental limitations due to polarizability. The electrostatic properties of atoms in the traditional MD simulations are represented by the partial charges of the atomic nuclei. The magnitudes and positions of the partial charges are treated as constants throughout the MD simulations. In reality, however, the magnitudes and distributions of charges in a biomolecule continuously change in response to the environment, so that the constant partial charge approximation is insufficient to describe the polarizability of the biomolecules. To bridge the gap between the constant partial charge and polarizability, various theoretical models have been applied to major FF groups. For example, the inducible dipoles model was adapted in Amber FF02 for the polarization of proteins and nucleic acids, while fluctuating charges and Drude oscillators models were applied to CHARMM FF for the polarization of proteins, nucleic acids, lipids, and carbohydrates.

As described above, it is important for users to understand that FFs are not perfect in describing the biomolecular behavior and unexpected limitations can be encountered as the MD simulations are executed for longer time scales or applied to noncanonical structures. In addition, users need to check for the most recent versions of the force field to choose the most reliable force field parameters for their MD simulations.

3. Limitations of MD simulations.

The capability of MD simulations such as simulation time, the size of the biological system, and the frequency of sampling (i.e., time step size) is often limited by computational resources. For example, the MD simulations of a 0.09 ns long trajectory for a 24-base pair DNA fragment in vacuum (total 1530 atoms) took several weeks in the early 1980s [32]. However, due to the rapid development of high performance computing systems, recently, explicit solvent MD simulations of the mature HIV-1 capsid (total 63 million atoms) reached 100 ns within 3 weeks [33], and a 1 ms-long MD simulation of ubiquitin was performed by the Anton system [34].

As presented in Subheading 3.4, an implicit solvent approach to MD has recently become more accurate with respect to nucleic acid simulation [27–29, 31]. In our tests of the GB-HCT protocol, a DNA-based nanocube model experienced distortions larger than in the explicit solvent simulations, but generally maintained structural integrity (the initial design's base pairing), whereas RNA nanocubes showed a tendency to open some base pairs and helix stacking across the 5′–3′ strand ends that meet in the middle of helical cube edges (GB-HCT run with the default offset of 0.09 Å and the 0.13 Å recommended for nucleic acids). Our evaluation of the GB-neck2 protocol on RNA and DNA nanocubes showed overall stability, with RNA cubes still experiencing some of the GB-HCT problems around the 5′ and 3′ ends of individual strands comprising the cubes. However, equilibration steps employing higher values of the Langevin collision frequency parameter (gamma_ln) than those used in the production MD run, appear to mitigate this problem and stress the importance of tuning the equilibration parameters to the starting structures. One has to monitor the transitions between the last stage of equilibration and the beginning of the production MD run (visually and/or with RMSD calculations) for any abrupt changes to the shape (distortions) and dynamics of the solute and adjust the sample protocol to make this transition as smooth as possible. Effectively, the exact number of equilibration steps and the strength of the harmonic restraints in the last equilibration step may be different for high resolution X-ray crystallography-based input structures than for the more approximate initial 3D models of nanostructures. Similar equilibration modifications applied to the GB-HCT protocol did not help to improve its problems with the RNA cubes. Simulation speed improvements resulting from faster processor choice (GPU) and implicit solvent simulation are illustrated in the following example. For a sample RNA nanocube of approximately 10,000 atoms run with the Amber 14 PMEMD module improves from ~1.8 ns/day for the explicit solvent PME simulation (total system size of ~230,000 atoms, run on a 2.7 GHz 32 core Intel Xeon Processor), up to ~9.2 ns/day on a single NVIDIA Tesla K40 GPU, while the implicit solvent simulation (GB-neck2 protocol with Langevin thermostat) yields ~18.0 ns/day on the same single GPU. Thus, the implicit solvent method is definitely worth testing and comparing to the explicit solvent results for the same nanostructures. It is also worth tracking reports on parameter updates, as the issue of 5′–3′ ends fraying and comparisons to the explicit solvent simulation results are targeted by the Amber development team as the next goal to improve in nucleic acid implicit solvent simulations [31]

Although the capability of MD simulations has significantly increased due to the improvements in high performance computing systems, there are still limitations with respect to the investigation of many important biological systems. For instance, Mg^{2+} ions play an important role in biological processes, such as in DNA repair mechanisms [35, 36], dimerization of HIV-1 [37], and self-assembly of RNA nanoparticles [4–6]. However, since the dehydration of Mg^{2+} ions and the binding formation of $Mg^{2+}\cdots O$ happen on a microsecond time scale, Mg^{2+} ion dynamics not only occur beyond regular MD simulation time scales but also require significant conformational sampling. Using enhanced sampling methods, these difficulties can sometimes be overcome. Transition path sampling (TPS) [38] is an example of an enhanced sampling method. TPS is the mixture of MD and Monte Carlo simulations, which allows collecting sufficient numbers of samples for two local minimum states separated by an energy barrier. In addition, coupled with umbrella sampling, TPS also provides the free energy profile of molecular transitions. TPS has been successfully applied to several biological systems including Mg^{2+} ion dynamics [36]. Besides TPS, there are several other enhanced sampling methods, such as replica exchange molecular dynamics (REMD) [39] simulations, Gaussian Accelerated Molecular Dynamics (GaMD) [40], and locally enhanced sampling (LES) [41].

Besides the sampling issue, all atom MD simulations still have limits on the size of biomolecular systems they can handle. For example, to develop biomedical applications using RNA nanoparticles, it is necessary to investigate the interactions between the RNA nanoparticles, the delivery agent and the cell membrane. However, the size of these systems including water and ions can be beyond the capability of computational resources. To overcome the computational limitations on large scale systems such as membranes, chromatin fibers, and viruses, coarse-grained approaches have been developed. In a coarse-grained approach, each nucleotide in an RNA can be represented by a few beads, so that the total number of atoms in the RNA nanostructure is dramatically reduced [11]. A special type of course-grained approach, the elastic network model, is an efficient method to investigate the dynamics of biomolecular systems. It was able to successfully characterize the RNA nanocube under feasible computing time scales [42]. Please refer to **Note 4** for a more detailed discussion of the elastic network model features. In addition, multiscale computational approaches using quantum mechanical and molecular mechanical (QM/MM) methods, all atom MD simulations, and coarse-grained approaches can provide significant information for biomolecular systems, including RNA nanostructures.

4. Elastic Network Model as an alternative or a complement to Molecular Dynamics Simulations

In many cases large scale (and lower frequency) collective motions of molecules, i.e., motion of entire domains of a molecule relative to each other, rather than high frequency local atomic motions, are of most interest. Molecular dynamics simulations by their nature predict both kinds of motions, and it is possible to extract the large scale, low frequency motions from MD trajectories by performing principal component (PC) analysis. However, in order to predict the low frequency collective motions one does not need to rely exclusively on molecular dynamics simulations, which, because of their complex force fields operating at the atomic level resolution, are computationally intensive and relatively slow. Orders of magnitude faster than MD, coarse-grained alternatives include variants of the elastic network models (ENM). An example of an ENM is the Anisotropic Network Model (ANM), in which a molecule, assumed to be at equilibrium, is represented as a network of nodes (masses) connected by springs within a cutoff distance treated as one of the adjustable parameters. Springs provide the potential energy for the calculation of fluctuations of the nodes around their starting positions [43–47]. The ANM model can be used to calculate the directionality, frequency, and relative magnitude of motions of each node within the network [48]. More precisely, ANM predicts normal modes of motion. Eigenvectors, called normal mode shapes, and eigenvalues, corresponding to the inverse squared frequencies of motion are calculated for each mode. Multiple studies demonstrated good agreement between the experimentally observed collective motion of molecules (large scale, rather than atomic level), and a small subset of the lowest frequency normal modes of motion predicted by ANM simulations [42, 49–53]. An overlap of ANM-predicted normal mode motions with the principal components of motions derived from MD simulations for the same biomolecules has also been demonstrated, showing intersection of the two methods [47, 54]. A general conclusion emerging from multiple studies is that the overall shape of a molecule (as captured by a network formed by connected nodes representing a molecule) at its equilibrium point is the key factor determining the major modes of its motions [55, 56]. An example of an ANM implementation is a software package called MAVENs, available for download as Windows or Mac OS X binaries or source files from Sourceforge (https://sourceforge.net/projects/maven/). For more information on the project, see the information on the project website http://maven.sourceforge.net [57].

5. However, please note that while we show the energy minimization step calling the same binary as other phases of the protocol, it can be safer in general (please refer to the Amber manual for the discussion of the precission model used on GPUs) to call a CPU-based Amber binary module for energy minimization (PMEMD or SANDER).

Acknowledgments

This project has been funded in part with federal funds from the Frederick National Laboratory for Cancer Research, National Institutes of Health, under contract HHSN261200800001E for W.K. K. and E.B. This research was supported in part by the Intramural Research Program of the NIH, National Cancer Institute, Center for Cancer Research. The content of this publication does not necessarily reflect the views or policies of the Department of Health and Human Services, nor does the mention of trade names, commercial products, or organizations imply endorsement by the US Government.

References

1. Severcan I, Geary C, Verzemnieks E, Chworos A, Jaeger L (2009) Square-shaped RNA particles from different RNA folds. Nano Lett 9 (3):1270–1277
2. Yingling YG, Shapiro BA (2007) Computational design of an RNA hexagonal nanoring and an RNA nanotube. Nano Lett 7 (8):2328–2334
3. Afonin KA, Bindewald E, Yaghoubian AJ, Voss N, Jacovetty E, Shapiro BA, Jaeger L (2010) In vitro assembly of cubic RNA-based scaffolds designed in silico. Nat Nanotechnol 5 (9):676–682
4. Grabow WW, Zakrevsky P, Afonin KA, Chworos A, Shapiro BA, Jaeger L (2011) Self-assembling RNA nanorings based on RNAI/II inverse kissing complexes. Nano Lett 11(2):878–887
5. Afonin KA, Grabow WW, Walker FM, Bindewald E, Dobrovolskaia MA, Shapiro BA, Jaeger L (2011) Design and self-assembly of siRNA-functionalized RNA nanoparticles for use in automated nanomedicine. Nat Protoc 6 (12):2022–2034
6. Afonin KA, Viard M, Koyfman AY, Martins AN, Kasprzak WK, Panigaj M, Desai R, Santhanam A, Grabow WW, Jaeger L, Heldman E, Reiser J, Chiu W, Freed EO, Shapiro BA (2014) Multifunctional RNA nanoparticles. Nano Lett 14(10):5662–5671
7. Martinez HM, Maizel JV Jr, Shapiro BA (2008) RNA2D3D: a program for generating, viewing, and comparing 3-dimensional models of RNA. J Biomol Struct Dyn 25(6):669–683
8. Bindewald E, Grunewald C, Boyle B, O'Connor M, Shapiro BA (2008) Computational strategies for the automated design of RNA nanoscale structures from building blocks using NanoTiler. J Mol Graph Model 27 (3):299–308
9. Kasprzak W, Bindewald E, Kim TJ, Jaeger L, Shapiro BA (2011) Use of RNA structure flexibility data in nanostructure modeling. Methods 54(2):239–250
10. Paliy M, Melnik R, Shapiro BA (2009) Molecular dynamics study of the RNA ring nanostructure: a phenomenon of self-stabilization. Phys Biol 6(4):046003
11. Paliy M, Melnik R, Shapiro BA (2010) Coarse-graining RNA nanostructures for molecular dynamics simulations. Phys Biol 7(3):036001
12. Kim T, Afonin KA, Viard M, Koyfman AY, Sparks S, Heldman E, Grinberg S, Linder C, Blumenthal RP, Shapiro BA (2013) In silico, in vitro, and in vivo studies indicate the potential use of bolaamphiphiles for therapeutic siRNAs delivery. Mol Ther Nucleic Acids 2:e80
13. Gupta K, Afonin KA, Viard M, Herrero V, Kasprzak W, Kagiampakis I, Kim T, Koyfman AY, Puri A, Stepler M, Sappe A, KewalRamani VN, Grinberg S, Linder C, Heldman E, Blumenthal R, Shapiro BA (2015) Bolaamphiphiles as carriers for siRNA delivery: from chemical syntheses to practical applications. J Control Release 213:142–151
14. Case DA et al (2014) AMBER 14, University of California, San Francisco
15. Phillips JC, Braun R, Wang W, Gumbart J, Tajkhorshid E, Villa E, Chipot C, Skeel RD, Kale L, Schulten K (2005) Scalable molecular dynamics with NAMD. J Comput Chem 26 (16):1781–1802
16. Berendsen HJC, van der Spoel D, Drunen R (1995) GROMACS: a message-passing parallel molecular dynamics implementation. Comp Phys Comm 91:43–56
17. Pronk S, Páll S, Schulz R, Larsson P, Bjelkmar P, Apostolov R, Shirts MR, Smith JC, Kasson PM, van der Spoel D, Hess B, Lindhal E (2013) GROMACS 4.5: a high-throughput and highly parallel open source molecular simulation toolkit. Bioinformatics 29(7): 845–854
18. Brooks BR, Bruccoleri RE, Olafson BD, States DJ, Swaminathan S, Karplus M (1983) CHARMM: a program for macromolecular energy, minimization, and dynamics calculations. J Comput Chem 4(2):187–217
19. Brooks BR et al (2009) CHARMM: the biomolecular simulation program. J Comput Chem 30(10):1545–1614
20. Zgrabova M, Otyepka M, Sponer J, Mladek A, Banas P, Cheatham TE 3rd, Jurecka P (2011) Refinement of the Cornell et al. nucleic acids force filed based on reference quantum chemical calculations of glycosidic torsion profiles. J Chem Theory Comput 7(9):2886–2902

21. Darden T, York D, Pedersen LG (1993) Particle mesh Ewald: an N-log(N) method for Ewald sums in large systems. J Chem Phys 98 (12):10089–10092
22. Essmann U, Perera L, Berkowitz ML, Darden T, Lee H, Pedersen LG (1995) A smooth particle mesh EWALD method. J Chem Phys 103 (19):8577–8593
23. Berendsen HJC, Postma JPM, van Gunsteren WF, DiNola A, Haak JR (1984) Molecular dynamics with coupling to an external bath. J Chem Phys 81(8):3684–3690
24. Srinivasan J, Cheatham TE 3rd, Cieplak P, Kollman PA, Case DA (1998) Continuum solvent studies of the stability of DNA, RNA, and phosphoramidate-DNA helices. J Am Chem Soc 120:9401–9409
25. Afonin KA, Kasprzak WK, Bindewald E, Kireeva M, Viard M, Kashlev M, Shapiro BA (2014) In silico design and enzymatic synthesis of functional RNA nanoparticles. Acc Chem Res 47:1731–1741
26. Onufriev A, Bashford D, Case DA (2004) Exploring protein native states and large-scale conformational changes with a modified generalized born model. Proteins 55 (2):383––394
27. Hawkins GD, Cramer CJ, Truhlar DG (1995) Pairwise solute descreening of solute charges from a dielectric medium. Chem Phys Lett 246 (1):122––129
28. Tsui V, Case DA (2000) Theory and applications of the generalized Born solvation model in macromolecular simulations. Biopolymers 56(4):275–291
29. Gaillard T, Case DA (2011) Evaluation of DNA force fields in implicit solvation. J Chem Theory Comput 7(10):3181–3198
30. Liu Y, Haddadian E, Sosnick TR, Freed KF, Gong H (2013) A novel implicit solvent model for simulating the molecular dynamics of RNA. Biophys J 105(5):1248–1257
31. Nguyen H, Perez A, Bermeo S, Simmerling C (2015) Refinement of generalized born implicit solvation parameters for nucleic acids and their complexes with proteins. J Chem Theory Comput 11(8):3714–3728
32. Levitt L (1983) Computer simulation of DNA double-helix dynamics. Cold Spring Harb Symp Quant Biol 47:251–275
33. Zhao G et al (2013) Mature HIV-1 capsid structure by cryo-electron microscopy and all-atom molecular dynamics. Nature 497:643–646
34. Pianaa S, Lindorff-Larsena K, Shawa DE (2013) Atomic-level description of ubiquitin folding. Proc Natl Acad Sci U S A 110:5915–5920
35. Freudenthal BD, Beard WA, Perera L, Shock DD, Kim T, Schlick T, Wilson SH (2015) Uncovering the polymerase-induced cytotoxicity of an oxidized nucleotide. Nature 517:635–639
36. Kim T, Freudenthal BD, Beard WA, Wilson SH, Schlick T (2016) Insertion of oxidized nucleotide triggers rapid DNA polymerase opening. Nucleic Acids Res 44:4409–4424
37. Kim T, Shapiro B (2013) The role of sodium and magnesium ion concentration in HIV-1 subtype-A and subtype-B kissing loop dimerization structures. J Biomol Struct Dyn 31 (5):495–510
38. Bolhuis PG, Chandler D, Dellago C, Geissler PL (2002) Transition path sampling: throwing ropes over rough mountain passes, in the dark. Annu Rev Phys Chem 53:291–318
39. Mitsutake A, Sugita Y, Okamoto Y (2001) Generalized-ensemble algorithms for molecular simulations of biopolymers. Biopolymers 60:96–123
40. Miao Y, Feher VA, McCammon JA (2015) Gaussian accelerated molecular dynamics: unconstrained enhanced sampling and free energy calculation. J Chem Theory Comput 11(8):3584–3595
41. Elber R, Karplus M (1990) Enhanced sampling in molecular dynamics. Use of the time dependent Hartree approximation for a simulation of carbon monoxide diffusion through myoglobin. J Am Chem Soc 112:9161–9175
42. Afonin KA, Kasprzak W, Bindewald E, Puppala PS, Diehl AR, Hall KT, Kim TJ, Zimmermann MT, Jernigan RL, Jaeger L, Shapiro BA (2014) Computational and experimental characterization of RNA cubic nanoscaffolds. Methods 67 (2):256–265
43. Bahar I, Atilgan AR, Erman B (1997) Direct evaluation of thermal fluctuations in proteins using a single-parameter harmonic potential. Fold Des 2(3):173–181
44. Bahar I, Erman B, Haliloglu T, Jernigan RL (1997) Efficient characterization of collective motions and inter residue correlations in proteins by low-resolution simulations. Biochemistry 36(44):13512–13523
45. Bahar I, Rader AJ (2005) Coarse-grained normal mode analysis in structural biology. Curr Opin Struct Biol 15(5):586–592
46. Sherwood P, Brooks BR, Sansom MS (2008) Multiscale methods for macromolecular simulations. Curr Opin Struct Biol 18(5):630–640
47. Yang L, Song G, Carriquiry A, Jernigan RL (2008) Close correspondence between the

motions from principal component analysis of multiple HIV-1 protease structures and elastic network modes. Structure 16(2):321–330
48. Atilgan AR, Durell SR, Jernigan RL et al (2001) Anisotropy of fluctuation dynamics of proteins with an elastic network model. Biophys J 80(1):505–515
49. Sen TZ, Feng Y, Garcia JV, Kloczkowski A, Jernigan RL (2006) The extent of cooperativity of protein motions observed with elastic network models is similar for atomic and coarser-grained models. J Chem Theory Comput 2(3):696–704
50. Wang Y, Jernigan RL (2005) Comparison of tRNA motions in the free and ribosomal bound structures. Biophys J 89(5):3399–3409
51. Yan A, Wang Y, Kloczkowski A, Jernigan RL (2008) Effects of protein subunits removal on the computed motions of partial 30S structures of the ribosome. J Chem Theory Comput 4 (10):1757–1767
52. Yang L, Song G, Jernigan RL (2007) How well can we understand large-scale protein motions using normal modes of elastic network models? Biophys J 93(3):920–929
53. Zimmermann MT, Jernigan RL (2014) Elastic network models capture the motions apparent within ensembles of RNA structures. RNA 20 (6):792–804
54. Bakan A, Bahar I (2011) Computational generation of inhibitor-bound conformers of p38 MAP kinase and comparison with experiments. Pac Symp Biocomput 181–192.
55. Doruker P, Jernigan RL, Bahar I (2002) Dynamics of large proteins through hierarchical levels of coarse-grained structures. J Comput Chem 23(1):119–127
56. Wang Y, Rader AJ, Bahar I, Jernigan RL (2004) Global ribosome motions revealed with elastic network model. J Struct Biol 147 (3):302–314
57. Zimmermann MT, Kloczkowski A, Jernigan RL (2011) MAVENs: motion analysis and visualization of elastic networks and structural ensembles. BMC Bioinformatics 12:264–270

Chapter 4

Rolling Circle Transcription for the Self-Assembly of Multimeric RNAi Structures and Its Applications in Nanomedicine

Mihue Jang and Hyung Jun Ahn

Abstract

The enzymatic process of rolling circle transcription (RCT) enables self-assembly of multimeric RNAi structures from a circular DNA template. The self-assembled RNAi structures can be manipulated easily by simple base pairing rules with short DNA fragments for constructing multifunctional nanoparticles in the field of nanomedicine. Here we describe the method to generate multifunctional RNAi nanoparticles applicable in nanomedicine.

Key words Rolling circle transcription, Multimeric RNAi structures, DNA–RNA hybrids, Base pairing, Delivery of siRNA

1 Introduction

RNA nanoparticles harboring multiple therapeutic molecules, such as RNAi or miRNA, have great potential to be a new generation of drug for efficient cancer therapy [1, 2]. In particular, RNAi is an endogenous regulatory process that can inhibit expression of any gene associated with diverse human disease [3–5]. A variety of RNAi therapeutics are currently undergoing clinical trials [6]. Many challenges exist, however, in clinical practice, and there is a need to develop an efficient RNAi delivery system. Here we introduce an RNAi delivery platform system for the efficient knockdown of target genes in tumor tissues [7]. We were able to generate a new siRNA carrier with multimeric RNAi hairpins through rolling circle transcription and base pairing of DNA–RNA (Figs. 1 and 2). We obtained multiple tandem copies of RNAi hairpins synthesized by an enzymatic process of rolling circle transcription, that self-assemble into RNA microsponge-like particles [8]. Multimeric RNAi hairpins can convert to siRNA duplexes through enzymatic processing by Dicer after cellular uptake. Next, we achieved packing

Eckart Bindewald and Bruce A. Shapiro (eds.), *RNA Nanostructures: Methods and Protocols*, Methods in Molecular Biology, vol. 1632, DOI 10.1007/978-1-4939-7138-1_4, © Springer Science+Business Media LLC 2017

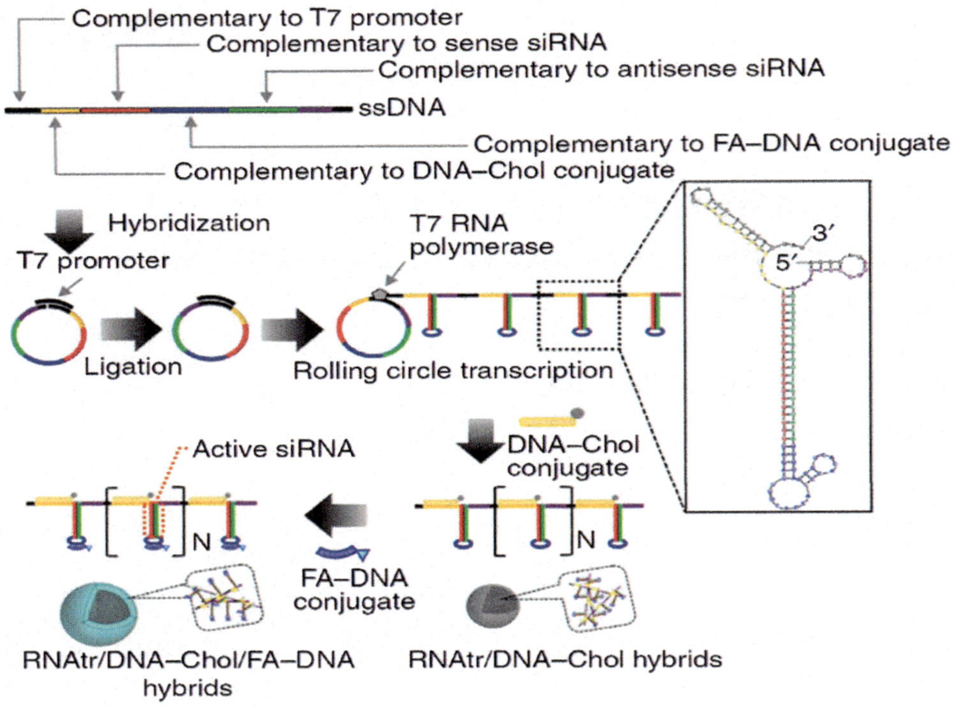

Fig. 1 A schematic design of multimeric RNAi nanoparticles via RCT and base pairing. DNA template encodes complementary sense and antisense of siRNA duplex. Multimeric RNAi hairpins (RNAtr) synthesized by rolling circle transcription can be self-assembed into highly condensed nanoparticles through the RNA/DNA base pairing with DNA–Chol conjugates. Sequential hybridization of FA–DNA conjugates on condensed RNAtr results in RNAtr/DNA–Chol/FA–DNA hybrids with self-assembled nanoparticles targeted to tumors. RNAi hairpins can be processed by Dicer enzyme and loaded into the RNA-induced silencing complex (RISC), yielding funtional siRNA duplex. The *box* shows the simulated hairpin of single unit of RNA transcripts (Figures reproduced with permission from ref. 6, ©2015 NPG)

of RNAi transcripts/DNA–cholesterol (DNA–Chol) hybrids into condensed RNAi nanoparticles without the aid of polycationic reagents through base pairing of DNA–RNA with DNA–Chol conjugates (Fig. 3). RNAi transcripts can be hybridized with DNA–Chol conjugates through linker regions in the RNAi transcripts, and can be highly condensed due to the hydrophobicity of cholesterol [9]. Several things should be considered, such as the length, Tm value, and structure of oligonucleotides, when designing the linker regions for RNA/DNA base pairing. Next, the self-assembled RNAi transcripts/DNA–Chol/folate–DNA (FA–DNA) nanoparticles are formed upon sequential hybridization with FA–DNA conjugates. Those nanoparticles presenting folate can target cancer sites where folate receptors are highly expressed (Fig. 3). Thus we have obtained a new RNAi carrier which can be efficiently adopted for systemic delivery of siRNA in cancer therapy.

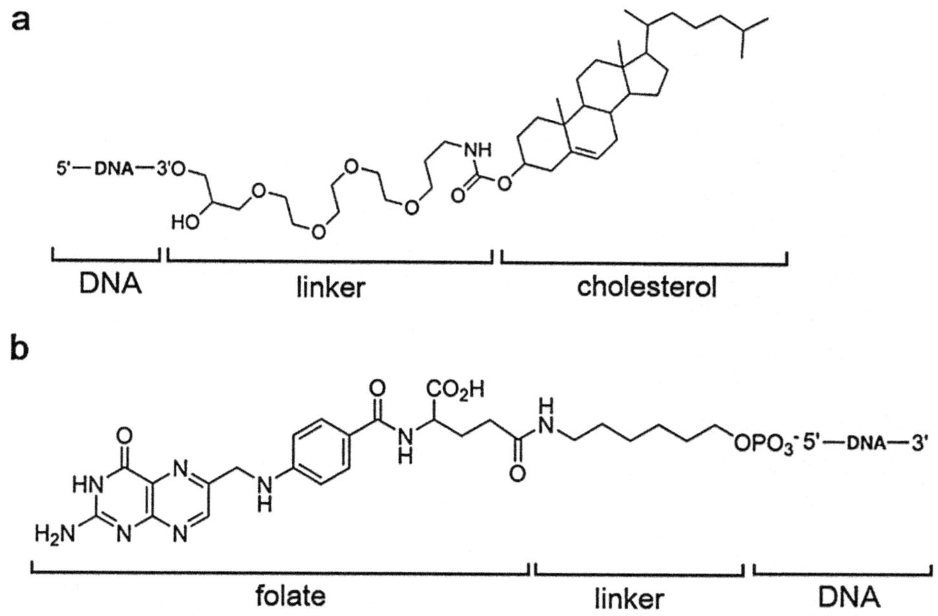

Fig. 2 Chemical structures showing the conjugates of DNA–Chol (**a**) and FA–DNA (**b**). Through RNA/DNA base pairing, DNA–Chol conjugates are hybridized with the RNA transcripts and FA–DNA conjugates with RNAtr/DNA–Chol result in RNAtr/DNA–Chol/FA–DNA hybrids (Figures reproduced with permission from ref. 6, ©2015 NPG)

2 Materials

2.1 Circular DNA Template

1. Design nucleotide sequence of single-stranded DNA (ssDNA) template encoding multimeric RNAi via RCT (*see* **Note 1** and Fig. 1).
2. 1× annealing solution: 50 mM Tris–HCl (pH 7.8), 10 mM $MgCl_2$, 10 mM DTT, 1 mM ATP. Store at −20 °C.
3. 1× ligation buffer: 50 mM Tris–HCl (pH 7.8), 10 mM $MgCl_2$, 10 mM DTT, 5 mM ATP. Store at −20 °C (*see* **Note 2**).

2.2 Rolling Circle Transcription

1. All reagents are stored at −20 °C.
2. Bacteriophage T7 RNA polymerase.
3. rNTP mixture solution.
4. 10× reaction buffer: 40 mM Tris–HCl (pH 7.9), 100 mM $MgCl_2$, 10 mM DTT, 2 mM spermidine.
5. RNase inhibitor.
6. DNase I restriction enzyme.

2.3 Agarose Gel Electrophoresis

1. 50× TAE buffer: Weigh 242 g Tris base and transfer it to 1 L beaker. Add about 600 mL distilled water. Mix until all Tris base dissolves completely. Then add 57.1 mL glacial acetic acid and 100 mL of 0.5 M EDTA (pH 8.0) (*see* **Note 3**). Mix the

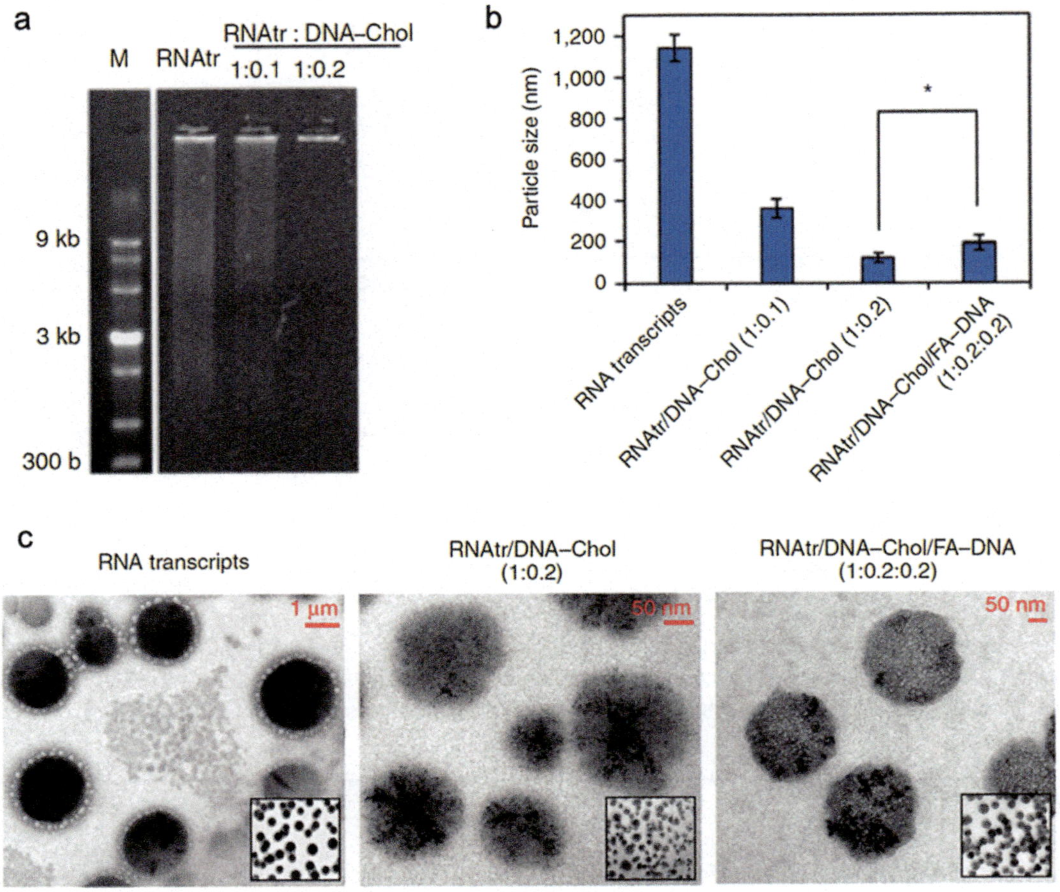

Fig. 3 Characterization of RNA/DNA hybrids. (**a**) Gel retardation assay of RNAi transcripts and RNAi transcripts/DNA–Chol hybrids at the indicated weight ratio. The RNAtr/DNA–Chol hybrids did not move down and stayed in the well at a weight ratio of 1:0.2. (**b**) Size distribution analyzed by DLS. (**c**) Transmission electron microscopy images (Figures reproduced with permission from ref. 6, ©2015 NPG)

solution again and adjust the pH to 8.3. Make up to 1 L with distilled water and store at room temperature. Sterilize the solution by autoclaving (*see* **Note 4**).

2. Agarose gel: prepare agarose and measure out agarose depending on the size of bands (*see* **Note 5**).
3. Nucleic acid staining dye (e.g., ethidium bromide (EtBr) or SYBR Gold).
4. Electrophoresis unit.

2.4 DNA–RNA Hybrids for Condensation of RCT Products

1. Cholesterol-conjugated DNA fragments (DNA–Chol conjugates) (*see* **Note 6**). Dissolve the DNA fragments in nuclease-free water and store at 4 °C (*see* **Note 7**).
2. Hybridization buffer: 50 mM Tris–HCl (pH 7.8), 10 mM $MgCl_2$, 10 mM DTT, 1 mM ATP. Store at −20 °C.

3. Amicon Ultra 0.5 mL Centrifugal filter, 3K MWCO, or equivalent.
4. 1 M $MgCl_2$ stock solution in distilled water stored at room temperature.

2.5 DNA–RNA Hybrids for Folate Receptor Targeting

1. Conjugation buffer: 100 mM MES (2-(*N*-morpholino)ethanesulfonic acid), 500 mM NaCl, pH 6.0. Store at room temperature.
2. 20 mM folate: dissolve 2.2 mg folate in conjugation buffer and store at −20 °C (*see* **Note 8**).
3. 20 mM EDC: 1-ethyl-3-(3-dimethylaminopropyl)carbodiimide hydrochloride. Dissolve 3.8 mg EDC in conjugation buffer and store at −20 °C.
4. 20 mM Sulfo-NHS: *N*-hydroxysulfosuccinimide. Dissolve 4 mg Sulfo-NHS in conjugation buffer and store at −20 °C.
5. 5′-amine-modified DNA fragments (*see* **Note 9**).
6. Hybridization buffer: 50 mM Tris–HCl (pH 7.8), 10 mM $MgCl_2$, 10 mM DTT, 1 mM ATP. Store at −20 °C.
7. DNA cleanup kit (e.g., PrepEase DNA Clean-Up Kit, Affymetrix).
8. RNA cleanup kit (e.g., RNeasy MinElute Cleanup kit, Qiagen).

3 Methods

3.1 Preparation of Circular DNA Template

1. Add T7 promoter primer solution to linear ssDNA template solution at a molar ratio of 1:1 and mix with 1× annealing solution. Make up to 20 μL with nuclease-free water in a PCR tube (*see* **Note 10**).
2. Heat mixed solution at 95 °C for 2 min and slowly cool down to 25 °C such that a 10 °C temperature reduction corresponds to a time interval of 10 min (*see* **Note 11**).
3. Add 5 U of T4 ligase to circular DNA solution containing 1× ligation buffer and incubate at 16 °C for 12 h (*see* **Note 12**).
4. Incubate the ligated solution at 65 °C for 10 min to inactivate T4 DNA ligase.
5. Purify the successfully ligated products using a DNA cleanup kit (*see* **Note 13**).
6. Check the ligated products on TAE agarose gel (*see* **Note 14**).

3.2 Rolling Circle Transcription for Generation of Polymerized RNA Transcripts

1. Mix 5 U of T7 RNA polymerase, 2 μL of 1× reaction buffer, 2 mM rNTP mixture solution, 1 U of RNase inhibitor, and 0.5 μM ligated DNA hybrids. Make up to 20 μL with nuclease-free water in e-tube.
2. Incubate reaction solutions at 37 °C for 1 h (*see* **Note 15**).
3. Treat with 2 U of DNase I at 37 °C for 20 min to remove the DNA template/T7 promoter hybrids.
4. Incubate the reaction mixture at 90 °C for 5 min (*see* **Note 16**).
5. Purify the reaction products by using an RNA cleanup kit or ethanol precipitation method.
6. Measure the concentration of RCT products using an RNA quantification assay kit (*see* **Note 17**).
7. Characterize the resulting RCT products on agarose gel electrophoresis (*see* **Note 18**).

3.3 Agarose Gel Analysis

1. Prepare the agarose and measure out agarose depending on sizes of bands.
2. Pour agarose into microwavable flask with 1× TAE buffer.
3. Microwave for 1–2 min to completely dissolve agarose.
4. Let agarose solution cool down.
5. Add nucleic acid staining dye in gel solution (*see* **Note 19**).
6. Pour the agarose solution into a gel tray with the well comb in place.
7. Dry agarose gel and run the gel in 1× TAE running buffer at 100 V for 0.5–1 h.

3.4 DNA–RNA Hybrids for Condensation of RCT Products

1. Mix DNA–Chol conjugate solutions with the RNA transcript solution in hybridization buffer at an indicated weight ratio (*see* Fig. 3 and **Note 20**).
2. Add 10 mM $MgCl_2$ to a final concentration of 1 mM and make up to 20 μL with nuclease-free water (*see* **Note 21**).
3. Heat the mixed solution at 65 °C for 10 min and slowly cool down to 4 °C by each 10 °C reduction with an interval of 10 min (*see* **Note 22**).
4. Remove free $MgCl_2$ by using an ultracentrifugal filter (*see* **Note 23**).

3.5 DNA–RNA Hybrids for Cancer-Specific Targeting

1. Mix 10 μL of folate solution, 12.6 μL of EDC solution, and 35 μL of Sulfo-NHS solution up to 500 μL with conjugation buffer (*see* **Note 24**).
2. Gently vortex the reaction mixture at room temperature for 15 min.
3. Add 1.4 μL of β-mercaptoethanol to deactivate excess EDC.

4. Add 44 μg DNA (*see* **Note 25** and Fig. 2) and gently vortex for 1.5–3 h at room temperature (*see* **Note 26**).
5. Purify the folate–DNA (FA–DNA) conjugates using a centrifugal filter (3K) (*see* **Note 27**).
6. Quantify the conjugation ratio of folate on DNA by UV spectrophotometry (*see* **Note 28**).
7. Mix FA–DNA conjugates with RNA transcripts/DNA–Chol hybrids at a weight ratio of 1:0.2 in hybridization buffer.
8. Add 10 mM $MgCl_2$ and make up to 20 μL with nuclease-free water.
9. Heat the mixed solution at 65 °C for 10 min and slowly cool down to 4 °C by each 10 °C reduction with an interval of 10 min.
10. Remove free $MgCl_2$ using an ultracentrifugal filter at 5000 × *g* with several times.

4 Notes

1. ssDNA should contain a phosphate group at the 5′ terminus of DNA to facilitate ligation. Complementary sequence of T7 promoter should be contained at each end of ssDNA template for generation of circular DNA. Use of a software tool can be better to predict a secondary RNA structure transcribed from DNA template for generating multimeric RNAi hairpins. The sequence design tool assigns sequences of the DNA template to minimize unintended secondary interactions. Note that Dicer processing requires a minimum stem length of ~21 nt [10]. The complete monomer unit for RNAi hairpins is shown in Fig. 1.
2. The ATP in the ligase buffer is essential for the ligation reaction, but is broken down by repeated freeze–thaw cycles. Make aliquots of ligase buffer from the fresh stock.
3. Dilute with distilled water to make final 1× working solution from 50× TAE buffer. 0.5 M EDTA solution should be autoclaved to avoid contamination. Autoclaving inactivates most enzymes and kills most microorganisms.
4. We recommend to adjust the solution pH at room temperature since pH is dependent on temperature. TAE buffer can be stored for several months. TAE buffer has lower buffering capacity then TBE, and therefore its use should be avoided for extended electrophoresis.
5. Adjust the amount of agarose to % g/100 mL TAE.
6. We use 3′-cholesterol-modified DNA (DNA–Chol) conjugates for condensation of the RNA transcripts. The nucleic acid

sequence of DNA–Chol conjugates should be designed to be complemented with linker regions of polymerized RNA transcripts through RNA/DNA base pairing.

7. Before use, heat the solution at 50 °C for 30 min when aggregates are formed, since cholesterol molecules can be easily aggregated due to the hydrophobicity of molecules.
8. Store as aliquots in tightly sealed vials at −20 °C. Generally, folate solution will be useable for up to 1 month. We recommend that you allow aliquots to equilibrate to room temperature for at least 1 h.
9. The 5′-amine group of the DNA fragments can be conjugated with a folate via an EDC/Sulfo-NHS cross-linking reaction involving a carboxyl group. The nucleic acid sequences of DNA fragments should be designed to be complemented with loop regions of polymerized RNA transcripts through RNA–DNA base pairing.
10. Dissolve ssDNA template and T7 promoter primer in nuclease-free water, respectively. Store at −20 °C. We use 0.3 μM of T7 promoter primer and ssDNA template, respectively.
11. The heating step is required for denaturation and the cooling step is required for sequence-specific base pairing of the ssDNA template with the primer. We use a PCR machine for precisely controlled temperature. We find those settings to be the best to anneal, but one can consider increasing the time intervals when there is weak base pairing. The agarose gel assay can be needed to test annealing efficiency.
12. A ligase connects the nick of circular DNA/T7 promoter hybrids. Additional 1 mM EDTA is helpful when ligation is weak.
13. This is optional. We also recommend the use of T4 polymerase to remove linear analogs or nicked hybrids as an alternative method.
14. We use 2–3% agarose gel to check linear hybrids and ligated hybrids.
15. T7 RNA polymerase can act progressively and continuously on the circular DNA hybrids generating multiple copies of tandem RNA units [8]. The size of the resulting RCT products can be controlled by reaction time or concentration of RNA polymerase [11]. As the RNA strands are transcribed continuously, they reach fiber-like, then sheet-like and finally microsponge-like RNA particles.
16. Enzymes present in the reaction mixture can be inactivated by heating.

17. We recommend an RNA quantification kit to make RNA quantification easy and accurate. The assay is highly selective for RNA over double-stranded DNA.
18. We use 1% agarose gel due to high molecular weights of the resulting RCT products.
19. Nucleic acid staining dye allows you to visualize the nucleic acids under UV light. We use SYBR Gold instead of EtBr, because EtBr is a known mutagen [12].
20. We observed a significant reduction of particle size at a weight ratio of 1:0.2 (Fig. 3). RNA/DNA–Chol hybrids self-assemble into highly packed nanoparticles due to the acquired amphiphilicity.
21. Higher salt concentration or higher temperature facilitates the hybridization rate of short oligonucleotides [13–15]. Therefore, supply additional $MgCl_2$ when hybridization rate is low.
22. Heat at 65 °C to denature secondary structure of RNA transcripts as an intial step. We use a PCR machine for precisely controlled temperature.
23. Repeate several times with distilled water to completely remove the free reagents.
24. Use EDC and Sulfo-NHS at molar ratio of 1:2.5.
25. Use folate and DNA at a weight ratio of 1:2. One can add excess of DNA if cross-linking between folate and DNA is not sufficient.
26. We recommend the use of a rotary vortex.
27. Repeat several times with nuclease-free water.
28. The ratio of folate to DNA on FA–DNA conjugates can be determined by measuring extinction coefficients of pure folate and DNA at two different wavelengths, 260 and 363 nm [7, 16].

References

1. Shu Y, Pi F, Sharma A, Rajabi M, Haque F, Shu D, Leggas M, Evers BM, Guo P (2014) Stable RNA nanoparticles as potential new generation drugs for cancer therapy. Adv Drug Deliv Rev 66:74–89
2. Burnett JC, Rossi JJ (2012) RNA-based therapeutics: current progress and future prospects. Chem Biol 19:60–71
3. Schiffelers RM, Ansari A, Xu J, Zhou Q, Tang Q, Storm G, Molema G, Lu PY, Scaria PV, Woodle MC (2004) Cancer siRNA therapy by tumor selective delivery with ligand-targeted sterically stabilized nanoparticle. Nucleic Acids Res 32:e149
4. Yin JQ, Gao J, Shao R, Tian WN, Wang J, Wan Y (2003) siRNA agents inhibit oncogene expression and attenuate human tumor cell growth. J Exp Ther Oncol 3:194–204
5. Jiang M, Milner J (2002) Selective silencing of viral gene expression in HPV-positive human cervical carcinoma cells treated with siRNA, a primer of RNA interference. Oncogene 21:6041–6048
6. Bobbin ML, Rossi JJ (2016) RNA Interference (RNAi)-Based therapeutics: delivering on the promise? Annu Rev Pharmacol Toxicol 56:103–122
7. Jang M, Kim JH, Nam HY, Kwon IC, Ahn HJ (2015) Design of a platform technology for systemic delivery of siRNA to tumours using rolling circle transcription. Nat Commun 6:7930

8. Lee JB, Hong J, Bonner DK, Poon Z, Hammond PT (2012) Self-assembled RNA interference microsponges for efficient siRNA delivery. Nat Mater 11:316–322
9. Daly TA, Wang M, Regen SL (2011) The origin of cholesterol's condensing effect. Langmuir 27:2159–2161
10. Ge Q, Ilves H, Dallas A, Kumar P, Shorenstein J, Kazakov SA, Johnston BH (2010) Minimal-length short hairpin RNAs: the relationship of structure and RNAi activity. RNA 16:106–117
11. Han D, Park Y, Nam H, Lee JB (2014) Enzymatic size control of RNA particles using complementary rolling circle transcription (cRCT) method for efficient siRNA production. Chem Commun (Camb) 50:11665–11667
12. Meyer RR, Simpson MV (1969) DNA biosynthesis in mitochondria. Differential inhibition of mitochondrial and nuclear DNA polymerases by the mutagenic dyes ethidium bromide and acriflavin. Biochem Biophys Res Commun 34:238–244
13. Kuhn H, Demidov VV, Coull JM, Fiandaca MJ, Gildea BD, Frank-Kamenetskii MD (2002) Hybridization of DNA and PNA molecular beacons to single-stranded and double-stranded DNA targets. J Am Chem Soc 124:1097–1103
14. Straus NA, Bonner TI (1972) Temperature dependence of RNA-DNA hybridization kinetics. Biochim Biophys Acta 277:87–95
15. Podder SK (1971) Co-operative non-enzymic base recognition. A kinetic study of interaction between GpGpGpC and GpCpCpC and of self-association of GpGpGpC. Eur J Biochem 22:467–477
16. Vlashi E, Kelderhouse LE, Sturgis JE, Low PS (2013) Effect of folate-targeted nanoparticle size on their rates of penetration into solid tumors. ACS Nano 7:8573–8582

Chapter 5

Computational Prediction of the Immunomodulatory Potential of RNA Sequences

Gandharva Nagpal, Kumardeep Chaudhary, Sandeep Kumar Dhanda, and Gajendra Pal Singh Raghava

Abstract

Advances in the knowledge of various roles played by non-coding RNAs have stimulated the application of RNA molecules as therapeutics. Among these molecules, miRNA, siRNA, and CRISPR-Cas9 associated gRNA have been identified as the most potent RNA molecule classes with diverse therapeutic applications. One of the major limitations of RNA-based therapeutics is immunotoxicity of RNA molecules as it may induce the innate immune system. In contrast, RNA molecules that are potent immunostimulators are strong candidates for use in vaccine adjuvants. Thus, it is important to understand the immunotoxic or immunostimulatory potential of these RNA molecules. The experimental techniques for determining immunostimulatory potential of siRNAs are time- and resource-consuming. To overcome this limitation, recently our group has developed a web-based server "imRNA" for predicting the immunomodulatory potential of RNA sequences. This server integrates a number of modules that allow users to perform various tasks including (1) generation of RNA analogs with reduced immunotoxicity, (2) identification of highly immunostimulatory regions in RNA sequence, and (3) virtual screening. This server may also assist users in the identification of minimum mutations required in a given RNA sequence to minimize its immunomodulatory potential that is required for designing RNA-based therapeutics. Besides, the server can be used for designing RNA-based vaccine adjuvants as it may assist users in the identification of mutations required for increasing immunomodulatory potential of a given RNA sequence. In summary, this chapter describes major applications of the "imRNA" server in designing RNA-based therapeutics and vaccine adjuvants (http://www.imtech.res.in/raghava/imrna/).

Key words Immunomodulatory RNA, RNA immunotoxicity, imRNA, Prediction, TLR 7, Adjuvant, Machine learning, SVM

1 Introduction

Various types of noncoding RNA molecules have shown promising potential for use in therapeutics. Among these, small-interfering RNA (siRNA) and micro-RNA (miRNA) are the major categories of RNA that have advanced into clinical trials for their translation into approved drugs. These two classes spearhead the RNA interference antisense technology that mainly works by lowering gene

Eckart Bindewald and Bruce A. Shapiro (eds.), *RNA Nanostructures: Methods and Protocols*, Methods in Molecular Biology, vol. 1632, DOI 10.1007/978-1-4939-7138-1_5,

expression caused by inhibition of mRNA translation [1]. Other mechanisms of antisense therapy include interference in RNA splicing, competing for binding with cognate proteins, disruption of ribozyme folding, and impeding telomerase activity involving RNA molecules [2]. The redirecting of a bacterial immunological pathway known as CRISPR-Cas9 (clustered regularly interspaced short palindromic repeats—CRISPR-associated genes), for genome engineering, is the latest paradigm in the application of RNA molecules as therapeutics. The CRISPR-Cas9 system can guide the enzymatic machinery to specific locations within any genome using RNA sequences complementary to the location in the genome, thus being capable of inducing site-specific cleavage and joining [3].

The emerging field of synthetic biology has demonstrated the potential of biopolymers like DNA, RNA, and proteins to provide biomaterials for medical applications. The polymeric and self-assembling nature of RNA allows programmed coalescence of RNA strands into complex nanostructures. These nanostructures have diverse applications like biocompatible materials for tissue engineering and vehicles for targeted drug delivery [4]. Khisamutdinov et al. have demonstrated that by manipulating the shape of RNA nanovehicles carrying a DNA adjuvant, the immunostimulatory effects can be enhanced for the same adjuvant molecule [5]. A U-rich synthetic RNA molecule itself has been shown to be a safe and potent adjuvant for a vaccine against cancer [6].

Thus, the activation of the mammalian innate immune system via RNA molecules can either be an advantage or disadvantage depending on the condition whether the therapy is enhanced or attenuated by the immune responses. In both the cases, the immunomodulatory potential of the RNA molecule is an important consideration affecting the choice of RNA sequences that could be developed as therapeutics. Hence, determination of the immunomodulatory potential of RNA sequences becomes an integral aspect in rational design of RNA-based therapy.

The innate immune system can recognize specific molecular patterns present in the microorganisms through pattern recognition receptors (PRRs). Toll-like receptors (TLRs) constitute a major family of PRRs, of which TLR 7 recognizes single-stranded oligoribonucleotides (ORNs). This recognition is further propagated by the innate immune system through cytokine induction. Figure 1 is an outline of cytokine induction by TLR 7 as a representative example of the toll-like receptor family of PRRs depicting how microbial single-stranded RNA activates TLR 7/8 and leads to activation of immune cells involved in both innate as well as adaptive immunity.

Experimental approaches for determining the TLR 7-mediated immunostimulatory effect of RNA largely involves the estimation of plasma level of cytokines like IFN-α and TNF-α [7, 8]. These approaches require highly sensitive instrumentation kits for

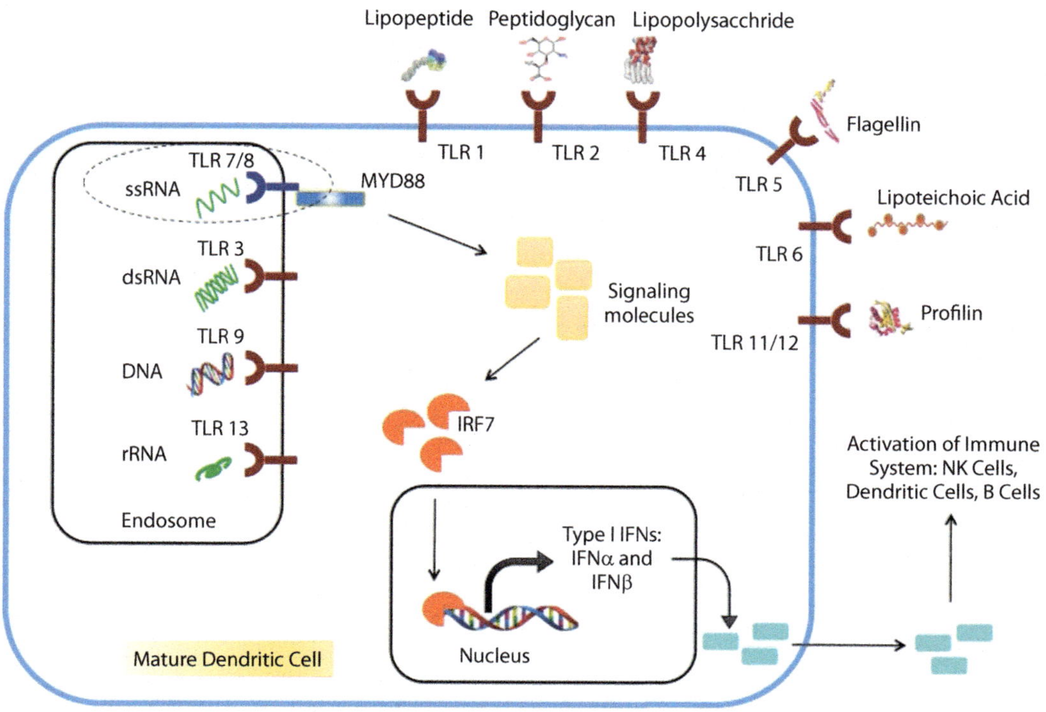

Fig. 1 Immune system activation by pattern recognition receptors particularly mediated by single-stranded RNA bound toll-like receptor 7/8. *TLR* toll-like receptor, *ssRNA* single-stranded RNA, *dsRNA* double-stranded RNA, *rRNA* ribosomal RNA, *MYD88* myeloid differentiation primary response gene 88, *IRF7* interferon regulatory transcription factor 7, *IFN* interferon, *NK Cells* natural killer cells, *B Cells* B lymphocytes

determining the immunostimulatory activity. In such a scenario, an in silico method for estimating the immunomodulatory potential of RNA sequences developed from the knowledge of experimentally known immunostimulatory RNA sequences could be highly useful for selecting the sequences of therapeutic RNA for both enhancing or eliminating the immunomodulatory effect depending on the need. The immunological side effects of RNA molecules were compiled previously by Olejniczak et al. in the form of a database named RNAimmuno [9]. A computational method to estimate the immunomodulatory potential of RNA sequences has not been developed hitherto.

The present chapter explains the functionality of imRNA [10], a web-based in silico method to predict the immunomodulatory potential of the RNA query sequences. The prediction models were developed using the Support Vector Machine learning methods on the single-stranded RNA sequences experimentally shown to be immunomodulatory (*see* **Notes 1** and **2**). Chaudhary et al. have discussed the details of the methodology used to develop the prediction models [10]. Here, various modules and their usage have been discussed elaborately in detail that may help a user to obtain desirable RNA sequences with respect to the estimated immunomodulatory potential. The primary emphasis of this web tool is the reduction of the immunotoxic side effect of therapeutic RNAs. However, the increased immunomodulatory potential of RNA sequences finds application in vaccines where these RNA molecules could act as potential vaccine adjuvants leading to the activation of innate immunity. Thus, imRNA houses tools to increase the immunomodulatory potential as well, thereby facilitating the design of potential vaccine adjuvants.

2 Methods

2.1 imRNA Schema

The web server "imRNA" developed for the prediction of immunomodulatory potential of RNA sequences is a method developed using the machine learning method—Support Vector Machines. Accordingly, the method relies on a scoring scheme for which a threshold value is obtained after optimization. In the case of imRNA, the score written as "IMscore" assigned to any RNA query sequence by the prediction models ranges between 0 and 9. The "IMscore," derived from the SVM score range used to optimize the prediction models, represents the probability of a query sequence being immunomodulatory or non-immunomodulatory. The higher the "IMscore," the higher is the probability of the query sequence to be an immunomodulator (*see* **Note 3**). On the other hand, a higher "IMscore" cutoff will decrease the coverage or inclusion of actual immunomodulators into the predicted positives. For increasing the coverage, the threshold probability ("IMscore") needs to be relaxed to a lower value amounting to a compromise on specificity (increased number of false positives). The prediction model has been optimized with the threshold "IMscore" of 4.5 that means if the RNA query sequence obtains a score equal to or above 4.5 by the model, it would be judged to be an "immunomodulatory" (positive) sequence and "nonimmunomodulatory" (negative) in case the score calculated by the model is below 4.5. The web server provides an additional facility for the user to change the threshold from 4.5 to other values. This effectively means that the user can alter the criterion (threshold probability) of deciding whether the input RNA sequence is immunomodulatory or not. The user can make the criterion more stringent by increasing the threshold in the case where the user is getting numerous positive hits among the query sequences submitted and wishes to restrict the search to the best few sequences having the potential of immunomodulation. On the other hand, when the user obtains only a few or no positive hits among the query sequences, it would be useful to relax the criterion by reducing the threshold from 4.5 to lower values such that the sequences having score very close to but below 4.5 would be considered to be immunomodulatory, thus suggesting the best candidates to proceed with further scientific investigation.

The prediction model used in imRNA was developed on a dataset containing 602 experimentally validated immunomodulatory sequences (positives) as well as 520 negatives [10]. The length of the dataset sequences ranged from 17 to 27. Hence, in principle, the prediction models have been trained to predict the immunomodulatory score only for the query sequences having length 17–27. In the web interface, the user is advised to hit the "Check Sequence" button in the form submission after inserting the query sequences wherever provided so as to ensure that the sequences have been given in the correct format. The "Check Sequence" button issues a message alert, reporting the correctness or incorrectness of the query sequence format with respect to three aspects—(1) whether the sequence has been put in FASTA format (in some forms as single sequence) or not, (2) in case of determination of immunomodulatory potential of the complete query sequence as a single entity the sequence should be in the length range 17–27 and (3) there should be no spaces or characters other than A, C, G, T, and U in the sequence. Here it is worthwhile to mention that the web server form submission accepts both T and U in the RNA sequence as both U and T are ultimately assumed to be uridines (U). Thus, ensuring the correctness of input is essential for the server scripts to perform as per the assumptions made in the inception and development of the imRNA web tool. The next section discusses in detail how the user can submit queries in various ways and interpret the results generated by the server using illustrations of RNA molecule sequences that were experimentally evaluated for immunostimulatory potential or are therapeutic RNAs being clinically tested.

2.2 imRNA Web Server Tools

For accommodating various tools within a single platform while preserving the user-friendly arrangement, the tools were categorized into three sections—(1) Vaccine Adjuvants, (2) siRNA Immunotoxicity, and (3) Additional Tools. Though many tools are essentially calculating the immunomodulatory score generated by the SVM-based prediction model and annotating the query sequences to be immunomodulatory or nonimmunomodulatory on the basis of the user-selected threshold, the presentation of the results are framed differently for correlating the immunomodulatory estimation with the scientific perspective a user might be investigating. For example, consider the tools "Virtual Screening" and "RNA Sequences" in the menu "Vaccine Adjuvants." Both are calculating SVM scores for the RNA sequences of length 17–27 for designating them as immunomodulatory or nonimmunomodulatory. Running the "Virtual Screening" tool is the simple classification of query sequences into immunomodulatory and nonimmunomodulatory classes, while the usage of the "RNA Sequences" is more for identifying immunomodulatory "regions" or subsequences within an RNA sequence longer than the length range for which the prediction model has been trained.

2.2.1 siRNA Immunotoxicity

Despite rapid advances in the therapeutic applications of RNA molecules and discovery of a new class of RNAs promising high-end applications like genome editing (CRISPR-cas9 associated guide RNAs), small-interfering RNAs (siRNAs) continue to dominate RNA therapeutics. However, the immunotoxicity of siRNA is a major limitation in its therapeutic application. Hence, promising therapeutic potential of siRNAs calls for a separate evaluation module that may facilitate the simultaneous evaluation of siRNA efficacy and its immunotoxicity. imRNA provides a main menu "siRNA Immunotoxicity" to help a user in selecting an siRNA sequence with desired efficacy as well as desired immunomodulatory potential so as to select highly effective siRNAs along with minimal immunotoxicity. In contrast, some workers may wish to design siRNAs that are highly effective as well as inducers of innate immune activation such that the siRNA serves a dual therapeutic effect of immune reactivity in addition to gene silencing, especially in the case of cancers [11]. The siRNA efficacy predictions are made by the "desiRm" program [12] and indicated by the Dscore in the result table.

Screen siRNA Library

This tool has the basic functionality of the "Virtual Screening" module with a customization of predicting the siRNA efficacy in addition to the prediction of immunomodulatory potential. An interesting example to demonstrate the capabilities of this tool would be that of the siRNAs studied by Hornung et al. [13]. The study evaluated the level of cytokine IFN-α induction by four unmodified siRNAs designated siRNA_9.1, siRNA_9.2, siRNA_9.3 and siRNA_9.4 and found that the cytokine induction by siRNA_9.2 and siRNA_9.3 was significantly higher than that of siRNA_9.1 and siRNA_9.4. Figure 2 contains a snapshot of the submission page and the result page where the sense and antisense strand sequences of all the siRNAs were put as a query. Both the strands of siRNA_9.2 and siRNA_9.3 are reported to be immunomodulatory based on the IMscore while a majority of the siRNA_9.1 and siRNA_9.4 strands are nonimmunomodulatory. Additionally, the predicted siRNA efficacies of the siRNA_9.2 and siRNA_9.3, as indicated by the Dscore, are also very high (minimum 0.62 and maximum 0.99). Thus, the server predictions are in agreement with the experimental findings.

siRNA Analogs

The "siRNA Analogs" tool generates all possible substitution mutants of the single query sequence and predicts the immunomodulatory potential of the mutants. Oriented toward application in siRNA designing, this module predicts the siRNA efficacy of the mutants using the "desiRm" program presented in the form of Dscore in the result table. The facility of sorting all the columns including the IMscore and Dscore allows the user to select sequences having desirable altered immunomodulatory scores

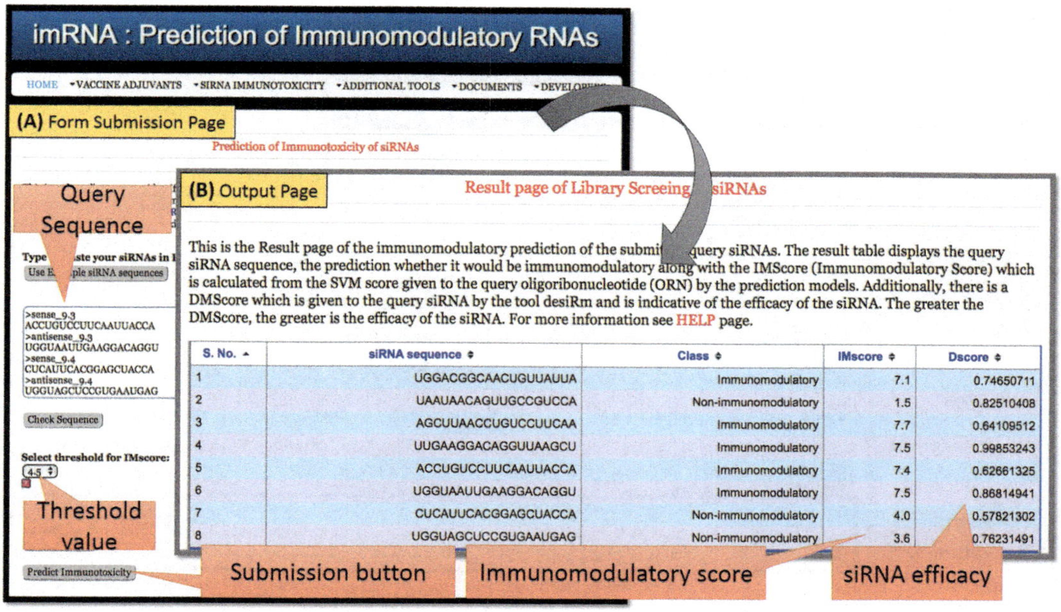

Fig. 2 The screenshots demonstrating the use of "siRNA toxicity" module of imRNA to predict the toxicity of siRNA. (*A*) The Form Submission page takes a list of siRNA sequences as input along with the user-defined threshold for the IMscore. It is automatically directed to output page after submission. (*B*) The output page is presented in tabular format to list each input sequence along with their IMscore and Dscore

along with desired altered siRNA efficacy in comparison to the parent siRNA. Thus, this module helps to improve the design of siRNA about the two of the most important functional aspects (Fig. 3).

2.2.2 Vaccine Adjuvants

This section equips the user with tools that help in identifying immunomodulatory sequences among a set of query sequences or a larger RNA sequence (*see* **Notes 4** and **5**) as well as designing sequences with increased immunomodulatory potential by mutating the residues within an identified immunomodulator (*see* **Note 6**).

Virtual Screening

The web page form in this tool has a text area for the user input query sequence(s) in FASTA format and a drop-down menu for selecting the SVM score selection. The "Use Example Sequence" button inserts example sequences in the required format, the "Check Sequence" button potentially raises alerts in cases of incorrectly formatted input data. The "Clear this form" button empties the text area and the "Predict" button submits the user query to the server to obtain the results (Fig. 4). For the sake of clarity, the example sequence of siRNA PF655 licensed to Quark Pharmaceuticals for treating ocular disease [14], might be interesting as this siRNA is currently subject to clinical trials. When the PF655 sequence 5′-"GUGCCAACCUGAUG-CAGCU" is submitted as one of the query sequences, the result

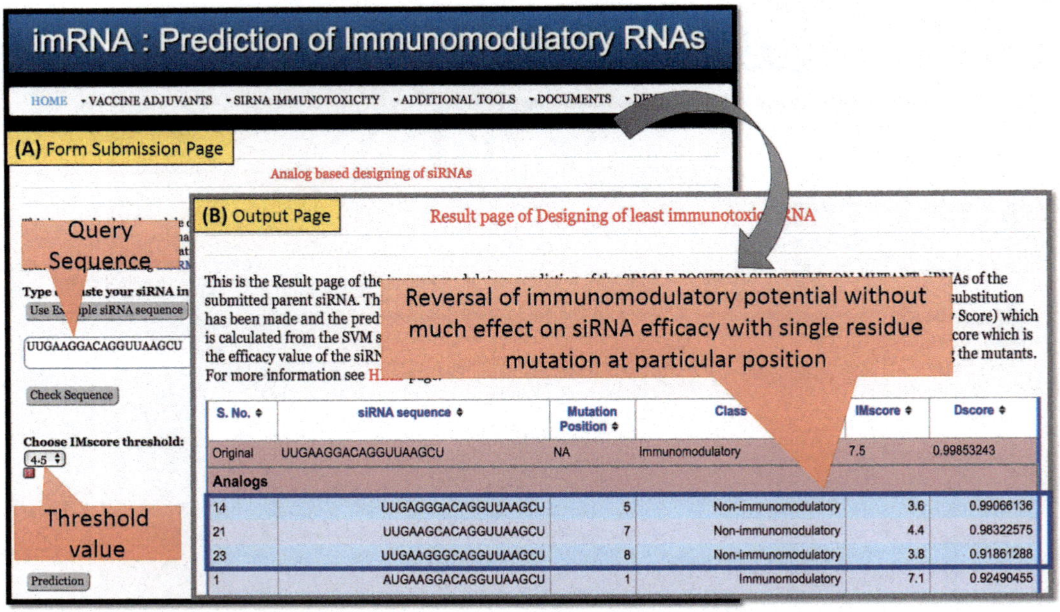

Fig. 3 The graphically annotated screenshots of "siRNA Analogs" module to find the best analog for a given siRNA sequence. (*A*) The input page takes a single siRNA sequence and generates all possible analogs at each position. (*B*) The result for each analog is displayed in tabular format

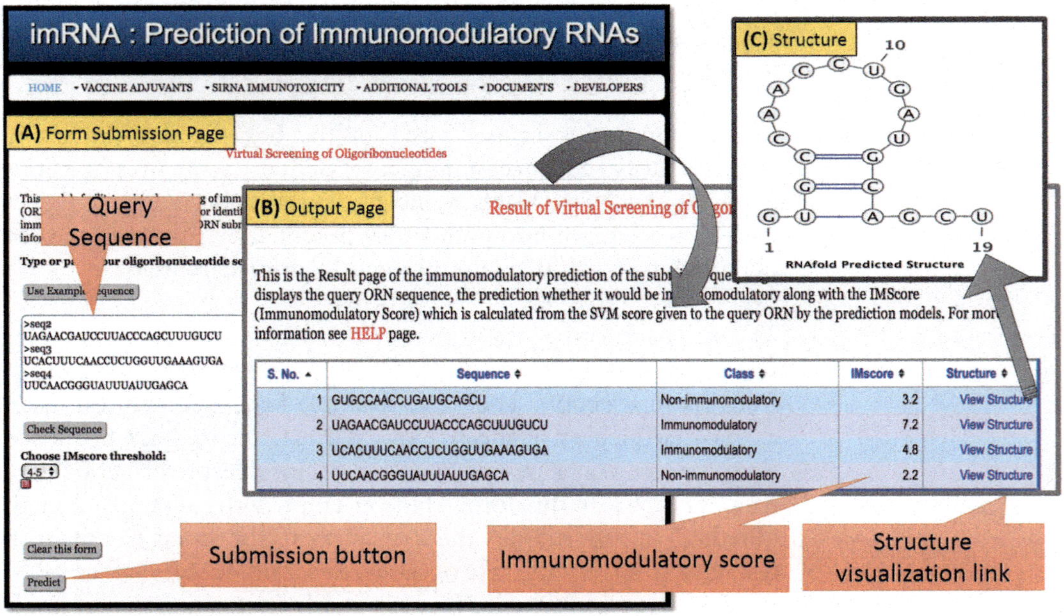

Fig. 4 The screenshots for "Virtual Screening" module. (*A*) The submission page takes multiple RNA sequences for predicting their immunomodulatory potential. (*B*) The output is displayed with sorting facility linked with each column header. (*C*) The structure of an RNA sequence has been visualized for illustration

page predicts PF655 to be nonimmunomodulatory. This is so because the SVM score of 3.2 is below the threshold value of 4.5 set as default on the submission page. The result table displays the predicted class (immunomodulatory or nonimmunomodulatory) and the SVM prediction score against each query sequence. The headers of the table can be clicked to sort the table according to the clicked column, thereby arranging the rows in increasing (ascending) order on the first click and decreasing (descending) order on the second click. The sortable table allows the user to compare the query sequences for their immunomodulatory potential that may help frame experiments for investigation. Additionally, the result table provides links to the predicted secondary structures corresponding to the query sequences in the last column through the clickable text "View Structure". Scientific literature highlights the occurrence and absence of certain types of nucleotides and motifs in the stem-loop structure of RNA molecules to which the immunomodulatory potential of the sequence might be attributed [10]. Thus, structure visualization might provide insight into the underlying principles of immunomodulation. The structure is predicted by the RNAFold program from TBI Vienna Package [15] and visualized using the VARNA applet [16].

Analogs of ORNs

Modifying a lead molecule for improving its therapeutic properties is a common practice in the area of drug discovery. In the case of RNA molecules, the most intuitive change to alter the properties would be to replace the base sequence in a stepwise manner. The "Analogs of ORNs" module caters to this need of virtually modifying the sequence of RNA molecules to identify sequence analogs of a parent query sequence with altered immunomodulatory potential. The module generates all possible single site substitutions in the parent sequence and provides the immunomodulatory potential of the mutants along with the mutation position in the result table. By sorting the "Class" column of the result table, the user can obtain top scoring immunomodulatory (first click) or nonimmunomodulatory mutants (second click). Figure 5 illustrates the substitution mutants of example sequence (5′ GUGCCAACCUGAUGCAGCU) of PF655 siRNA. If the user wishes to obtain a siRNA having lower immunomodulatory potential than PF655, the result page provides a nonimmunomodulatory mutant (5′ GUGCCCACCUGAUGCAGCU) obtained by mutating the sixth position from A to C that has the immunomodulatory score of 1.8 which is much lower than that of PF655 (3.2). On the other hand, mutating the 13th position in PF655 from U to C gives an immunomodulatory score of 5.0 which is also above the default threshold, thus providing PF655 mutant (5′ GUGCCAACCUGACGCAGCU) predicted to be immunomodulatory in contrast to PF655 which is predicted to be nonimmunomodulatory (IMscore 3.2). Both of these sequences can be obtained by simply

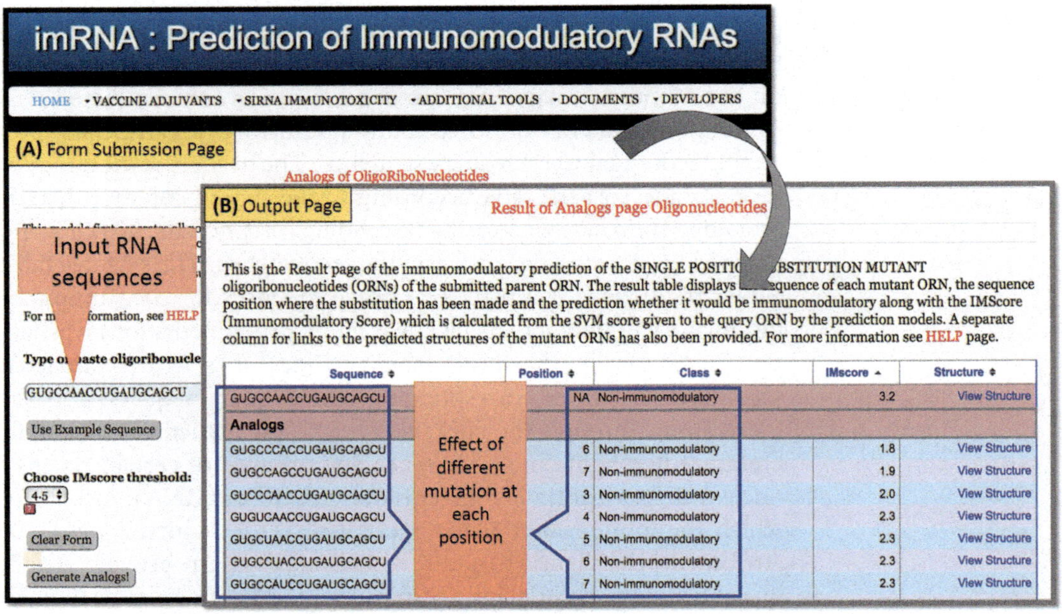

Fig. 5 The annotated screenshots for the screening of the best mutant for a given RNA sequence using "Analogs of ORNs" module. (*A*) The RNA sequence can be given as input for submission page. (*B*) The user can select the best mutant based on the given IMscore in the result page

clicking on the IMscore header in the result table. The default threshold of 4.5 provides only five mutants that are predicted to be immunomodulatory. If the user needs more than five mutants (say around ten mutants) as the sequence choice that are most likely to be immunomodulatory, the user simply needs to go back to the "Analogs of ORNs" form submission page and lower the threshold from 4.5 to 4 and submit the PF655 sequence as the query by hitting the "Generate Analogs!" button. The result table now provides nine sequences that are most likely to be immunomodulatory among the mutants. These nine sequences include four additional sequences apart from those predicted to be immunomodulatory with threshold 4.5. This happened because, in the previous case, the IMscores (4.2 and 4.4) of the newly found mutants were very close but less than the default threshold 4.5. By lowering the threshold to value 4, these sequences were then designated immunomodulatory as their IMscore was greater than 4. Similarly, increasing the threshold score to 5 leaves only one mutant (with U to C substitution at 13th position) as the single immunomodulatory sequence. Thus, increasing or decreasing the threshold values corresponds to making the search criterion stringent or lenient respectively so as to decrease or increase the number of positive hits (immunomodulatory sequences in this case) respectively. In a nutshell, the search by prediction of immunomodulatory sequences by the model is threshold dependent.

RNA Sequences

Often times a user might be interested in knowing and extracting the subsequences (regions) of a long RNA sequence that might be immunomodulatory. For instance, the Human Immunodeficiency Virus or the HIV is a virus infecting the helper T cells of the human immune system predominantly. Since HIV has an RNA genome, a researcher might be interested in finding those regions in the RNA genome of HIV that might serve as immunostimulator of the innate immunity. The "RNA Sequences" tool of imRNA may be helpful in such cases. Using the user-defined sequence length window in the form submission page, tool generates all possible sequences of the selected length through a scanning window method and predicts the IMscore for each possible sequence. Here the allowed window length is in the range of the dataset sequence lengths on which the prediction model was optimized. In the case of the example of the HIV RNA genome considered here, submitting this sequence (via clicking the "Use Example Sequence" button) with a default window length 27 and default threshold 4.5 yields a result page where the region with the highest IMscore of 6.4 is a subsequence of length 27 starting at position 846. As is the case for the other tools, varying the thresholds could yield a larger number of likely immunomodulatory regions and the most likely of these immunomodulatory regions. Another example, demonstrating the application of this module, is the mRNA transcript of the gene *RTP801* that is the target of the therapeutic siRNA PF655 which is currently undergoing a clinical trial for the treatment of ocular disease [14]. On submitting the mRNA transcript of *RTP801* with default values of threshold and window length, the result page assigns nonimmunomodulatory status based on the IMscore to the subsequences of length 27 starting at positions 472–480. Interestingly, the PF655 siRNA sequence also falls in this region. Through this example, one can rationally identify additional strategic regions at the 3′ end of *RTP801* mRNA that are nonimmunomodulatory and identify more siRNA-based drugs similar to PF655. Figure 6 is an illustration of identifying immunomodulatory or nonimmunomodulatory regions of *RTP801* mRNA.

2.2.3 Additional Tools

Motif analyses of the immunomodulatory sequences, using the MERCI program [17], revealed that certain sequence motifs are preferred in the immunomodulatory sequences as compared to the nonimmunomodulatory ones [10]. The occurrence of immunomodulatory motifs could be a direct indication that an RNA sequence could be immunomodulatory. Also, sequence similarity to an experimentally verified immunomodulator could also indicate the immunomodulatory capacity of an RNA query sequence. Hence, tools for detecting the occurrence of immunomodulatory motifs in a query sequence and for determining BLAST-based sequence similarity to experimentally verified immunomodulators

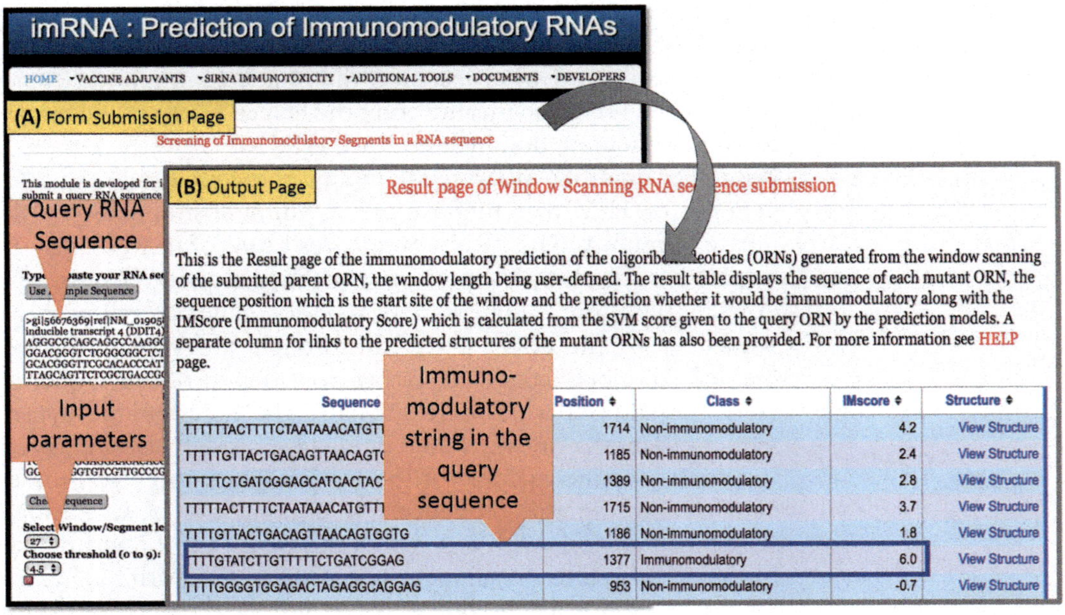

Fig. 6 The combined screenshots for the "RNA Sequence" module. (*A*) The module can scan the given RNA sequences for all the immunomodulatory stretches with user-defined window length. The window length can vary from 17 to 27 nucleotides. (*B*) The results for all the oligonucleotides will be presented in tabular format, displaying IMscore and link for structure visualization

are separately presented in the imRNA web server in the menu "Additional Tools."

Scan Motifs

The immunomodulatory motifs to be searched in the query sequence have been predetermined [10]. In the result page, the "Scan Motifs" tool displays all the immunomodulatory motifs found in each of the query sequences. Also, the starting position of each motif is highlighted in a different color compared to the rest of the sequence. Additionally, motifs found in the nonimmunomodulatory sequences taken in the dataset for developing the prediction models are also searched in the query sequences and reported. Considering the example of unmodified siRNA sequences used in the study by Hornung et al. [13] again, sequence siRNA_9.2 was found to be the most immunostimulatory as compared to the other sequences. On submitting antisense sequences of siRNA_9.2 and PF655, the antisense strand siRNA_9.2 does include an immunomodulatory motif, which is in agreement with the experimental findings of the study, while PF655 showed the occurrence of only the nonimmunomodulatory motifs. Figure 7 provides a snapshot of the example discussed here for "Scan Motifs."

BLAST Search

The form submission page of the "BLAST Search" tool takes a single query sequence and performs a similarity search with a database of experimentally verified immunomodulatory sequences

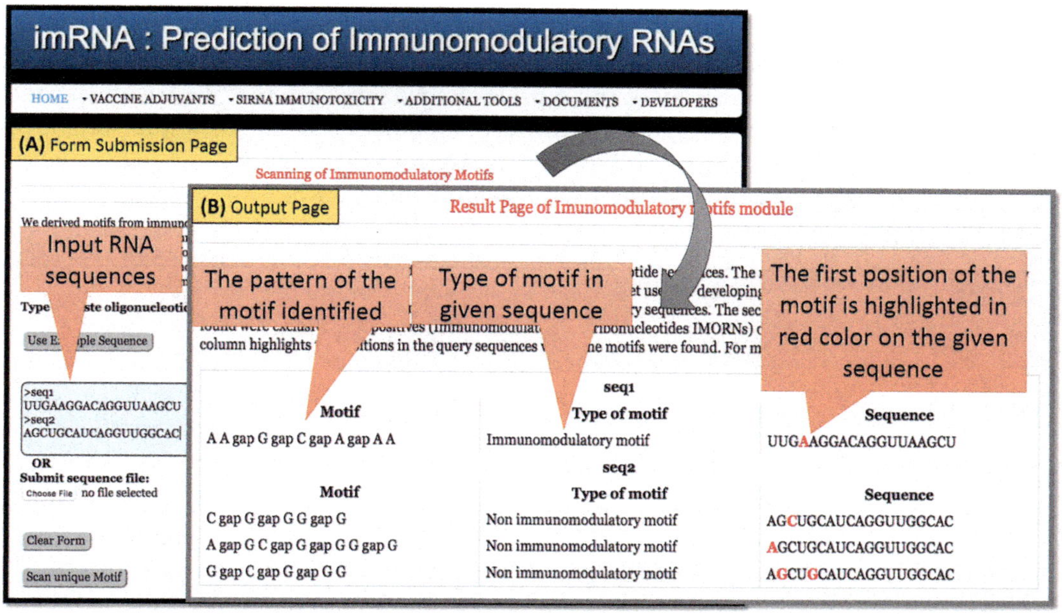

Fig. 7 The annotated graphical presentation of "Motif module" from imRNA server. (*A*) The module will search the immunomodulatory and nonimmunomodulatory motifs in the user-provided sequences. (*B*) The resultant motifs will be highlighted in the result page for the query sequences

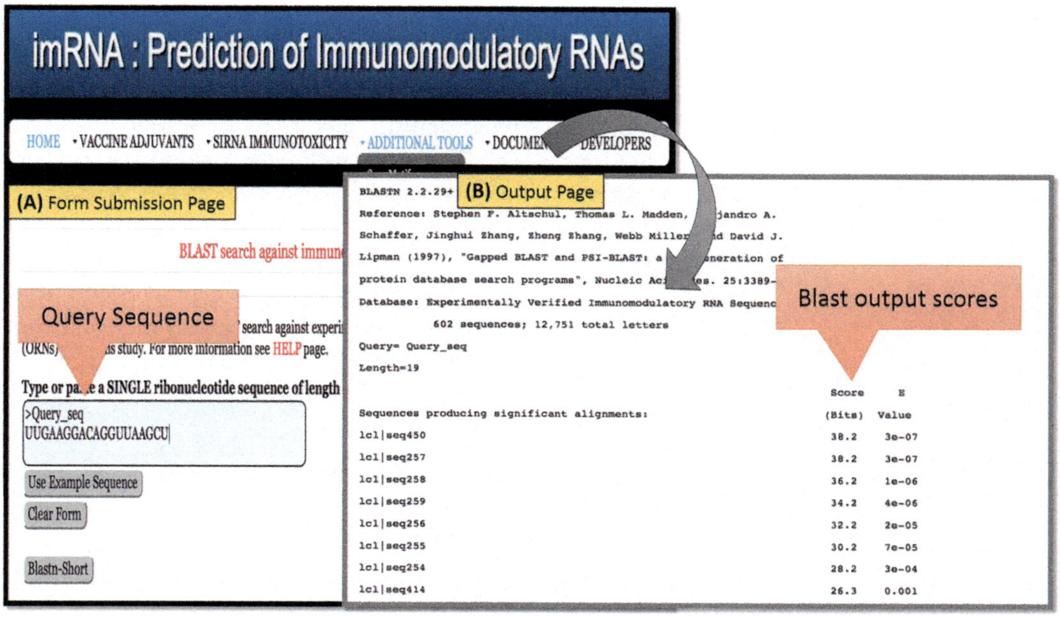

Fig. 8 The screenshot of the "BLAST Search" module from imRNA server. (*A*) The module will help the user to run BLAST against imRNA dataset for their sequences. (*B*) The result will be provided in typical BLAST output

using the "Blastn-short" tool optimized by NCBI for sequence alignment of sequences with length less than 30. The result page is a display of the output file generated by the "Blastn-short" tool providing the individual alignments along with the alignment scores and the corresponding *E*-values. Figure 8 is a representative example of the BLAST alignment of a RNA query sequence.

3 Notes

1. The server described in this chapter is developed for predicting the immunomodulatory potential of single strand RNA. Thus, the server is not suitable for predicting the immunomodulatory potential of double-stranded or chemically modified RNAs.
2. All models used in this server are trained and tested on single-stranded RNA oligonucleotides containing a minimum of 17 and a maximum of 27 nucleotides. Hence, the imRNA server is most efficient for predicting immunomodulatory potential of RNA oligonucleotides having length between 17 and 27 nucleotides.
3. IMscore represents the probability of the correct prediction of immunomodulatory oligonucleotides, which is represented as a score between 0 (weakest) and 9 (highest). In order to select highly reliable immunomodulatory oligonucleotides, the user should select oligonucleotides with the highest IMscore.
4. This server is best suited for predicting immunomodulatory RNA sequences having length less than or equal to 27. In order to identify the immunomodulatory potential of long RNA sequences, the user should use the "RNA Sequence" module of the server, which allows the user to identify immunomodulatory regions in long RNA sequences. Based on the number of immunomodulatory regions one may assign RNA sequence as immunomodulatory (immunotoxic) or nonimmunomodulatory (nonimmunotoxic). Be careful when using this module as this is an extension of our prediction module and the hypothesis is not validated.
5. The Virtual Screening module of imRNA is very useful for researchers as it allows users to identify immunomodulatory RNA sequences from a large pool of sequences.
6. In case user is interested in increasing or decreasing immunotoxicity or immunomodulatory potential, user should use "Analogs of ORNs." This module systematically creates all possible analogs in user's query sequence, then computes the immunomodulatory potential of each analog. This allows users to identify mutations responsible for drastically increasing or decreasing the immunomodulatory potential.

4 Limitations and Future Prospects

Mammalian systems have separate Pattern Recognition Receptors (PRRs) for the detection of various types of molecules originating from microorganisms. Among these, Toll-like Receptors 7/8 (TLR 7/8) recognize single-stranded RNA while TLR 3 recognizes double-stranded RNA molecules. In addition to these, several receptor families like RIG-like receptors, NOD-like receptors, etc. constitute the class of cytosolic receptors recognizing RNA molecules.

The method for prediction of RNA immunomodulatory potential described in this chapter has been developed based on the nucleotide sequences of the single-stranded RNA molecules, known to be recognized by the TLRs 7/8. Sequence-independent immunostimulation of the innate immune system has been shown to occur by virtue of double-strandedness via TLR 3 [18, 19] or due to presence of 5′-triphosphates via RIG-I and PKR receptors [20, 21]. Such cases of immunostimulation due to backbone modifications are beyond the scope of the current method. Thus, future improvements of the prediction of the immunomodulatory potential of RNAs might include consideration of backbone modification, localization of the RNA molecule in cellular compartments as well as the delivery vehicle used for transporting the therapeutic RNA into the cell.

In spite of the abovementioned shortcomings, "imRNA" is an introductory tool that enables in silico estimation and sequence-based estimated alteration of the immunomodulatory potential of RNA molecules. In the future, better prediction models may be developed using a larger training dataset. Consideration of the RNA structure in addition to nucleotide sequence might improve the performance of prediction. Finally, receptor-based docking methods or QSAR models might also prove to be significant advancements over the method described here.

References

1. Lam JKW, Chow MYT, Zhang Y, Leung SWS (2015) siRNA versus miRNA as therapeutics for gene silencing. Mol Ther Nucleic Acids 4: e252. doi:10.1038/mtna.2015.23
2. Goodchild J (2004) Oligonucleotide therapeutics: 25 years agrowing. Curr Opin Mol Ther 6:120–128
3. Hsu PD, Lander ES, Zhang F (2014) Development and applications of CRISPR-Cas9 for genome engineering. Cell 157:1262–1278. doi:10.1016/j.cell.2014.05.010
4. Li H, Lee T, Dziubla T et al (2015) RNA as a stable polymer to build controllable and defined nanostructures for material and biomedical applications. Nano Today 10:631–655. doi:10.1016/j.nantod.2015.09.003
5. Khisamutdinov EF, Li H, Jasinski DL et al (2014) Enhancing immunomodulation on innate immunity by shape transition among RNA triangle, square and pentagon nanovehicles. Nucleic Acids Res 42:9996–10004. doi:10.1093/nar/gku516
6. Heidenreich R, Jasny E, Kowalczyk A et al (2015) A novel RNA-based adjuvant combines strong immunostimulatory capacities with a

favorable safety profile. Int J Cancer 137:372–384. doi:10.1002/ijc.29402

7. Gantier MP, Tong S, Behlke MA et al (2010) Rational design of immunostimulatory siRNAs. Mol Ther 18:785–795. doi:10.1038/mt.2010.4
8. Goodchild A, Nopper N, King A et al (2009) Sequence determinants of innate immune activation by short interfering RNAs. BMC Immunol 10:40. doi:10.1186/1471-2172-10-40
9. Olejniczak M, Galka-Marciniak P, Polak K et al (2012) RNAimmuno: a database of the nonspecific immunological effects of RNA interference and microRNA reagents. RNA 18:930–935. doi:10.1261/rna.025627.110
10. Chaudhary K, Nagpal G, Dhanda SK, Raghava GPS (2016) Prediction of Immunomodulatory potential of an RNA sequence for designing non-toxic siRNAs and RNA-based vaccine adjuvants. Sci Rep 6:20678. doi:10.1038/srep20678
11. Schlee M, Hornung V, Hartmann G (2006) siRNA and isRNA: two edges of one sword. Mol Ther 14:463–470. doi:10.1016/j.ymthe.2006.06.001
12. Ahmed F, Raghava GPS (2011) Designing of highly effective complementary and mismatch siRNAs for silencing a gene. PLoS One 6: e23443. doi:10.1371/journal.pone.0023443
13. Hornung V, Guenthner-Biller M, Bourquin C et al (2005) Sequence-specific potent induction of IFN-alpha by short interfering RNA in plasmacytoid dendritic cells through TLR7. Nat Med 11:263–270. doi:10.1038/nm1191
14. Lee DU, Huang W, Rittenhouse KD, Jessen B (2012) Retina expression and cross-species validation of gene silencing by PF-655, a small interfering RNA against RTP801 for the treatment of ocular disease. J Ocul Pharmacol Ther 28:222–230. doi:10.1089/jop.2011.0116
15. Lorenz R, Bernhart SH, Höner Zu Siederdissen C et al (2011) ViennaRNA package 2.0. Algorithms Mol Biol 6:26. doi:10.1186/1748-7188-6-26
16. Darty K, Denise A, Ponty Y (2009) VARNA: interactive drawing and editing of the RNA secondary structure. Bioinformatics 25:1974–1975. doi:10.1093/bioinformatics/btp250
17. Vens C, Rosso M-N, Danchin EGJ (2011) Identifying discriminative classification-based motifs in biological sequences. Bioinformatics 27:1231–1238. doi:10.1093/bioinformatics/btr110
18. Alexopoulou L, Holt AC, Medzhitov R, Flavell RA (2001) Recognition of double-stranded RNA and activation of NF-kappaB by Toll-like receptor 3. Nature 413:732–738. doi:10.1038/35099560
19. Kato H, Takeuchi O, Sato S et al (2006) Differential roles of MDA5 and RIG-I helicases in the recognition of RNA viruses. Nature 441:101–105. doi:10.1038/nature04734
20. Hornung V, Ellegast J, Kim S et al (2006) 5′-Triphosphate RNA is the ligand for RIG-I. Science 314:994–997. doi:10.1126/science.1132505
21. Nallagatla SR, Hwang J, Toroney R et al (2007) 5′-triphosphate-dependent activation of PKR by RNAs with short stem-loops. Science 318:1455–1458. doi:10.1126/science.1147347

Chapter 6

Cotranscriptional Production of Chemically Modified RNA Nanoparticles

Maria L. Kireeva, Kirill A. Afonin, Bruce A. Shapiro, and Mikhail Kashlev

Abstract

RNA nanoparticles consisting of multiple RNA strands of different sequences forming various three-dimensional structures emerge as promising carriers of siRNAs, RNA aptamers, and ribozymes. In vitro transcription of a mixture of dsDNA templates encoding all the subunits of the RNA nanoparticle may result in cotranscriptional self-assembly of the nanoparticle. Based on our experience with production of RNA nanorings, RNA nanocubes, and RNA three-way junctions, we propose a strategy for optimization of the cotranscriptional production of chemically modified ribonuclease-resistant RNA nanoparticles.

Key words In vitro transcription, T7 RNA polymerase, T7 R&DNA polymerase, 2′-F-dUTP, 2′-F-dCTP, RNA nanorings, RNA nanocubes, Cotranscriptional assembly

1 Introduction

RNA-based therapeutics and biotechnology tools are becoming increasingly popular. The most widely used applications, including siRNA (reviewed in [1, 2]) and gene targeting by CRISPR-Cas9 [3], typically rely on chemically synthesized short RNA oligonucleotides [4]. Other applications, such as transfection of mRNA [5, 6], and technologies based on assembly of the RNA nanoparticles [7–9], require the synthesis of a relatively large (nanomolar and higher) amounts of long RNA molecules. The production of RNAs longer than about 100 nt is usually achieved by in vitro transcription on synthetic or PCR-amplified DNA templates. This chapter describes experimental approaches for optimizing conditions for linear DNA templates transcription by bacteriophage T7 RNA polymerase (T7 RNAP) in order to achieve self-assembly of RNA nanoparticles functionalized with siRNA and RNA aptamers [10]. Cotranscriptional self-assembly has been performed for RNA nanorings (single and double rings, with six and 12 different RNA core structural strands respectively), nanocubes containing six and ten different RNA core structural strands [11–14] (*see*

Eckart Bindewald and Bruce A. Shapiro (eds.), *RNA Nanostructures: Methods and Protocols*, Methods in Molecular Biology, vol. 1632, DOI 10.1007/978-1-4939-7138-1_6, © Springer Science+Business Media LLC 2017

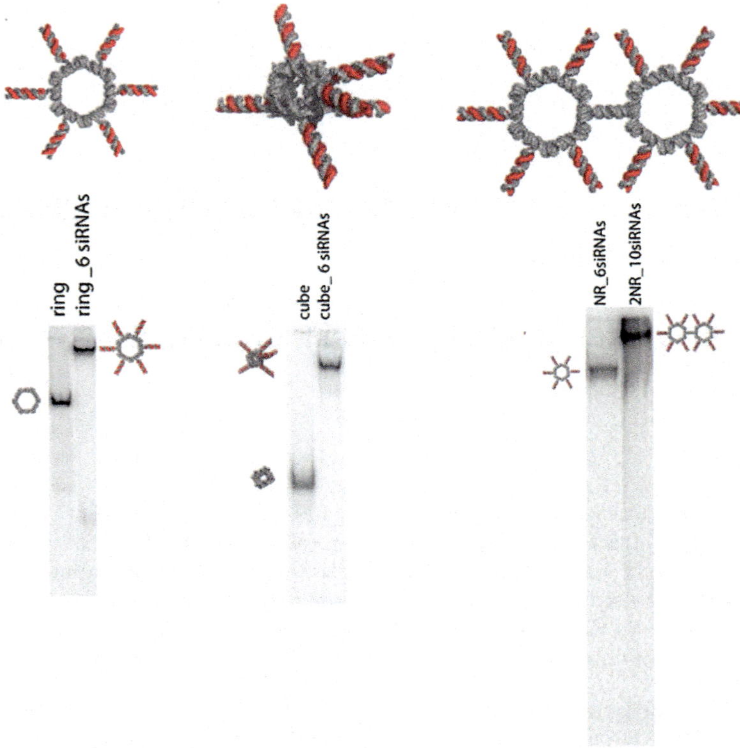

Fig. 1 Illustration of our previously cotranscriptionally assembled unmodified RNA nanoparticles (*ring*, *cube*, and *double ring*) with and without siRNA arms and their corresponding assembly gels (*see* [11, 13])

Fig. 1), and three-way junctions assembled from three RNA strands. Other examples of self-assembling RNAs were also shown to form cotranscriptionally [15]. Cotranscriptionally folded RNA nanorings were successfully used in delivering siRNA to human breast cancer cells [11, 12]. Recently, cotranscriptional assemblies of larger RNA structures have been also demonstrated [16].

While RNA nanoparticles efficiently self-assemble under the conditions in which transcription is performed [10, 11], the molar yield of the mature nanoparticle cannot exceed the yield of the RNA strand from the least efficiently transcribed template. Excessive transcription of one or a few DNA templates in the mix may lead to the appearance of subassembled particles and thus interfere with the efficient purification of the mature nanoparticle. Therefore, optimization of the transcription conditions to increase and balance the amounts of transcripts produced from each individual template represents one main methodological challenge for the cotranscriptional production of RNA nanoparticles. Some templates are not efficiently transcribed under a broad variety of conditions tested. For these cases we used the findings by Uhlenbeck and coauthors that the efficiency of transcription initiation and promoter escape by bacteriophage T7 RNAP does not only strictly require two GMP residues encoded at the 5′ end of the transcript [17], but also strongly depends on the nature of the next four to five residues [18]. Addition of the polypurine stretch at the 5′ end of the transcript encoded by the poorly transcribed templates significantly improved the RNA yield and enabled the production of equal amounts of RNA nanocube subunits and RNA three-way junctions, as described below in Subheadings 3.2 and 3.4.

Synthesis of chemically modified nanoparticles, which are resistant to nucleases [11], poses an additional technical difficulty. NTPs carrying chemical modifications, such as fluorine modification of the sugar ring (2′-F-dNTPs) structurally resemble dNTPs and, therefore, are less efficiently incorporated by T7 RNAP during initiation and elongation [19]. Substitution of Mg^{2+} with Mn^{2+} was first reported to enable dNTP incorporation by T7 RNAP, albeit with a low efficiency, by Wyatt and Walker [20]. The combination of 2.5 mM $MgCl_2$ and 2.5 mM $MnCl_2$ supported the full-length transcript synthesis when ATP or CTP were replaced by dATP or dCTP and enabled the incorporation of 2′-O-methyl ATP and 2′-O-methyl CTP [21]. The same concentrations of divalent cations were used to synthesize transcripts modified with 8-N_3AMP [22]. In our hands, the addition of 0.5 mM $MnCl_2$ along with 5 mM $MgCl_2$ to the transcription reaction sufficiently relaxes the substrate specificity of T7 RNAP to promote efficient incorporation of 2′-F-dUTP and 2′-F-dCTP and the production of fluorinated ribonuclease-resistant nanorings and nanocubes [11, 23]. Analyses of the $MnCl_2$ concentration effect on the yield of the regular and chemically modified RNAs is described in [11]. A similar approach to determine the optimal concentration of $MgCl_2$ in the presence of $MnCl_2$ is outlined below (in Subheading 3.2). Production of chemically modified RNAs has been reported with the error-prone Y639 mutant of T7 RNAP (T7 R&DNAP™, Epicentre), which poorly discriminates between rNTP and dNTP substrates [19]. The production of chemically modified nanoring subunits by wild type and mutant T7 RNAP variants is presented in Subheading 3.1.

Currently, it is challenging to provide a generic protocol for RNA expression and nanoparticle assembly that will work without substantial modifications for any newly designed scaffold. In our experience, each template set requires stepwise optimization of the reaction conditions performed as outlined in Fig. 2. The conditions developed for the nanoring cotranscriptional assembly [11] represent a good starting point for the reaction buffer composition and concentrations of T7 RNAP and template DNA. In addition to the parameters outlined in Fig. 2, the initial optimization of the reaction conditions might include substituting KCl with polyamines or potassium glutamate and Tris–HCl with HEPES, as recommended for in vitro transcription by *E. coli* RNAP [24]. Four protocols outlined below provide a framework for testing a variety of the

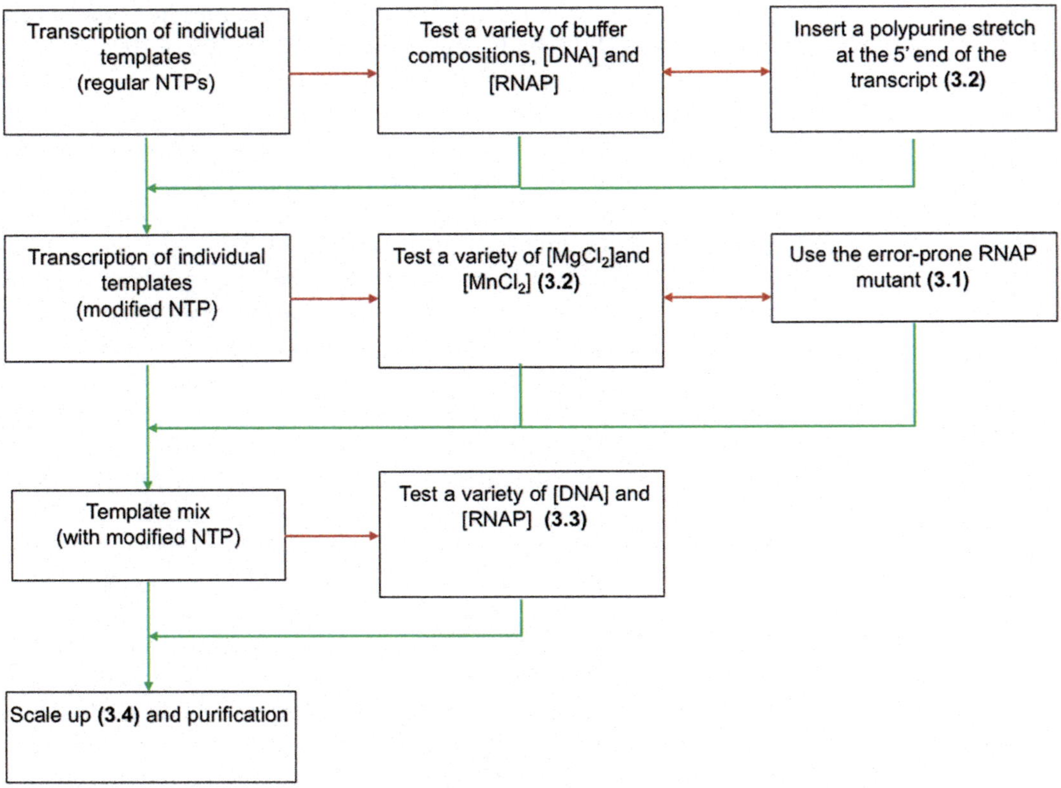

Fig. 2 Workflow for establishing reaction conditions supporting the cotranscriptional assembly of an RNA nanoparticle. The *green arrows* indicate the satisfactory outcomes of the optimization steps. The *red arrows* point toward the steps to be taken to improve the RNA yield. The numbers in *bold* indicate the Subheadings from *Methods*

reaction parameters and choosing the conditions resulting in the best yield of the nanoparticle of your choice. Subsequent up-scaling of the reaction is straightforward and involves the proportional increase of all the reagent volumes [11].

2 Materials

2.1 Components for the In Vitro Transcription and RNA Nanoparticle Purification

1. 10× transcription buffer (TB): 200 mM Tris–HCl, pH 7.9, 400 mM KCl, 50 mM $MgCl_2$ (*see* **Note 1**).
2. 10 mM $MnCl_2$ in H_2O (*see* **Note 2**).
3. 50, 150, 250 mM $MgCl_2$.
4. DTT: 0.1 mM 1,4-dithiothreitol.
5. RNasin: ribonuclease inhibitor (isolated from human placenta) supplied in 20 mM HEPES–KOH (pH 7.6), 50 mM KCl, 8 mM DTT, 50% (v/v) glycerol (Promega).
6. T7 RNAP (200 units/μL, TM910K, Epicentre).

7. T7R&DNA polymerase (50 units/μL, D7P9205K, Epicentre).
8. ATP, GTP, CTP, and UTP (GE Healthcare).
9. Purified ATP, GTP, and CTP (*see* **Note 3**).
10. 2′-F-dUTP and 2′-F-dCTP (Epicentre).
11. 4 NTP stock: 1 mM solution of each ATP, GTP, CTP, and UTP (use the commercial NTPs).
12. ACGF-U stock: 1 mM solution of each ATP, GTP, CTP, and 2′-F-dUTP in water (*see* **Note 4**).
13. AUCF-C stock: 1 mM solution of each ATP, GTP, UTP, and 2′-F-dCTP (*see* **Note 4**).
14. α-[^{32}P] GTP (3000 Ci/mmol, PerkinElmer BLU506H) (*see* **Note 5**).
15. Acetylated bovine serum albumin (BSA): 20 mg/mL.
16. RNase H from *Escherichia coli* (Sigma or Promega).
17. Centrifugal filter devices (microcons) for volumes up to 500 μL with a 100 kDa molecular weight cutoff (Amicon Ultra-0.5; Millipore).

2.2 Components for the Polyacrylamide gel Electrophoresis (PAGE) of RNA

1. Nondenaturing (native) PAGE buffer stock (5×): Dissolve 27.5 g of boric acid and 104 g of Trizma base in water and adjust to a final volume of 1 L.
2. 1 M $Mg(OAc)_2$.
3. Native PAGE running buffer (1×): combine 200 mL of the 5× running buffer and 2 mL 1 M $Mg(OAc)_2$ and adjust to a final volume of 1 L with water. Cool down the running buffer after the preparation. It can be stored at 4 °C for several days.
4. Native gel loading buffer: combine 200 μL 5× native PAGE buffer stock, 2 μL 1 M $Mg(OAc)_2$, 0.5 g glycerol, and 10 μg of each bromophenol blue and xylene cyanol, and adjust to 1 mL with water. Vortex thoroughly.
5. Acrylamide solution for nondenaturing PAGE: 7% acrylamide–bis-acrylamide 29:1.
6. Denaturing gel-loading solution: 7 M urea, 50 mM EDTA, 0.001% bromophenol blue and xylene cyanol.
7. 5× Tris–borate–EDTA buffer (5× TBE): Dissolve 220 g of boric acid, 432 g of Trizma base, 160 mL 0.5 M EDTA in water and adjust to a final volume of 8 L.
8. Denaturing acrylamide solution: 20% acrylamide–bis-acrylamide 19:1, 7 M urea, 1× TBE. The solution should be filtered through 0.22 μ filter.
9. 10% ammonium persulfate (APS).

10. N,N,N′,N′-tetramethylethylenediamine (TEMED).
11. 20 × 20 cm electrophoretic glass plates.
12. 0.8 mm comb and spacers (for the nondenaturing PAGE).
13. 0.4 mm comb and spacers (for the denaturing PAGE).
14. Vertical electrophoretic unit (Thermo Scientific or similar).
15. Storage Phosphor screen and a Phosphorimager.

2.3 Software

1. Image Processing Software Such as ImageQuant or ImageJ.

3 Methods

3.1 Comparison of the Wild Type T7 RNAP and the Mutant T7 R&DNAP for the Production of Chemically Modified RNA

1. Prepare a master mix: 6 μL of 10× TB, 3 μL of 10 mM $MnCl_2$, 39 μL of 0.02 μM template DNA, 3 μL of α-[^{32}P] GTP.
2. Split the master mix to two 25.5 μL aliquots. Add 3 μL of 1 mM 4 NTP stock to the first aliquot and 3 μL of 1 mM ACGF-U stock to the second aliquot (*see* **Note 6**).
3. Take out two 9.5 μL aliquots from each mix. Start the reaction by adding 0.5 μL of T7 RNAP or T7 R&DNAP. Incubate at 37 °C for 60 min.
4. Prepare the denaturing gel:
   (a) Prewarm 25 mL of the denaturing acrylamide solution to room temperature in a 50 mL disposable screw-cap tube, add 250 μL of 10% APS, mix by inverting the tube, add 25 μL of TEMED, and mix again.
   (b) Cast the gel using the 0.4 mm-thick spacers and a comb.
   (c) Let the polymerization proceed for at least 30 min. It is not necessary to prerun the gel before the samples are loaded.
5. Stop the transcription reactions by adding 10 μL of denaturing gel loading buffer. Incubate at 95 °C for 2 min.
6. Load up to 10 μL of sample per well. Perform electrophoresis in 1× TBE for 50 min at 50 W constant power. The bromophenol blue dye should migrate 3/4 of the gel length.
7. Transfer the gel to a sheet of the developed X-ray film or leave it on the glass plate, and wrap with a plastic film.
8. Expose to a Phosphor screen and scan using Phosphor imager. The result of the described experiment performed using the DNA template encoding the strand F of a basic (nonfunctionalized) nanoring [11] is shown in Fig. 3 (*see* **Note 7**).

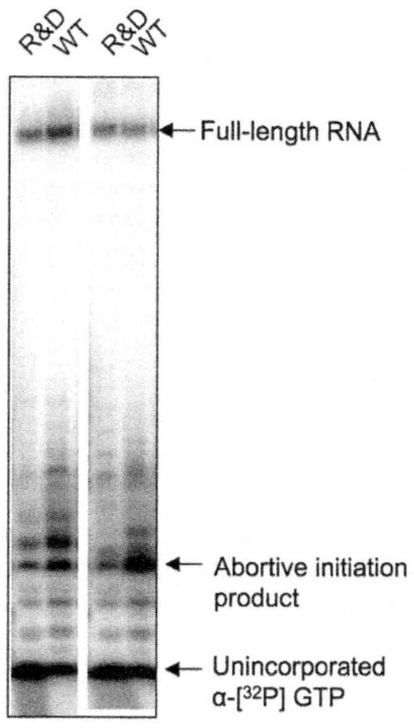

Fig. 3 Transcription by the wild type T7 RNAP (WT) and the Y639F T7 RNAP mutant (R&D) in the presence of regular NTPs and with UTP substituted by 2′-F-dUTP (ACGF-U mix) (Subheading 3.1)

3.2 Optimization of the Transcript Yield by Modification of the Sequence at the RNA 5′ End

The following protocol suggests a framework for the analysis of the effect of a 5′-end extension of the nonfunctionalized RNA nanocube strand A sequence [13] (5′GGCAACUUUGAUCCCUCGGUU UAGCGCCGGCCUUUUCUCCCACACUUUCACG3′) with a purine-rich stretch (5′GGGAAAGGAAGAGC3′) obtaining a new template encoding the modified RNA strand A [23]. Additionally, the protocol includes optimization of [$MgCl_2$] and incorporation of 2′-F-dNTPs.

1. Prepare two master mixes, one with the original template, and one with the modified template, planning for twelve 10 μL reactions in each: 15 μL of 10× TB, 7.5 μL of 10 mM DTT, 1.5 μL of RNasin, 3 μL of T7 RNAP, 3 μL of 0.4 μM template, 7.5 μL of 10 mM $MnCl_2$, 82.5 μL of H_2O.
2. Prepare three tubes with 5 μL each of 2 mM four NTPs, ACGF-U and ACGF-C mixes. Add 5 μL of α-[^{32}P] GTP to each tube to obtain the labeled NTP mixes.
3. Prepare four tubes containing 30 μL of each master mix. Add 3.75 μL of H_2O or 50, 150, or 250 mM $MgCl_2$ to the tubes to test transcription in 5, 10, 20, and 30 mM $MgCl_2$.
4. Take out three 9 μL aliquots from each of the eight tubes containing two different templates and four various concentrations of $MgCl_2$. Start transcription by adding the labeled NTP mixes to each of the three tubes. Incubate at 37 °C for 30 min.

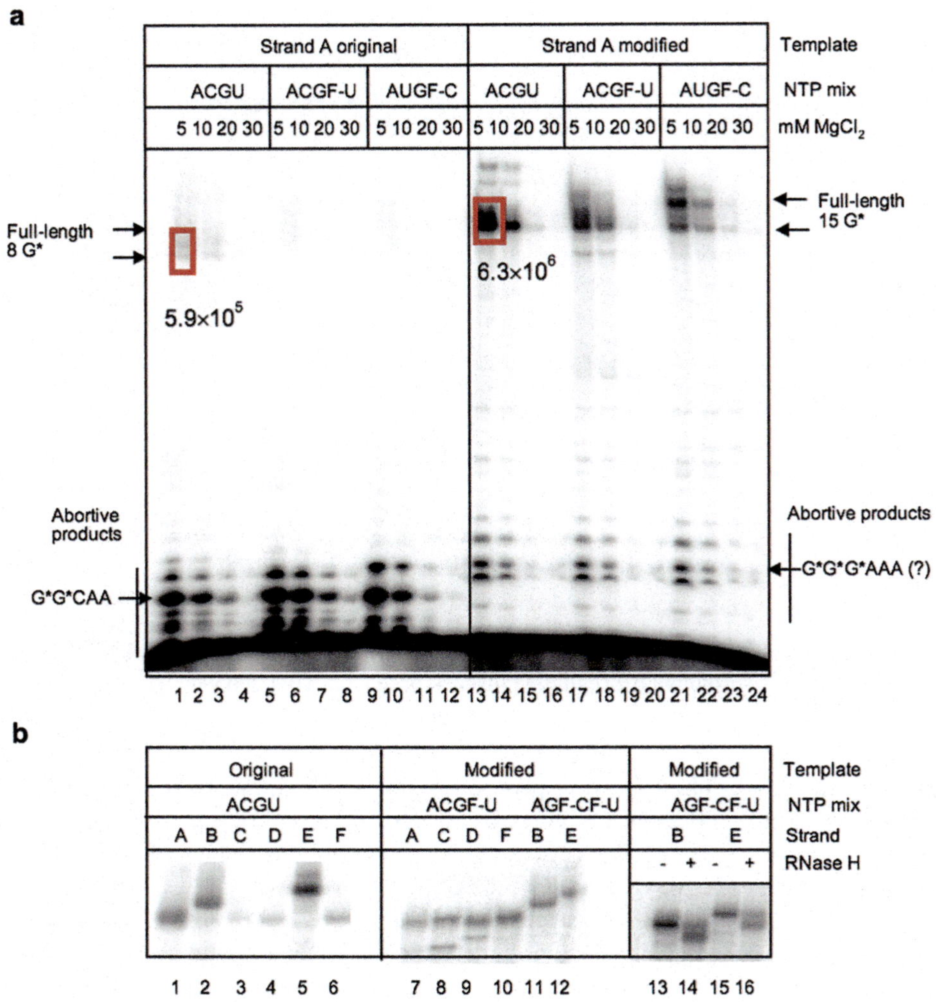

Fig. 4 Effects of the transcript sequence and $MgCl_2$ on transcription. Panel (**a**) demonstrates that insertion of a polypurine stretch at the 5′end of the transcript substantially promotes transcription by T7 RNAP and that increased $MgCl_2$ inhibits transcription on both the original and extended templates. The experiment is described in Subheading 3.2, **steps 1–6**. Panel (**b**) illustrates the outcome of transcription performed in the presence of 5 mM $MgCl_2$ and 0.5 mM $MnCl_2$ in the presence of NTP mixes indicated on top of the gel on the original (*lanes 1–6*) and extended (*lanes 7–16*) templates. The original nanocube subunit sequence is restored by the post-transcriptional treatment with RNase H (Subheading 3.2, **step 8**)

5. Analyze the reaction products by denaturing PAGE (*see* Subheading 3.1, **steps 4–8**).
6. Quantify the RNA yield using image processing software, taking into account the different number of the radioactively labeled GMP residues in the transcripts obtained from different templates (Fig. 4a) (*see* **Note 8**).
7. The balancing of nanocube subunit yields by the 5′ extension is shown in Fig. 4b (compare the lanes 1–6 and lanes 7–12).

8. If necessary, remove the 5′ purine-rich stretch from the transcript to restore the original RNA sequence, using the DNA oligo-guided degradation by RNase H.
   (a) After transcription is completed, transfer the reaction tube to room temperature.
   (b) Add 1 μM final of DNA oligonucleotide complementary to the purine-rich stretch (5′GCTCTTCCTTTCCC3′) and 0.01 unit/μL RNase H (Sigma).
   (c) Incubate for 30 min.
   (d) Determine the efficiency of digestion (*see* **Note 9**) by denaturing PAGE (*see* Subheading 3.1, **steps 4–8**) (Fig. 4b, lanes 13–16).

3.3 Optimization of the T7 RNAP and Template DNA Concentrations in the Cotranscriptional Assembly of the RNA Nanoring Containing 2′-Fluorinated UMPs

This protocol represents an example of fine-tuning the cotranscriptional assembly conditions after the best conditions for the individual subunit expression in equally high amounts have been established.

1. Prepare the master mix: 7.5 μL of 10× TB, 53.5 μL of H_2O, 3.25 μL of 10 mM $MnCl_2$, 7.5 μL of ACGF-U, 3.25 μL of α-[^{32}P] GTP, 0.75 μL of 100 mM DTT, and 0.75 μL of RNasin.
2. Mix six templates, adjusting the concentration to 4 μM each.
3. Prepare three thin-walled PCR tubes with 6.5 μL, one with 14 μL, and one with 29 μL of the master mix. Add 0.5 μL of the template mix to each tube.
4. Start the reaction by adding 0.25 and 1 μL of T7 RNAP to the first two tubes and 0.5 μL of T7 RNAP to the remaining tubes. Incubate for 3 h at 37 °C in a thermocycler to diminish evaporation.
5. During the incubation, prepare the native gel: prewarm 75 mL of the native acrylamide solution to room temperature in a beaker, add 500 μL of 10% APS, carefully mix with a glass rod or a disposable pipette, add 30 μL of TEMED, and mix again. Cast the gel using the 0.8 mm-thick spacers and a comb. Let the polymerization proceed for at least 30 min. Native PAGE should be performed at 4 °C. It is necessary to prerun the gel for 15 min at 25 W before the samples are loaded.
6. Take out 7 μL aliquots from each tube and add 7 μL of the native gel loading buffer to each. Freeze the remaining transcription mix at −80 °C for the future use or proceed with purification (*see* Subheading 3.3, **step 9** below).
7. Load the samples and perform PAGE for 4 h at 25 W at 4 °C. Leave the gel on one plate, and wrap with the plastic film.

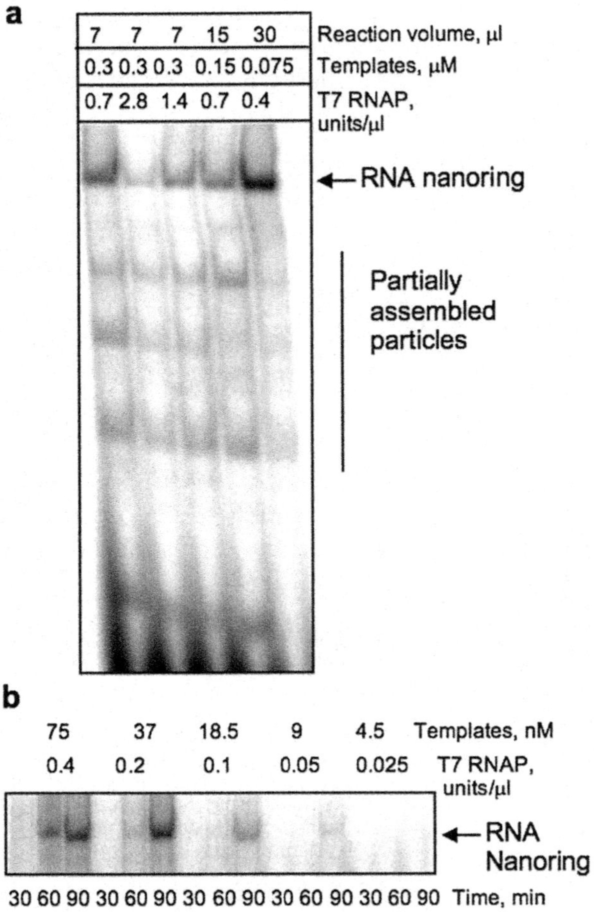

Fig. 5 High concentration of T7 RNAP is inhibitory to transcription and nanoparticle assembly. Panel (**a**) illustrates the experiment described in Subheading 3.3. Further dilution of the reaction decreases the nanoparticle yield as shown in panel (**b**)

8. Expose to a Phosphor screen for 30 min and scan using Phosphor imager. Results showing that a high concentration of T7 RNAP is inhibitory to transcription are shown in Fig. 5.
9. Purify the remaining nanorings from unincorporated NTPs using centrifugal filter devices (microcons).
   (a) Dilute the transcription mix with 450 μL of 1× native PAGE running buffer containing 0.1 mg/mL acetylated BSA (*see* **Note 10**) and transfer to a microcon chamber.
   (b) Spin for 8–10 min at 20,000 × *g*. Transfer the chamber to a fresh collection tube, and discard the used collection tube with the radioactive flow-through.
   (c) Repeat two more times.

(d) Insert the chamber upside down into a fresh collection tube, and spin down briefly on a tabletop centrifuge to collect the retentate containing the purified nanoring.

(e) Take out 0.7 μL and add to 10 μL of the native gel loading buffer.

(f) Analyze by native PAGE as described in Subheading 3.3, **steps 5** and 7.

3.4 Large-Scale Cotranscriptional Production of a Nanoparticle Carrying Three Separate siRNA Precursors Connected by Three-Way Junctions [25] and Removal of the 5′-end Extensions by RNase H Treatment

Three-way junctions are compact RNA tertiary motifs connecting RNA helices. The helices may carry RNA-based functionalities. Fig. 6a depicts RNA sequences designed to obtain the nanoparticle carrying three different asymmetric Dicer substrates (tested in our previous work [26] against eGFP, GSTP1-1, and HIV protease) connected by 3 three-way junctions. The three RNA strands start from 5′-AC-3′, 5′-AA-3′, and 5′-GA-3′ dinucleotides, which makes the templates encoding these RNAs poor substrates for T7 RNAP [17]. To support high and balanced yields of all three RNAs for cotranscriptional assembly, we added a purine-rich stretch (*see* Subheading 3.2) to the 5′-ends of all RNA strands. The cotranscriptional assembly of the three-way junction RNA nanoparticle, and removal of the extra RNA using RNase H is outlined below.

1. Prepare the master mix: 10 μL of 10× TB, 71 μL of H_2O, 5 μL of 10 mM $MnCl_2$, 1 μL of 25 mM 4 NTPs, 3 μL of α-[^{32}P] GTP, 1 μL of 100 mM DTT, and 1 μL of RNasin.
2. Prepare the mix of the template DNA fragments containing 0.2 μM of each of the three templates.
3. Add 2 μL of the template mix to the master mix.
4. Start transcription by adding 5 μL of T7 RNAP. Incubate for 3 h at 37 °C.
5. Bring the transcription reaction to room temperature.
6. Add 1 μL of 100 μM DNA oligo for RNase H (Fig. 6a).
7. Take out 1 μL from the transcription reaction, add to 10 μL of native gel loading buffer, and place on ice ("No RNase H" control).
8. Split the transcription reaction to two 50 μL aliquots. Add 5 μL of RNase H (Sigma) to one aliquot and 5 μL of RNase H (Promega) to another aliquot. Incubate for 30 min at room temperature.
9. Take out 1 μL from each of the two reactions and add to 10 μL of native gel loading buffer.
10. Analyze the samples, including the "No RNase H" control, by native PAGE (*see* Subheading 3.3, **steps 5** and 7). Fig. 6b shows that treatment with RNase H results in an increase of electrophoretic mobility of the RNA nanoparticle due to removal of the single strand RNA regions by RNase H. Two batches of RNase H tested in the experiment work equally well.

a

```
Strand 1
5′ - ACCCUGAAGUUCAUCUGCACCACCGGCAUUGGGGCAGUGCCUUCACAUAGUCAUCCUUGC
Strand 2
5′ - AAGGAUGACUAUGUGAAGGCACUGCCUCUAACCUCUUCUAAUACUGUAUCAUCUGCUCCU
Strand 3
5′ - GAGCAGAUGAUACAGUAUUAGAAGAGGGACCGUGCCGGUGGUGCAGAUGAACUUCAGGGUCA

Strand 1 extended
5′ - gggaaaggaagagcACCCUGAAGUUCAUCUGCACCACCGGCAUUGGGGCAGUGCCUUCACAUAGUCAUCCUUGC
Strand 2 extended
5′ - gggaaaggaagagcAAGGAUGACUAUGUGAAGGCACUGCCUCUAACCUCUUCUAAUACUGUAUCAUCUGCUCCU
Strand 3 extended
5′ - gggaaaggaagagcGAGCAGAUGAUACAGUAUUAGAAGAGGGACCGUGCCGGUGGUGCAGAUGAACUUCAGGGUCA

DNA oligo for RNase H
3′ - ccctttccttctcg - 5′
```

b

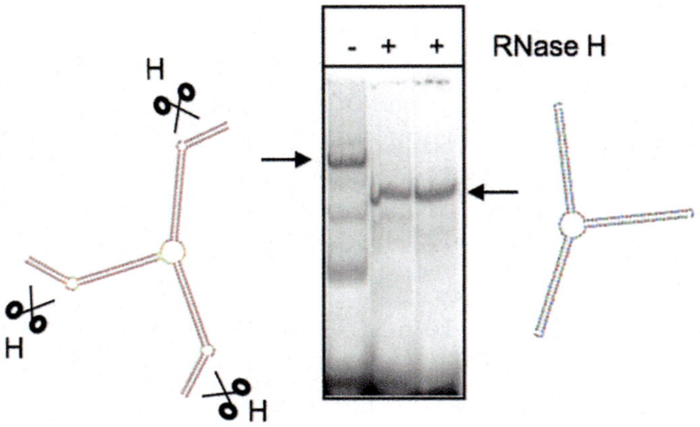

Fig. 6 Cotranscriptional assembly of a three-way junction carrying three different asymmetric Dicer substrates (tested in our previous work [26]) indicated by *black*, *red*, and *green* colors, respectively in panel (**a**). The three-way junction motif is colored in *blue*. The 5′-end extensions of the RNA strands are shown in *orange*, and the complementary sequence of the ssDNA that targets the RNase H-mediated degradation of the ssRNA starting sequence extensions are shown on the bottom of panel (**a**). The relative mobilities of the RNA nanoparticles before and after the treatment with RNase H analyzed on native-PAGE are illustrated in panel (**b**)

4 Notes

1. We have been routinely using regular ddH_2O, instead of DEPC-treated or commercially produced RNase-free water without ill effect on the quality of RNA.
2. $MnCl_2$ precipitates in TB within several hours. Add $MnCl_2$ immediately before transcription is started.
3. The NTP purification rationale and procedure are described in detail in [27].

4. Use the purified nonmodified NTPs to avoid contamination of the 2′-F-NTP with its unmodified analog.
5. Exercise caution while handling radioactive material, always use the appropriate shielding and strictly follow your institutional rules about radioactive material handling and disposal. Use extra care while discarding buffers from the bottom chamber of the gel electrophoresis apparatus.
6. When planning to run *x* reactions, make the mix sufficient for at least $x + 1$ reactions to account for the pipetting errors.
7. The concentration of T7 R&DNAP is four times lower than the concentration of the wild type T7 RNAP. Taking into account the difference in the concentrations of the wild type and the mutant T7 RNAPs, it is clear that T7 R&DNAP is much more efficient (*per unit* of the enzyme) to incorporate 2′-F-dUTP in the transcription conditions described here. However, the price of R&DNAP *per microliter* of the commercially available enzyme solution is still at least two times higher than the price of the wild type T7 RNAP. Considering that the yields of chemically modified RNA obtained with equal *volumes* of wild type T7 RNAP and R&DNAP in Fig. 3 are nearly identical, we have chosen the wild type T7 RNAP for a better price-to-performance ratio in our experiments. T7 R&DNAP should definitely be considered if less expensive approaches to increase the yield of the chemically modified RNA from a certain template prove unsuccessful.
8. For example, the signal from unmodified full-length transcription product obtained from the original template at 5 mM $MgCl_2$ (Fig. 4a, lane 1) is 11 times lower than the signal from the corresponding sample obtained in the modified template encoding the transcript containing a purine-rich stretch at the 5′ end (Fig. 4a, lane 11). However, the RNA yield on the modified template is only 5.7 times higher, because the original RNA strand contains 8 GMP residues, and the modified one contains 15 GMP residues. The decrease in the amount of the abortive products on the modified template is more dramatic than it appears, because the prominent abortive products on the original template carry only two GMP residues, while relatively minor abortive products appearing during transcription of the modified template carry three or more GMPs.
9. Mn^{2+} in millimolar concentrations has been reported to inhibit RNase H from *E. coli* [28]. If the efficiency of the reaction is low, desalt the RNA the using centrifuge filter units with a 10 kDa MWCO (*see* Subheading 3.3, **step 8**).
10. Addition of BSA is absolutely essential, because it prevents binding of the nanoring to the microcon membrane. A purification attempt in the absence of BSA resulted in a complete

absence of the nanoring in the eluate and accumulation of the radioactive signal on the membrane.

Acknowledgments

We thank Lorena Parlea and Paul Zakrevsky for fruitful discussions, and Josh Turek-Herman for proofreading the manuscript. The contents of this publication do not necessarily reveal the views or policies of the Department of Health and Human Services, nor does the mention of trade names, commercial product, or organizations imply endorsement by the US Government. This work was supported by the Intramural Research Program of the National Institutes of Health, Center for Cancer Research, the National Cancer Institute.

References

1. Castanotto D, Rossi JJ (2009) The promises and pitfalls of RNA-interference-based therapeutics. Nature 457(7228):426–433. doi:10.1038/nature07758
2. Zuckerman JE, Davis ME (2015) Clinical experiences with systemically administered siRNA-based therapeutics in cancer. Nat Rev Drug Discov 14(12):843–856. doi:10.1038/nrd4685
3. Cho SW, Kim S, Kim JM, Kim JS (2013) Targeted genome engineering in human cells with the Cas9 RNA-guided endonuclease. Nat Biotechnol 31(3):230–232. doi:10.1038/nbt.2507
4. Kelley ML, Strezoska Z, He K, Vermeulen A, Smith A (2016) Versatility of chemically synthesized guide RNAs for CRISPR-Cas9 genome editing. J Biotechnol 233:74–83. doi:10.1016/j.jbiotec.2016.06.011
5. Li L, Allen C, Shivakumar R, Peshwa MV (2013) Large volume flow electroporation of mRNA: clinical scale process. Methods Mol Biol 969:127–138. doi:10.1007/978-1-62703-260-5_9
6. Preskey D, Allison TF, Jones M, Mamchaoui K, Unger C (2016) Synthetically modified mRNA for efficient and fast human iPS cell generation and direct transdifferentiation to myoblasts. Biochem Biophys Res Commun 473(3):743–751. doi:10.1016/j.bbrc.2015.09.102
7. Grabow WW, Jaeger L (2014) RNA self-assembly and RNA nanotechnology. Acc Chem Res 47(6):1871–1880. doi:10.1021/ar500076k
8. Haque F, Guo P (2015) Overview of methods in RNA nanotechnology: synthesis, purification, and characterization of RNA nanoparticles. Methods Mol Biol 1297:1–19. doi:10.1007/978-1-4939-2562-9_1
9. Shukla GC, Haque F, Tor Y, Wilhelmsson LM, Toulme JJ, Isambert H, Guo P, Rossi JJ, Tenenbaum SA, Shapiro BA (2011) A boost for the emerging field of RNA nanotechnology. ACS Nano 5(5):3405–3418. doi:10.1021/nn200989r
10. Afonin KA, Kasprzak WK, Bindewald E, Kireeva M, Viard M, Kashlev M, Shapiro BA (2014) In silico design and enzymatic synthesis of functional RNA nanoparticles. Acc Chem Res 47(6):1731–1741. doi:10.1021/ar400329z
11. Afonin KA, Kireeva M, Grabow WW, Kashlev M, Jaeger L, Shapiro BA (2012) Co-transcriptional assembly of chemically modified RNA nanoparticles functionalized with siRNAs. Nano Lett 12(10):5192–5195. doi:10.1021/nl302302e
12. Afonin KA, Grabow WW, Walker FM, Bindewald E, Dobrovolskaia MA, Shapiro BA, Jaeger L (2011) Design and self-assembly of siRNA-functionalized RNA nanoparticles for use in automated nanomedicine. Nat Protoc 6(12):2022–2034. doi:10.1038/nprot.2011.418
13. Afonin KA, Bindewald E, Yaghoubian AJ, Voss N, Jacovetty E, Shapiro BA, Jaeger L (2010) In vitro assembly of cubic RNA-based scaffolds designed in silico. Nat Nanotechnol 5(9):676–682. doi:10.1038/nnano.2010.160

14. Afonin KA, Viard M, Tedbury P, Bindewald E, Parlea L, Howington M, Valdman M, Johns-Boehme A, Brainerd C, Freed EO, Shapiro BA (2016) The use of minimal RNA toeholds to trigger the activation of multiple functionalities. Nano Lett 16(3):1746–1753. doi:10.1021/acs.nanolett.5b04676

15. Afonin KA, Lin YP, Calkins ER, Jaeger L (2012) Attenuation of loop-receptor interactions with pseudoknot formation. Nucleic Acids Res 40(5):2168–2180. doi:10.1093/nar/gkr926

16. Geary C, Rothemund PW, Andersen ES (2014) RNA nanostructures. A single-stranded architecture for cotranscriptional folding of RNA nanostructures. Science 345(6198):799–804. doi:10.1126/science.1253920

17. Milligan JF, Uhlenbeck OC (1989) Synthesis of small RNAs using T7 RNA polymerase. Methods Enzymol 180:51–62

18. Milligan JF, Groebe DR, Witherell GW, Uhlenbeck OC (1987) Oligoribonucleotide synthesis using T7 RNA polymerase and synthetic DNA templates. Nucleic Acids Res 15(21):8783–8798

19. Sousa R, Padilla R (1995) A mutant T7 RNA polymerase as a DNA polymerase. EMBO J 14(18):4609–4621

20. Wyatt JR, Walker GT (1989) Deoxynucleotide-containing oligoribonucleotide duplexes: stability and susceptibility to RNase V1 and RNase H. Nucleic Acids Res 17(19):7833–7842

21. Conrad F, Hanne A, Gaur RK, Krupp G (1995) Enzymatic synthesis of 2′-modified nucleic acids: identification of important phosphate and ribose moieties in RNase P substrates. Nucleic Acids Res 23(11):1845–1853

22. Gopalakrishna S, Gusti V, Nair S, Sahar S, Gaur RK (2004) Template-dependent incorporation of 8-N3AMP into RNA with bacteriophage T7 RNA polymerase. RNA 10(11):1820–1830. doi:10.1261/rna.5222504

23. Afonin KA, Viard M, Kagiampakis I, Case CL, Dobrovolskaia MA, Hofmann J, Vrzak A, Kireeva M, Kasprzak WK, KewalRamani VN, Shapiro BA (2015) Triggering of RNA interference with RNA-RNA, RNA-DNA, and DNA-RNA nanoparticles. ACS Nano 9(1):251–259. doi:10.1021/nn504508s

24. Holmes SF, Foster JE, Erie DA (2003) Kinetics of multisubunit RNA polymerases: experimental methods and data analysis. Methods Enzymol 371:71–81. doi:10.1016/S0076-6879(03)71005-2

25. Parlea L, Bindewald E, Sharan R, Bartlett N, Moriarty D, Oliver J, Afonin KA, Shapiro BA (2016) Ring catalog: a resource for designing self-assembling RNA nanostructures. Methods 103:128–137. doi:10.1016/j.ymeth.2016.04.016

26. Afonin KA, Viard M, Martins AN, Lockett SJ, Maciag AE, Freed EO, Heldman E, Jaeger L, Blumenthal R, Shapiro BA (2013) Activation of different split functionalities on re-association of RNA-DNA hybrids. Nat Nanotechnol 8(4):296–304. doi:10.1038/nnano.2013.44

27. Lubkowska L, Kireeva ML (2015) Direct competition assay for transcription fidelity. Methods Mol Biol 1276:153–164. doi:10.1007/978-1-4939-2392-2_8

28. Goedken ER, Marqusee S (2001) Co-crystal of *Escherichia coli* RNase HI with $Mn2+$ ions reveals two divalent metals bound in the active site. J Biol Chem 276(10):7266–7271. doi:10.1074/jbc.M009626200

Chapter 7

Supported Fluid Lipid Bilayer as a Scaffold to Direct Assembly of RNA Nanostructures

Aleksandra P. Dabkowska, Agnes Michanek, Luc Jaeger, Arkadiusz Chworos, Tommy Nylander, and Emma Sparr

Abstract

RNA architectonics offers the possibility to design and assemble RNA into specific shapes, such as nanoscale 3D solids or nanogrids. Combining the minute size of these programmable shapes with precise positioning on a surface further enhances their potential as building blocks in nanotechnology and nanomedicine. Here we describe a bottom-up approach to direct the arrangement of nucleic acid nanostructures by using a supported fluid lipid bilayer as a surface scaffold. The strong attractive electrostatic interactions between RNA polyanions and cationic lipids promote RNA adsorption and self-assembly. Protocol steps for the characterization of assembled RNA complexes via several complementary methods (QCM-D, ellipsometry, confocal fluorescence microscopy, AFM) are also provided. Due to their tunable nature, lipid bilayers can be used to organize RNA laterally on the micrometer scale and thus facilitate the building of more complex 3D structures. The bilayer-based approach can be extended to other programmable RNA or DNA objects to construct intricate structures, such as 2D grids or 3D cages, with high precision.

Key words RNA architectonics, Nanostructure, 2D ordering, Supported lipid bilayer, Ellipsometry, QCM-D, Confocal fluorescence microscopy, AFM

1 Introduction

RNA and DNA molecules can be used as building blocks to create bespoke nanoscale objects. The specificity of the interactions between complementary bases enables directed and selective self-assembly of complex nano-sized architectures with well-defined properties. By taking advantage of RNA architectonics, it is possible to design intricate RNA shapes, including RNA polyhedrons, [1] cubes, [2] fibers, [3, 4] squares [5], and nanorings. [6] These shapes can be further connected into programmable arrays and nanogrids [5, 7]. Complementary nucleic acid strands can also hybridize to link together materials such as particles [8, 9] or proteins in a programmable way, thus directing the assembly of nano-objects with high precision in terms of shape and organization.

Eckart Bindewald and Bruce A. Shapiro (eds.), *RNA Nanostructures: Methods and Protocols*, Methods in Molecular Biology, vol. 1632, DOI 10.1007/978-1-4939-7138-1_7, © Springer Science+Business Media LLC 2017

Directing the arrangement of nucleic acid nanostructures at interfaces opens up new possibilities to use these biomolecules to construct functional nanostructured surfaces with applications in, for example, bioanalytical and diagnostic devices. However, there are still limitations to building up larger and more complex structures. At present, most self-assembly strategies for RNA or DNA 3D nanostructures are performed in bulk solution after which the preassembled objects are deposited onto a solid support when so desired [1, 4, 7, 10].

The additional potential of mediating and directing the assembly of nucleic acid objects on the surface has recently been exploited by using fluid lipid bilayers as scaffolds [11–14]. By using a (partly) cationic lipid bilayer as a scaffold for the negatively charged RNA assembly, there is no need for chemical modification of the RNA molecules or the surface to achieve controlled, nanoscale ordering over millimeter-scale surface regions.

1.1 Outline of the Method

In this protocol, we describe how to assemble a layer of 3D RNA structures from RNA building blocks on a fluid lipid bilayer (Fig. 1). The RNA building blocks consist of two complementary tectosquares that are programmed to connect to each other via tail–tail interactions to create a 3D polyhedron. The final 3D RNA polyhedron has the geometry of a nonuniform square antiprism.

To begin the assembly, first a positively charged lipid bilayer is formed on a silica surface, which can be a silicon wafer or a cover glass. Next, the first RNA tectosquares, referred to as *tectosquare 1*, are added and adsorb to the bilayer surface. The adsorption is driven by electrostatic attraction between the oppositely charged

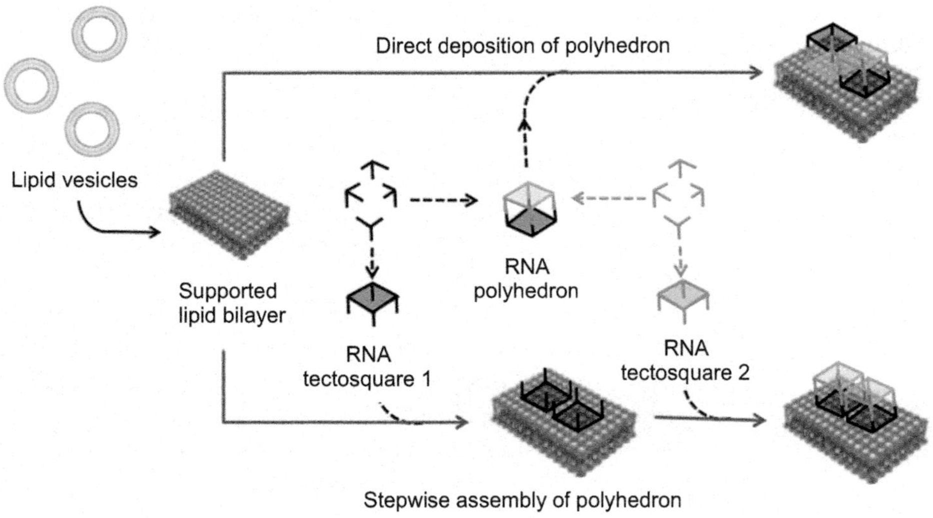

Fig. 1 Schematic showing the directed assembly of RNA polyhedron structures using supported fluid lipid bilayer as a scaffold to guide the building up of a polyhedron from RNA components

bilayer and RNA, which results in a uniform layer of RNA on the lipid bilayer. The second RNA tectosquares (*tectosquare 2*) are then added. As each *tectosquare 2* is complementary to *tectosquare 1*, these bind together to create a layer of oriented 3D RNA polyhedrons. The formation of the 3D layer is controlled by sequence specific interactions, surface charge, supported bilayer fluidity, and changes in the salt composition and concentration [12]. Assembling the structures at the confinement of a surface allows us to orient the RNA polyhedrons. As many properties of bilayer surfaces are tunable (for example, molecular composition, charge density, and fluidity), the bilayers can be used to order the RNA nano-objects laterally on the micrometer scale and thus act as a foundation for building up more complex 3D structures.

The process of building up 3D RNA on a surface can be studied with many surface-sensitive techniques (Fig. 2). Here we describe time-resolved characterization of each step in the RNA assembly using simultaneous quartz-crystal microbalance with dissipation (QCM-D) and ellipsometry measurements. QCM-D and

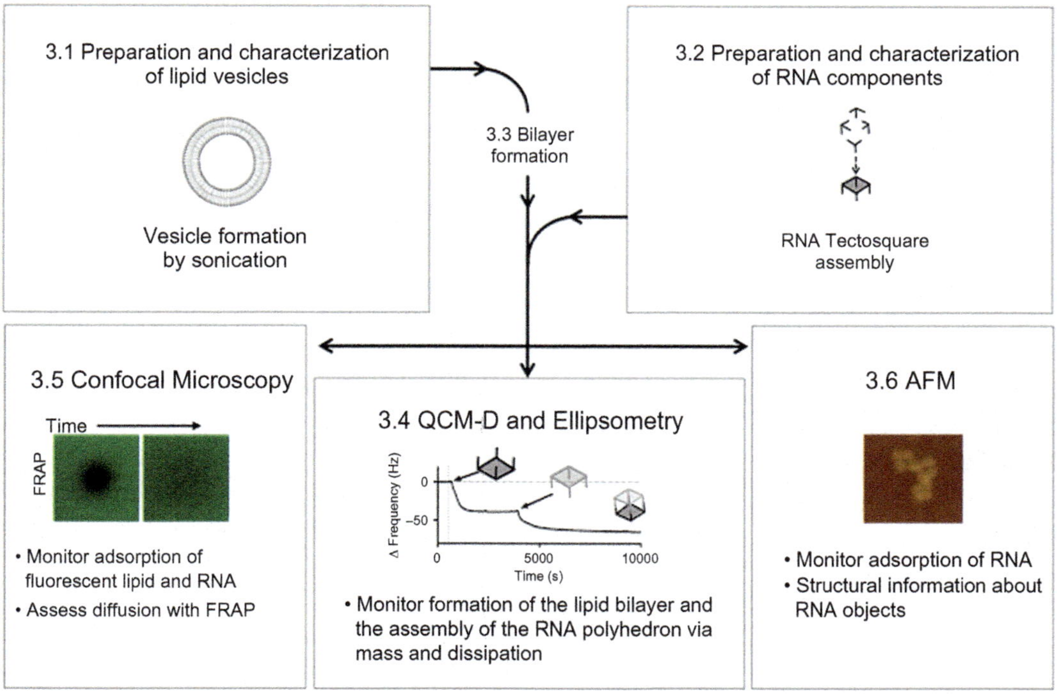

Fig. 2 Workflow for characterization of direct assembly of RNA nanostructures on supported fluid lipid bilayer. Following the basic characterization of the RNA and lipid components (Subheadings 3.1 and 3.2), different experimental approaches are pursued to obtain real-time information about the formation of the lipid bilayer and the subsequent assembly of the polyhedron. Time-resolved characterization of each step in the RNA assembly is performed using simultaneous QCM-D and ellipsometry measurements, confocal fluorescence microcopy, and AFM. The characterization techniques can be used to explore the effects of various solution conditions (e.g., counterion concentration) and surface properties (e.g., changing the bilayer composition)

ellipsometry are complementary as they provide information about the mechanical and optical properties of the surface layer, respectively. Combining these techniques imparts in situ, real-time information about the formation of the lipid bilayer and the subsequent assembly of the polyhedrons. Via two independent protocols described here, it is also possible to visualize the surface nanostructures formed on bilayers using confocal fluorescence microscopy and atomic force microscopy (AFM). The mobility in the deposited layer is also investigated by means of fluorescence recovery after photobleaching (FRAP).

As a practical example, we describe the formation of 3D RNA polyhedrons on positively charged 1,2-dioleoyl-3-trimethylammonium-propane (DOTAP) bilayers. However, the proposed protocol can easily be optimized to build up many different nucleic acid structures such as 2D grid arrays or RNA 3D nanoparticles on surfaces.

2 Materials

Prepare all solutions using water ultrapurified to a sensitivity of >18 MΩ cm at 25 °C (Milli-Q® Ultrapure water purification system, Millipore, Massachusetts, USA). As RNA is sensitive to nuclease degradation, clean all glassware as follows: soak in 10% hydrochloric acid for several hours, wash ten times with Milli-Q water, and sterilize at 200 °C for 8 h. Autoclave all aqueous solutions/buffers (121 °C for 20 min).

Follow all manufacturers' safety guidelines and storage recommendations when working with chemicals and equipment. Follow all waste disposal regulations when disposing of excess and used materials.

2.1 Preparation and Characterization of Lipid Vesicles

1. *Bilayer formation buffer:* 10 mM HEPES, pH 7. Weigh out 2.381 g of HEPES (4-(2-hydroxyethyl)piperazine-1-ethanesulfonic acid sodium salt; ≥99.5% purity) into a 1-L beaker. Add water to a volume 900 mL and mix well. Adjust to pH 7 with HCl. Make up to 1 L with water. Store at 4 °C. Filter through a syringe filter (Anotop 25 Plus, 0.2 μm, Whatman) prior to use.
2. *Stock lipid solutions:* Weigh out appropriate amounts of each lipid, DOTAP (18:1 TAP, 1,2-dioleoyl-3-trimethylammonium-propane) and RhPE (1,2-dioleoyl-*sn*-glycero-3-phosphoethanolamine-N-(lissamine rhodamine B sulfonyl) (ammonium salt)) both with 99% purity, into a clean glass vial with a hard plastic cap lined with Teflon. Use a glass Hamilton syringe to add solvent mixture of chloroform–methanol (9:1 (vol/vol), both 99.8% purity) to dissolve lipids and make 10 mg/mL stock solution (*see* **Note 1**).

3. 100 μL volumetric glass syringe (e.g., Hamilton® syringe).
4. Probe tip sonicator (e.g., Vibracell tip sonicator, Sonics & materials Ltd., USA).
5. Benchtop centrifuge.
6. Dynamic light scattering (DLS) instrument (e.g., Zetasizer Nano Z, Malvern Instruments Ltd., Worcestershire, UK).

2.2 Preparation and Characterization of RNA Components

1. *RNA tectosquares:* Prepare according to the protocol described in Severcan et al. [1] and store in 20 μL aliquots at −80 °C. Essentially, RNA molecules (size 120–180 nt) are prepared in run-off transcription reaction on dsDNA templates using T7 RNA polymerase, supplemented with inorganic pyrophosphatase and RNase blocker. A crude transcription mixture can be purified using polyacrylamide gel electrophoresis (PAGE) resulting in pure RNA monomers. RNA tectosquares and other particles can be assembled from RNA monomers in *Association buffer 1* and stored for further analysis.
2. *Association buffer 1 (0.2 mM Mg^{2+}):* 10 mM Tris–borate, 50 mM KCl, and 0.2 mM $Mg(OAc)_2$, pH 8.2. Add 100 mL of water to a 1-L beaker. Weigh 1.21 g of Tris (Tris(hydroxymethyl)aminomethane (Tris) of ultra pure grade), 0.62 g of boric acid (approx. 99% purity), 3.72 g of KCl (($\geq$99.5% purity), and 0.043 g of $Mg(OAc)_2$ (99% purity). Transfer each to the beaker while mixing. Add water up to 900 mL. Store at room temperature (*see* **Note 2**).
3. *Association buffer 2 (15 mM Mg^{2+}):* 10 mM Tris–borate, 50 mM KCl, and 0.2 mM $Mg(OAc)_2$, pH 8.2. Add 100 mL of water to a 1-L beaker. Weigh 1.21 g of Tris (Tris(hydroxymethyl)aminomethane (Tris) of ultra pure grade), 0.62 g of boric acid (approx. 99% purity), 3.72 g of KCl (($\geq$99.5% purity), and 3.216 g of $Mg(OAc)_2$ (99% purity). Transfer each to the beaker while mixing. Add water up to 900 mL. Store at room temperature.
4. GelStar nucleic acid gel stain (10,000× concentrate from Lonza Bioscience) (*see* **Note 3**).
5. Mica (muscovite mica grade V-1).
6. Nitrogen.
7. AFM system (for instance XE-100 Park system) used in non-contact mode in air. For that typical cantilevers with medium spring constant 10–50 N/m oscillating in resonance frequency 200–350 kHz can be used (for the Park system the PPP-NCHR: $k = 42$ N/m: $f = 330$ kHz is recommended). Alternatively samples can be imaged under solution (*Association buffer 2*) using soft AFM probes with a spring constant 0.1–0.8 N/m and a resonance frequency of 10–40 kHz.

2.3 Simultaneous QCM-D and Ellipsometry Measurements

1. Q Sense E4 system from Q Sense (Gothenburg, Sweden) with a Q-Sense Ellipsometry Module comprising a temperature-controlled flow cell and Q-sense software for measurement as well as Q-tools software for analysis. The advantage of combining ellipsometer and QCM-D is that the ellipsometer measures the dry mass and the QCM-D the wet mass, giving information of the absolute amount and the hydration of the layer.
2. QCM crystals coated with a 50-nm thick SiO_2 layer (QSX 303, Q-Sense AB). The QCM crystals are quartz crystals with a fundamental resonant frequency of 4.95 MHz and are covered with a 50 nm thick SiO_2 layer on a thin gold surface layer connected to the two electrodes.
3. Null-ellipsometer (for example Optrel Multiskop ellipsometer from Optrel, Berlin, Germany with a Nd:YAG, $\lambda = 5320$ Å laser) [15]. The angle of incidence is adjusted to 65° to have the light beam pass perpendicular through the windows of the simultaneous QCM-D ellipsometry module. The measurement cell should allow the exchange of solution at about the same relative rate as the QCM-D measurement cell to be able to directly compare the results from the two techniques. Alternatively, if a simultaneous QCM-D and elliopsometry setup is not available, use a standard ellipsometer for in situ measurements at the solid–liquid interface.
4. Peristaltic pump (e.g., Ismatec IPC-N 4) to control the flow of samples and buffers into the QCM-D measurement chamber as required during the experiment.

2.4 Confocal Fluorescence Microscopy

1. Confocal Fluorescence or TIRF Microscope. In this practical example, fluorescence microscopy images shown are recorded using a confocal laser scanning microscope (CLSM) Leica CLSM SP5 operated in the inverted mode (D60001) with a 60× glycerol-immersion objective. HeNe (543 nm) and Ar ion (488 nm) lasers are used to excite the Rhodamine-labeled (543 nm) and NBD-labeled (463 nm) lipids, respectively.
2. Hydrogen peroxide (H_2O_2; 30%), ammonium hydroxide (NH_4OH; 27%).
3. Hot plate.
4. Teflon® coverslip rack designed to vertically hold eight standard round or square coverslips (such as the Coverslip Mini-Rack (C-14784) from Molecular Probes).
5. Plasma cleaner (such as from Harrick Scientific, New York, USA).
6. Microscopy flow cells (such as the sticky-Slide $I^{0.4}$ Luer, ibidi GmbH, Martinsried, Germany).

3 Methods

Perform experiments at room temperature unless otherwise specified.

3.1 Preparation and Characterization of Lipid Vesicles

1. Using a Hamilton syringe, volumetrically dispense 300 μL of DOTAP lipid stock (10 mg/mL solution in chloroform from **step 2** in Subheading 2.1) into a 4-mL glass vial (*see* **Note 4**).
2. Dry under a slow stream of nitrogen to remove the solvent to form a thin film on the glass surface.
3. Remove trace solvent by storing vial under vacuum for 6–16 h.
4. Add 3 mL of bilayer buffer to dry lipid film and allow vial to stand for 30 min.
5. Insert clean sonicator probe into the vesicle dispersion and sonicate with ten 30-s pulses at 30% amplitude until the dispersion is clear.
6. Remove clear vesicle dispersion into centrifuge tubes and centrifuge at about 14,000 RCF for 10 min to remove any debris from the probe. Retain supernatant and discard pellet, if any.
7. Measure the vesicle size using dynamic light scattering (DLS) to check that the size is around or below 100 nm and the polydispersity index is below 0.2.
8. To form fluorescently labeled lipid bilayers, add 5 μL of fluorescent lipid stock (RhPE; 10 mg/mL solution in chloroform) into the vial in **step 1** and follow **steps 2–7**.

3.2 Preparation and Characterization of RNA Components

1. Thaw one aliquot of each *tectosquare 1* and *tectosquare 2* on ice.
2. Characterize the structure of each tectosquare by AFM imaging (Fig. 3):

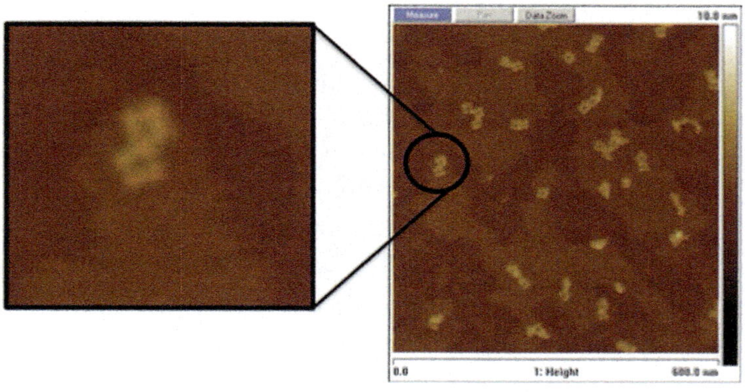

Fig. 3 AFM height imaging screenshot of RNA *tectosquare 1* on mica in air. The *scale bar* is 600 nm. The *inset* shows a close-up of two squares

(a) Add 1.8 μL of RNA tectosquare (0.8 mg/mL stock) to 48 μL of association buffer to achieve 200 nM RNA solution. Invert carefully and leave on ice.

(b) Use adhesive to attach a piece of mica to the AFM button.

(c) Cleave off the top layer of mica using tweezers.

(d) Rinse the surface with 50 μL of association buffer 2. Remove excess buffer from surface.

(e) Pipette 10 μL of RNA solution onto freshly cleaved mica. Incubate for 5 min (*see* **Note 5**).

(f) Rinse gently with 50 μL of buffer.

(g) Rinse with 500 μL of water.

(h) Dry until a gentle stream of nitrogen.

(i) Mount onto the AFM sample stage.

(j) Image the topography of the RNA on the surface using noncontact mode in air according to standard AFM procedures.

3. To fluorescently label the RNA tectosquares for microscopy studies, incubate RNA tectosquares (10 μg/mL) in association buffer 1 with the GelStar dye (excitation 493 nm, emission 527 nm) at its 1× nominal concentration on ice prior to microscopy (*see* **Note 6**).

3.3 Monitoring Lipid Bilayer Formation by QCM-D

1. Prior to use, clean the sensor crystal by soaking in 2% SDS solution for 1 h, rinsing with water, rinsing with ethanol, and drying under a stream of nitrogen and finally treating in a plasma cleaner for 5 min.
2. Insert cleaned QCM-D crystal into the measurement chamber of the Q-sense system (*see* **Note 7**).
3. Initiate the instrument software and set the temperature control to 25 °C.
4. Determine the resonance frequency and dissipation values in air for overtones 1, 3, 5, 7, 9, and 11.
5. Pump in HEPES buffer into the module via the peristaltic pump at a flow rate of 200 μL/min to fully exchange the solution in the measurement chamber (minimum 2 mL depending on the length of tubing) (*see* **Note 8**).
6. Determine the resonance frequency and dissipation values in buffer for overtones 1, 3, 5, 7, 9, and 11.
7. Begin the QCM-D measurement while flowing buffer for 10 min. Check that the frequency and dissipation values stabilize (*see* **Note 9**).

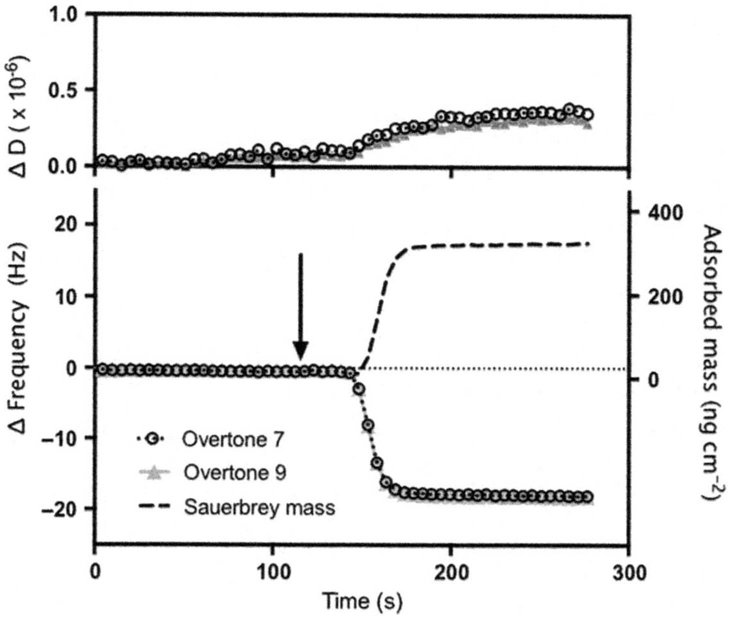

Fig. 4 Representative QCM-D measurement of supported bilayer formation on a silica surface showing changes in frequency (*bottom panel*) and dissipation (*top panel*). The *arrow* indicates the time of injection of DOTAP vesicles. Two overtones (7 and 9) are shown. The adsorbed mass as a function of time is calculated using the Sauerbrey expression and shown as a *dashed line* (*right y-axis*)

8. Once the values have stabilized, stop the buffer flow, restart the measurements, and measure for 3 min to obtain a stable baseline.
9. Carefully move the tubing from the buffer solution to the vesicle dispersion and flow the vesicles solution (0.3–0.5 mg/mL) at a flow rate of 100 μL/min. Continue flowing until the bilayer has formed and the frequency and dissipation values have stabilized.
10. Pump in HEPES buffer at a flow rate of 200 μL/min for at least 10 min to rinse away excess vesicles.
11. Check that the formation of the bilayer progressed properly by looking at the change in frequency and dissipation over time (Fig. 4). For the DOTAP bilayer used here, the final changes for frequency and dissipation should be -20 ± 1 Hz and $<0.5 \times 10^{-6}$, respectively (*see* **Note 10**).
12. Use the Qtools software to calculate the wet mass of the lipid film adsorbed to the silica surface using the Sauerbrey expression:

$$\Delta m = \frac{C}{o_n} \Delta f$$

where $C \approx 17.7$ ng Hz^{-1} cm^{-2} for a 5 MHz crystal and o_n is the overtone number [16]. For the case of DOTAP, the mass of the deposited bilayers would be expected to be 380–400 ng/cm^2 from the Sauerbrey expression (*see* **Note 11**).

3.4 Self-Assembly of RNA Polyhedron Monitored by QCM-D and Ellipsometry

1. Equilibrate the cell with bare surfaces and determine the optical properties of the bare substrate with using the ellipsometer and the frequency and dissipation values using QCM-D.
2. Flow in association buffer 1 at 200 μL/min for at least 10 min into the measurement chambers containing the DOTAP bilayer formed in Subheading 3.1.
3. Wait for the frequency and dissipation values to stabilize.
4. Once the values have stabilized, stop the buffer flow, use the QSoft software to determine the resonance frequency and dissipation values in buffer for overtones 1, 3, 5, 7, 9, and 11.
5. Restart the measurement and measure for 5 min to obtain a baseline.
6. Carefully move the tubing from the association buffer to the solution with the first RNA building block (*tectosquare 1*; 10 μg/mL). Inject the RNA at a flow rate of 10 μL/min. Continue flowing until the frequency and dissipation values as well as the readings from the ellipsometer have stabilized (*see* **Note 12**).
7. Rinse by flowing the association buffer 1 at a flow rate of 100 μL/min (*see* **Note 13**).
8. Exchange buffer to association buffer 2 using a flow rate of 100 μL/min. Continue flowing until the frequency and dissipation values have stabilized.
9. This time carefully move the tubing from the association buffer 2 to the solution with the second RNA building block (*tectosquare 2*; 10 μg/mL). Inject the RNA at a flow rate of 10 μL/min. Continue flowing until the frequency and dissipation values have stabilized (Fig. 5).
10. Use the Q-tools software to model the wet mass of the RNA adsorbed at each stage using a Voigt-based model for each adsorbed layer (*see* **Note 14**).

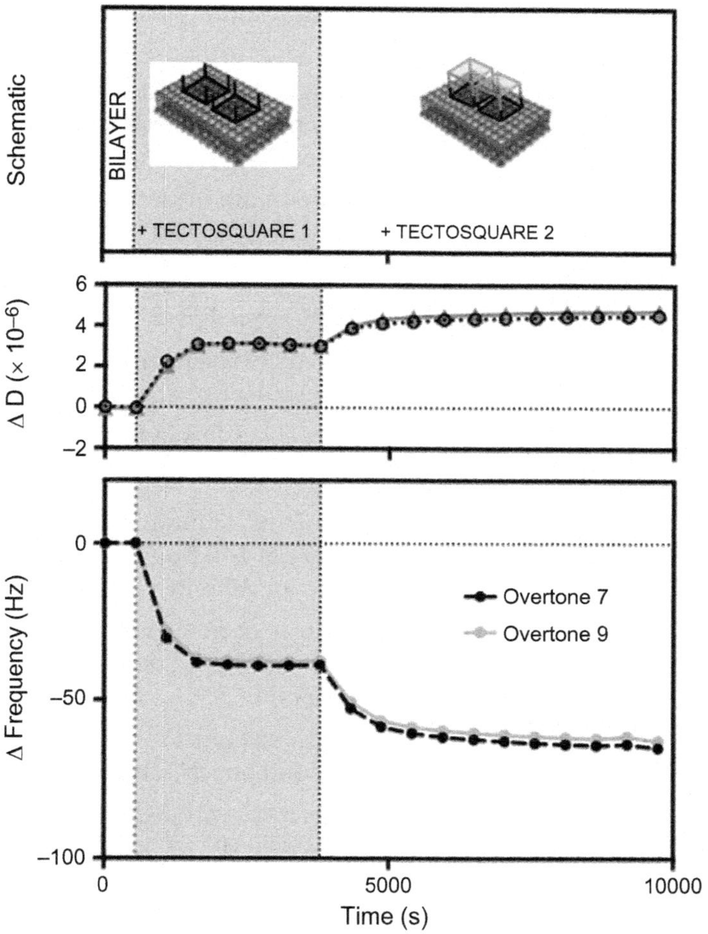

Fig. 5 Representative QCM-D measurement showing the stepwise assembly of an RNA polyhedron on a lipid bilayer from the sequential addition of RNA *tectosquare 1* (*grey panel*) and *tectosquare 2* (*white panel*). Two overtones (7 and 9) are shown. The *schematic* shows the sequential addition of each RNA component. First, the addition of *tectosquares 1* results in the formation of the lower RNA layer (shown over *grey* background). Second, the addition of *tectosquares 2* results in the binding of each *tectosquaree 2* to a *tectosquare 1* to form a full layer of RNA polyhedrons. The timecourse of the formation of each layer is characterized by a decrease in frequency (*lower panel*) due to the increased mass on the surface as well as a concomitant increase in dissipation (*middle panel*) due to the protruding 3D structure of the RNA polyhedrons

3.5 Self-Assembly of RNA Polyhedron Monitored by Confocal Fluorescence Microscopy

1. Clean microscope coverslips using a version of the RCA 1 [17] cleaning method:
   (a) In a 500-mL beaker, prepare the RCA bath by pouring in 80 mL of water, 10 mL of 30% hydrogen peroxide (H_2O_2), 10 mL of 27% ammonium hydroxide to achieve a 5:1:1 (H_2O–H_2O_2–NH_4OH) volume ratio.

(b) Heat solution on a hot plate to 70 °C (*see* **Note 15**).

(c) Place coverslips carefully into the Teflon coverslip rack designed to vertically hold eight standard round or square coverslips (such as the Coverslip Mini-Rack (C-14784) from Molecular Probes) (*see* **Note 16**).

(d) Careful dip in the coverslip rack to fully cover the coverslips. Leave at 70 °C for 5–15 min.

(e) Rinse well in water (*see* **Note 17**).

(f) Blow coverslips dry with nitrogen (*see* **Note 18**).

(g) Place coverslips in plasma cleaner and treat for 5 min in reduced air plasma of 0.02 mbar.

(h) Use coverslips immediately after plasma treatment.

2. Mount clean coverslip onto a microscopy flow cell.
3. Inject 0.5 mL of fluorescent lipid vesicles (0.3–0.5 mg/mL) at a flow rate of 100 μL/min into each channel of the flow cell and incubate for 30 min.
4. Rinse each channel with 10 mL of water at 200 μL/min, followed by 2 mL of association buffer 1 at a flow rate of 200 μL/min.
5. Place flow cell onto a temperature-controlled microscope stage (25 °C) and attach tubing to the peristaltic pump.
6. Image bilayers with 543 nm laser line to excite the Rhodamine-labeled DOTAP.
7. Perform fluorescence recovery after photobleaching (FRAP) on the bare bilayers by bleaching a spot with a diameter of 5 μm using the highest intensity of the laser, according to the microscope guidelines. Right after the bleaching step, record time-lapse images in short intervals (< 1 s) to monitor the fluorescence recovery. Repeat in different regions to ensure that bilayer is continuous and homogeneous (in the case of microscopy cells used here, the surface area available for bilayer formation is around 0.6 cm^2).
8. After ascertaining that the lipid bilayer is of good quality via FRAP in **step** 7, carefully inject the first RNA building block (*tectosquare 1*; 10 μg/mL) at a flow rate of 10 μL/min.
9. Take time-lapse images of the Rhodamine-labeled bilayers and the GelStar-labeled RNA, respectively. The excitation and emission maxima of GelStar stain are 493 nm and 527 nm (532 nm for RNA) respectively.
10. Once the adsorption has reached a plateau, rinse the channel with 10 mL of association buffer 1 at a flow rate of 200 μL/min.
11. Perform FRAP on the bilayers as in **step** 7.

12. Calculate the diffusion constants of the bilayers directly from the images using specialized software (FRAP_analysis [18] is an example of software that is freely available). FRAP can be performed with different concentrations of RNA to determine the effect of RNA concentration on the lipid diffusivity within the bilayer.

3.6 Self-Assembly of RNA Polyhedron Monitored by AFM

1. Use adhesive to attach a piece of mica to AFM button.
2. Cleave off top layer of mica using tweezers.
3. Mount surface onto microscope stage.
4. Add a drop of lipid solution onto the mica surface.
5. Incubate for 20 min.
6. Rinse with copious amounts of *Association buffer 2.*
7. Carefully add the RNA building blocks sequentially as described for Subheading 3.3, rinsing with copious amounts of *Association buffer 2* in between components.
8. Image the topography of the RNA on the surface using non-contact mode imaging under solution (*Association buffer 2*) according to standard AFM procedures.

4 Notes

1. Do not use syringes or containers made of polymeric material for storing organic solvents, as these will leach contaminants. Instead, use glass and stainless steel. Seal the vials well and store in a cool, dry, and well-ventilated place for up to a month. The concentration of lipids increases whenever the vial is opened due to solvent evaporation. Minimize this problem by working quickly in a cool fume cupboard and by keeping note of any volume changes that occur in the stock solutions. If the desired lipid solution is a mixture of two or more lipids, then the mixture should be made immediately from fresh stock solutions. Storing lipid mixtures will reduce discrepancy between desired and actual component ratios, which could result from potentially different evaporation of separate stock solutions. Work with chloroform and methanol should always be done in a fume cupboard and following all of the manufacturer's safety precautions.
2. If storing the buffer, it is advisable to make a 5× concentrated stock and dilute as needed.
3. Follow all precautions when working with the stain, which is a potential teratogen, as well as with the DMSO-containing solutions, as DMSO is a known skin penetration enhancer.

4. All steps involving solvents must be performed in the fume-hood, using the appropriate personal protection.
5. Do not allow the drop to dry.
6. The GelStar dye does not bind to DOTAP bilayer at the concentrations used here, but it is important to consider possible artifacts due to fluorophore interactions with the bilayer if other lipids are used. It is therefore important to run control experiments to rule out inadvertent interactions.
7. The crystal should be inserted immediately after plasma treatment.
8. All solutions pumped into the QCM-D flow chambers should be degassed to prevent bubble formation and equilibrated to the measurement temperature to minimize thermal drift that could affect measurements.
9. If the values do not stabilize, check for any leaks, air bubbles, cleanliness issues or improper placement of the crystal.
10. QCM-D is a sensitive tool for measuring the real-time kinetics of bilayer formation. As the bilayer formation kinetics are highly reproducible under the same experimental conditions for vesicles of the same lipid composition, concentration and buffer, the frequency and dissipation changes over time can be used as a signal signature to check for good bilayer formation [19–21].
11. The Sauerbrey equation is a good approximation for thin, acoustically rigid and evenly distributed adsorbed films formed in biomolecular systems [22]. However, even acoustically rigid films may trap solvent, and this additional mass will also be sensed in the QCM-D measurements. Therefore, the Sauerbrey-derived mass only corresponds well to the molecular mass of the film when the amount of coupled water is low. This is the case for planar supported lipid bilayers, and therefore the Saurebrey equation can be used to determine the adsorbed mass.
12. The solutions of the RNA polyhedrons should be freshly prepared. In order to reduce the lag time, it is possible to increase the flow rate to 100 μL/min for the first minute of flow such that the nucleic acids reach the measurement cell faster.
13. Check that no desorption occurs as observed by no change in the frequency. The experimental approach relies on a robust adsorption. Therefore, minimal RNA desorption (i.e., not significant within errors of the measurements) should be noted upon rinsing with buffer. As a control experiment it is possible to monitor the adsorption of RNA molecules to the bare SiO_2 surface, which should show no adsorption due to repulsion

since the RNA is negatively charged and the SiO_2 surface is slightly negatively charged at neutral pH.

14. For structures that extend from the surface, such as the RNA polyhedrons, the Sauerbrey expression is not applicable to calculate the adsorbed mass [22]. This limitation of the Sauerbrey model occurs because the adsorbed material may extend into the solution in such a way that the amount of hydrodynamically coupled water becomes significant. Secondly, the conversion of ΔF into wet mass fails because of the large dissipation induced. The amplitude difference, Ψ, and the phase shift, Δ when light is reflected on a surface is measured with ellipsometry [23, 24]. From these values the thickness and refractive index of the film formed at the surface are determined using an iterative process and based on the optical characterization of the bare substrate. Adsorbed amount (Γ), the dry mass, can be obtained according to de Feijter et al [25].

$$\Gamma = d_f \frac{n_f - n_0}{dn/dc}$$

where d_f is film thickness, n_f is the film refractive index, and n_0 is the refractive index of the solvent. The refractive index increment dn/dc has been determined to be 0.148 cm^3/g for the lipid [26] and 0.168 cm^3/g for the RNA adsorbed to the lipid bilayer [27].

15. Perform in a fume hood. Wear goggles, resistant gloves, and protective labwear. Never leave hot plates unattended.
16. It is best to use a rack made of Teflon material such that it can withstand strong acids and bases and is resistant to heat.
17. Collect waste and follow all instructions for waste removal.
18. At this stage, the coverslip surfaces should be hydrophilic and therefore should wet easily.

References

1. Severcan I, Geary C, Chworos A et al (2010) A polyhedron made of tRNAs. Nat Chem 2:772–779. doi:10.1038/nchem.733
2. K a A, Bindewald E, Yaghoubian AJ et al (2010) In vitro assembly of cubic RNA-based scaffolds designed in silico. Nat Nanotechnol 5:676–682. doi:10.1038/nnano.2010.160
3. Nasalean L, Baudrey S, Leontis NB, Jaeger L (2006) Controlling RNA self-assembly to form filaments. Nucleic Acids Res 34:1381–1392. doi:10.1093/nar/gkl008
4. Geary C, Chworos A, Jaeger L (2011) Promoting RNA helical stacking via A-minor junctions. Nucleic Acids Res 39:1066–1080. doi:10.1093/nar/gkq748
5. Severcan I, Geary C, Verzemnieks E et al (2009) Square-shaped RNA particles from different RNA folds. Nano Lett 9:1270–1277. doi:10.1021/nl900261h
6. Grabow WW, Zakrevsky P, Afonin KA et al (2011) Self-assembling RNA nanorings based on RNAI/II inverse kissing complexes. Nano Lett 11:878–887. doi:10.1021/nl104271s

7. Chworos A, Severcan I, Koyfman AY et al (2004) Building programmable jigsaw puzzles with RNA. Science 306:2068–2072. doi:10.1126/science.1104686
8. Alivisatos AP, Johnsson KP, Peng X et al (1996) Organization of "nanocrystal molecules" using DNA. Nature 382:609–611. doi:10.1038/382609a0
9. Mirkin C, Letsinger RL, Mucic RC, Storhoff JJ (1996) A DNA-based method for rationally assembling nanoparticles into macroscopic materials. Nature 382:607–609
10. Aldaye FA, Sleiman HF (2007) Dynamic DNA templates for discrete gold nanoparticle assemblies: control of geometry, modularity, write/erase and structural switching. J Am Chem Soc 129:4130–4131. doi:10.1021/ja070017i
11. Michanek A, Björklund M, Nylander T, Sparr E (2012) ssRNA base pairing at a bilayer interface can be controlled by the acyl chain order. Soft Matter 8:10428. doi:10.1039/c2sm06700e
12. Dabkowska AP, Michanek A, Jaeger L et al (2015) Assembly of RNA nanostructures on supported lipid bilayers. Nanoscale 7:583–596. doi:10.1039/C4NR05968A
13. Suzuki Y, Endo M, Sugiyama H (2015) Lipid-bilayer-assisted two-dimensional self-assembly of DNA origami nanostructures. Nat Commun 6:8052. doi:10.1038/ncomms9052
14. Kocabey S, Kempter S, List J et al (2015) Membrane-assisted growth of DNA origami nanostructure arrays. ACS Nano 9:3530–3539. doi:10.1021/acsnano.5b00161
15. Harke M, Teppner R, Schulz O et al (1997) Description of a single modular optical setup for ellipsometry, surface plasmons, waveguide modes and their corresponding imaging techniques including Brewster angle microscopy. Rev Sci Instrum 68:3130
16. Vandoolaeghe P, Rennie AR, Campbell RA et al (2008) Adsorption of cubic liquid crystalline nanoparticles on model membranes. Soft Matter 4:2267. doi:10.1039/b801630e
17. Kern W, Puotinen D (1970) Cleaning solutions based on hydrogen peroxide for use in silicon semiconductor technology. RCA Rev 31:187–206
18. Jönsson P, Jonsson MP, Tegenfeldt JO, Höök F (2008) A method improving the accuracy of fluorescence recovery after photobleaching analysis. Biophys J 95:5334–5348. doi:10.1529/biophysj.108.134874
19. Keller CA, Kasemo B (1998) Surface specific kinetics of lipid vesicle adsorption measured with a quartz crystal microbalance. Biophys J 75:1397–1402. doi:10.1016/S0006-3495(98)74057-3
20. Richter R, Mukhopadhyay A, Brisson A (2003) Pathways of lipid vesicle deposition on solid surfaces: a combined QCM-D and AFM study. Biophys J 85:3035–3047. doi:10.1016/S0006-3495(03)74722-5
21. Cho NJ, Frank CW, Kasemo B, Hook F (2010) Quartz crystal microbalance with dissipation monitoring of supported lipid bilayers on various substrates. Nat Protoc 5:1096–1106. doi:10.1038/Nprot.2010.65
22. Kasemo B, Fant C, Sott K et al (2001) Variations in coupled water, viscoelastic properties, and film thickness of a Mefp-1 protein film during adsorption and cross-linking: a quartz crystal microbalance with dissipation monitoring, ellipsometry, and surface plasmon. Anal Chem 73:5796–5804
23. Azzam RMA, Bashara NM (1996) Ellipsometry and polarized light. Elsevier B.V, Amsterdam
24. Tiberg F, Landgren M (1993) Characterization of thin nonionic surfactant films at the silica/water interface by means of ellipsometry. Langmuir 9:927–932. doi:10.1021/la00028a009
25. de Feijter A, Benjamins J, Veer FA (1978) Ellipsometry as a tool to study the adsorption behavior of synthetic and biopolymers at the air–water interface. Biopolymers 17:1759. doi:10.1002/bip.1978.36017071
26. Ainalem M-L, Kristen N, Edler KJ et al (2010) DNA binding to zwitterionic model membranes. Langmuir 26:4965–4976. doi:10.1021/la9036327
27. Nieuwenhuysen P, De Voeght F, Clauwaert J (1981) The molecular weight of Artemia ribosomes, as determined from their refractive-index increment and light-scattering intensity. Biochem J 197:689

Chapter 8

Evaluation of Thermal Stability of RNA Nanoparticles by Temperature Gradient Gel Electrophoresis (TGGE) in Native Condition

Kheiria Benkato, Benjamin O'Brien, My N. Bui, Daniel L. Jasinski, Peixuan Guo, and Emil F. Khisamutdinov

Abstract

Temperature gradient gel electrophoresis (TGGE) is a powerful tool used to analyze the thermal stabilities of nucleic acids. While TGGE is a decades-old technique, it has recently gained favor in the field of RNA nanotechnology, notably in assessing the thermal stabilities of RNA nanoparticles (NPs). With TGGE, an electrical current and a linear temperature gradient are applied simultaneously to NP-loaded polyacrylamide gel, separating the negatively charged NPs based on their thermal behavior (a more stable RNA complex will remain intact through higher temperature ranges). The linear temperature gradient can be set either perpendicular or parallel to the electrical current, as either will make the NPs undergo a transition from native to denatured conformations. Often, the melting transition is influenced by sequence variations, secondary/tertiary structures, concentrations, and external factors such as the presence of a denaturing agent (e.g., urea), the presence of monovalent or divalent metal ions, and the pH of the solvent. In this chapter, we describe the experimental setup and the analysis of the thermal stability of RNA NPs in native conditions using a modified version of a commercially available TGGE system.

Key words Temperature gradient gel electrophoresis, TGGE, Melting temperature, RNA nanoparticle, pRNA, 3-way junction, 3WJ

1 Introduction

Thermodynamic stabilities of RNA NPs are characterized by the ability of an individual sequence to specifically recognize its complementary strand using the nucleotides' ability to engage in canonical (Watson–Crick, i.e., A–U and C–G) and noncanonical interactions, including base-stacking interactions. In general, the stability of nucleic acid molecules formed by the pairing of two strands can be assessed using several biophysical techniques, including spectrophotometric methods (e.g., UV melting) [1]. However, the melting curves acquired by measuring UV absorbance with respect to temperature often have multiple transitions, making it

Eckart Bindewald and Bruce A. Shapiro (eds.), *RNA Nanostructures: Methods and Protocols*, Methods in Molecular Biology, vol. 1632, DOI 10.1007/978-1-4939-7138-1_8, © Springer Science+Business Media LLC 2017

impossible to determine the melting point (T_M) specifically for a complex nanoparticle composed of more than two strands. One of the advantages of TGGE is that bands corresponding to the native conformation of a RNA complex can be visualized clearly on a gel matrix. This allows for a direct determination of the T_M value, assuming that there are only two states of RNA folding involved in the melting: the native form (intact RNA complex band detectable on the gel) and the denatured conformation (indeed *any* denatured conformation). Another advantage of TGGE is that it permits the screening of a library of RNA and DNA molecules in order to select the most stable conformation, as has been demonstrated by P. Bevilacqua's group [2, 3]. This feature makes it suitable for SELEX protocols [4]. In addition, TGGE is often used in conjunction with PCR to determine differences in sequence between similar sized DNA or RNA strands [5–7]. This is useful for detecting mutations in DNA or RNA sequences, as a suspected mutant strand can be compared to its original using differences in melting temperatures. The protocol is accurate for up to 50% of the sequence variations of strands containing ~500 base pairs. For example, given a sequence containing 100 mutant variations, the technique could detect upward of 50 [8]. TGGE can also be used to detect structural changes in DNA or RNA due to the addition of intercalator compounds such as ethidium bromide, daunomycin, and doxorubicin [9]. These compounds intercalate between DNA base pairs due to stacking interactions, altering the overall DNA double helix. Like the detection of mutations, one can compare a sample altered with an intercalator to the original strand or polymer and determine whether the intercalator was integrated by analyzing the shift in T_M.

Various TGGE protocols have been documented and the majority of them feature the use of denaturing agents like urea to destabilize the hydrogen forces of attraction in long DNA and RNA molecules [10–12]. These agents, however, decrease the thermal stability of nucleic acids, thereby reducing the melting temperature and rendering direct analysis of the T_M problematic. The protocol described herein utilizes only native conditions, avoiding use of denaturing agents.

TGGE can be implemented in a fashion that is either perpendicular or parallel to the electrical current (Fig. 1). In a perpendicular analysis, multiple aliquots of a single purified RNA NP are placed into each well corresponding to known temperatures gradually increasing from left to right (as shown in Fig. 1a). After electrophoresis is complete, the band corresponding to the RNA complex sample eventually undergoes thermal denaturation at a certain temperature. This allows for the assessment of the apparent T_M value of the NPs. As such, there are various RNA NPs whose T_M values have been evaluated: pRNA 3WJ [13] and RNA triangle, [14] RNA squares and pentagons, [15] RNA tecto-squares [16]

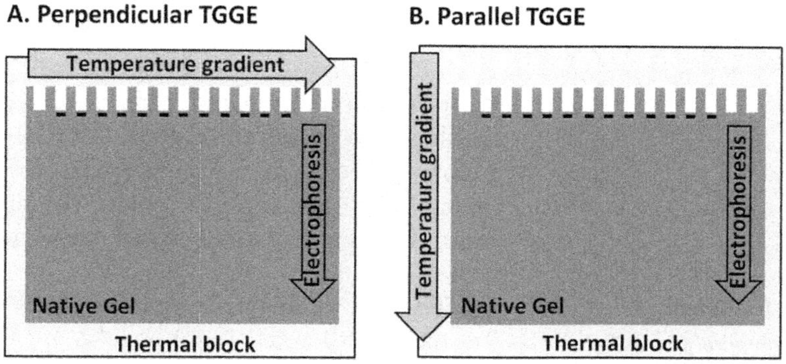

Fig. 1 Demonstration of (**a**) perpendicular and (**b**) parallel settings of temperature gradient gel electrophoresis

and nano-rings [17], RNA nano-cubes [18], and polyhedrons [19]. Moreover, perpendicular TGGE can be used to extract thermodynamic parameters of RNA nanoparticles. This has been demonstrated by P. Guo's group, where the T_M parameters of pRNA 3WJ NP have been analyzed as a function of concentration [20]. In contrast to perpendicular TGGE, parallel TGGE allows for the screening of multiple nanoparticles in high-throughput fashion. This is especially useful when comparing the thermal stabilities of a library of RNA nanoparticles at once. One example of such an experiment was the comparison the stabilities of 3WJ NPs composed of RNA, DNA, 2′-fluoro uridine and cytidine modified RNA strands (2′F-U/C RNA), respectively [13]. Multiple samples are run side-by-side through a temperature range that is gradually increased from top to bottom as the nucleic acid complexes continue to migrate from their wells (Fig. 1b). The heat-resistant RNA constructs migrate more slowly as the hydrogen and stacking interactions between participating strands form stable complexes. As a tendency, perpendicular analysis is more useful for examining the thermodynamic properties of a single NP, whereas parallel analysis is more useful for comparing the properties of several NPs.

The stability of RNA NPs is important for many aspects of application in nanomedicine, RNA synthetic biology, biomedical diagnostics, etc. This chapter focuses on the TGGE method to study the melting behavior of an RNA NP in native conditions using total RNA staining to visualize the bands. The method utilizes a commercially available TGGE system with modifications that are detailed in the methods section.

2 Materials

All solutions are prepared using double deionized water (dd H_2O) that has been passed through a filter (0.2 μm pore size). All reagents are of analytical grade, either purchased from a certified company or prepared in the laboratory. The reagents are typically stored at room temperature unless noted otherwise.

2.1 Native Polyacrylamide Gel Electrophoresis (PAGE).

1. Tris–borate–magnesium chloride (TBM, pH = 8.0) buffer; 1× buffer: 89 mM Tris base, 86 mM boric acid, and 10 mM $MgCl_2$.
2. Tris–magnesium saline (TMS) buffer, pH = 8.0; 1× buffer: 50 mM Tris–HCl, 100 mM NaCl, 10 mM $MgCl_2$.
3. Acrylamide–bis-acrylamide, 40% solution (mix ratio 29:1). Store at 4 °C. This reagent is available from several vendors, e.g., Thermo Fisher Scientific, Sigma-Aldrich. Alternatively, weigh 38.7 g of acrylamide and 1.3 g bis-acrylamide and transfer to a 150-mL graduated beaker containing about 50 mL of water. Mix for about 30 min using on a magnet stirrer and add dd H_2O up to 100 mL mark on the beaker. It is suggested to filter the resulting mixture through a 0.2 μm. The solution can be stored at 4 °C for at least 1 month if properly wrapped with aluminum foil (*see* **Note 1**).
4. Native gel loading buffer (6× concentrated): weigh 4 g sucrose, 25 mg bromophenol blue, and 25 mg of xylene cyanol and transfer to a graduate 15-mL falcon test tube. Bring the total volume up to 10 mL with dd H_2O.
5. Ammonium persulfate (APS): 10% solution in water (*see* **Note 2**).
6. TEMED: *N, N, N′, N′*-tetramethyl-ethylenediamine. Store at 4 °C (*see* **Note 3**).

2.2 Gel Electrophoresis System

In principle, any commercially available or custom-made TGGE can be used to unravel thermostability of RNA nanoparticles. As an example, a TGGE apparatus from Biometra GmbH with added modification was utilized in the current protocol. The gel casting using the original glass plates was inconvenient in our trials; therefore, we searched for alternative casting plates and electrophoresis systems that can compensate for the commercial recommended counterparts. Below is the suggested list of materials that could fit into the 9 × 9 cm area temperature block of the TGGE System.

1. Electrophoresis unit that fits into the temperature block of the utilized TGGE system. In this work, a SE250 Mighty Small II Mini Vertical Electrophoresis Unit has been used (**Hoefer, Inc.) has been used, including the following:**

(a) Lower buffer chamber.
(b) Upper buffer chamber/cooling core.
(c) Safety lid with high voltage leads.
(d) Casting cradle w/sealing gasket set.
(e) Casting clamp assemblies.
(f) Cams—4 pcs.
(g) Glass plates, rectangular, 10 × 8 cm.
(h) Alumina plates, notched, 10 × 8 cm.
(i) Spring clamps.
(j) Combs, 15 well (0.75 mm).
(k) T-spacers (0.75 mm).
(l) Gel seal.

2.3 Temperature Gradient Gel Electrophoresis

The protocol is based on the use of a 115 V TGGE System (e.g., Biometra GmbH) complete with:

(a) TGGE controller with integrated power supply, and control function of temperature and electrophoresis conditions.
(b) TGGE-electrophoresis unit with two removable buffer chambers, Peltier-element powered gradient block, and control cable.
(c) Buffer wicks.
(d) Cover films.
(e) Plastic clamps.

2.4 RNA Nanoparticles.

1. Gel-purified RNA nanoparticles. The concentration can be diluted to 50 ng/μL using 1× TMS buffer (or a buffer of choice). Alternatively, if the yield of RNA self-assembly is greater than 90%, the preassembled NPs can be prepared without purification.

2.5 Analysis

1. A gel imaging system.
2. Gel quantification software (e.g., Image J, Image Quant or equivalent).
3. Data analysis and graphing software (e.g., Origin, Excel or equivalent).

3 Methods

The customized TGGE procedure implies two general steps: (1) insertion of the RNA NP into the gel matrix using a standard vertical electrophoresis system and (2) transfer of the glass plate

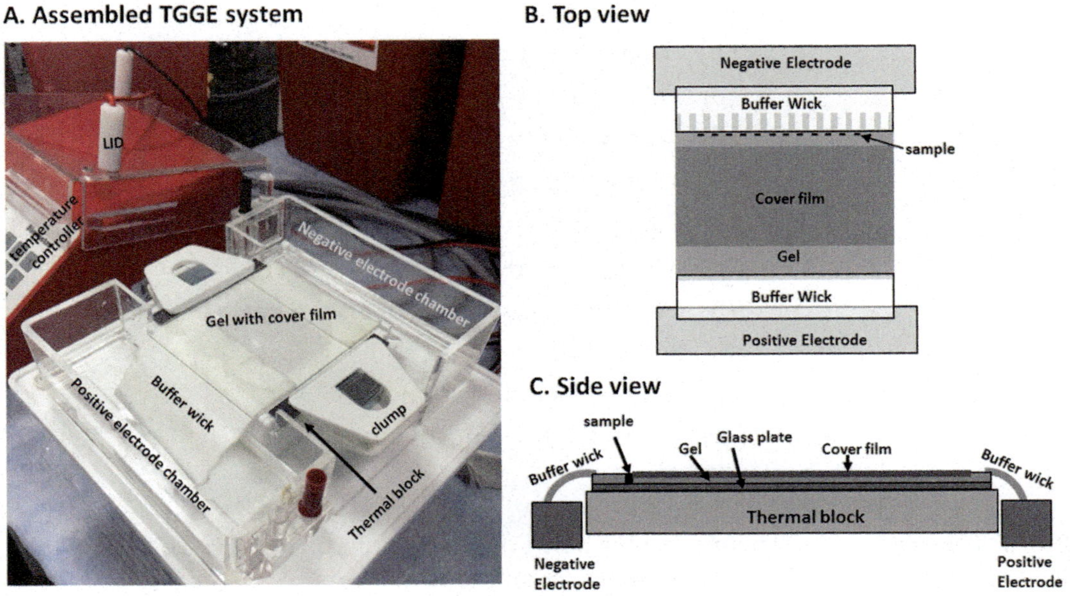

Fig. 2 A detailed view of customized TGGE system. (**a**) Photograph of complete assembly unit. Schematic top and side views with descriptions of components are shown in panels **b** and **c**, respectively

containing polyaclylamide (PAA) gel into the TGGE apparatus for evaluation of thermal stability. Figure 2 depicts the custom-assembled TGGE. The method primarily describes perpendicular TGGE but this protocol also applies for the parallel version as well.

3.1 RNA NP Insertion into 7% Polyacrylamide Gel Matrix at Native Conditions

1. Prepare and set up a vertical electrophoresis apparatus (see instructions for the electrophoresis unit, Subheading 2.2).
2. Prepare PAA mixture by mixing 0.875 mL of 40% PAA, 0.5 mL of 10 × TBM, and 3. 625 mL of dd H_2O. Add 60 μL of ammonium persulfate and 6 μL of TEMED.
3. Cast gel with a 8 cm × 10 cm × 0.75 cm gel cassette. Use 15-well combs.
4. Wait until the gel polymerizes for ~30 min, fill the inner and outer chambers with 1× TBM buffer, carefully remove 15-well comb and using a syringe wash formed wells (*see* **Note 4**).
5. Mix purified RNA nanoparticles with 6× native loading buffer. To increase the density of the RNA samples it is recommended to use 20 μL of RNA (50 ng/μL) and 10 μL of the 6× gel loading buffer.
6. Add 2 μL of the resulting mixture to the central 11 out of 15 wells and electrophorese at 70 V at room temperature until the sample has entered the gel (*see* **Note 5**).
7. After the electrophoresis is complete (~5–10 min depending on the NP size), disassemble glass plates with a spatula leaving

PAA gel on the small plate (white alumina plate, notched, available from GE Lifesciences http://www.gelifesciences.com/) together with T spacers (*see* **Note 6**).

3.2 Application of the PAA Gel with Prerun RNA NP to the TGGE Thermal Block

At this point the RNA NPs have entered the gel matrix and it can be safely transferred to the TGGE system. Prior to transferring, the back side of the aluminum plate should be thoroughly cleaned. Overall, the protocol is similar with respect to the description in the TGGE manual from Biometra (please refer to technical report.

http://www.biometra.de/files/themes/biometra_standard/pdfs/Manual_TGGE_Mini_Oct_2009.pdf). However, since the original plates are not in use, it is important to match the gels' wells with the temperature line # 1 (L1, closest to temperature T1) to line #6 (L6, closest to temperature T2) (as shown in Fig. 3) on the temperature block to obtain accurate temperature reading.

1. Fill buffer chambers with ~250 mL of 1× TBM buffer and place buffer wicks inside the chambers. Let buffer wicks soak for ~10 min.

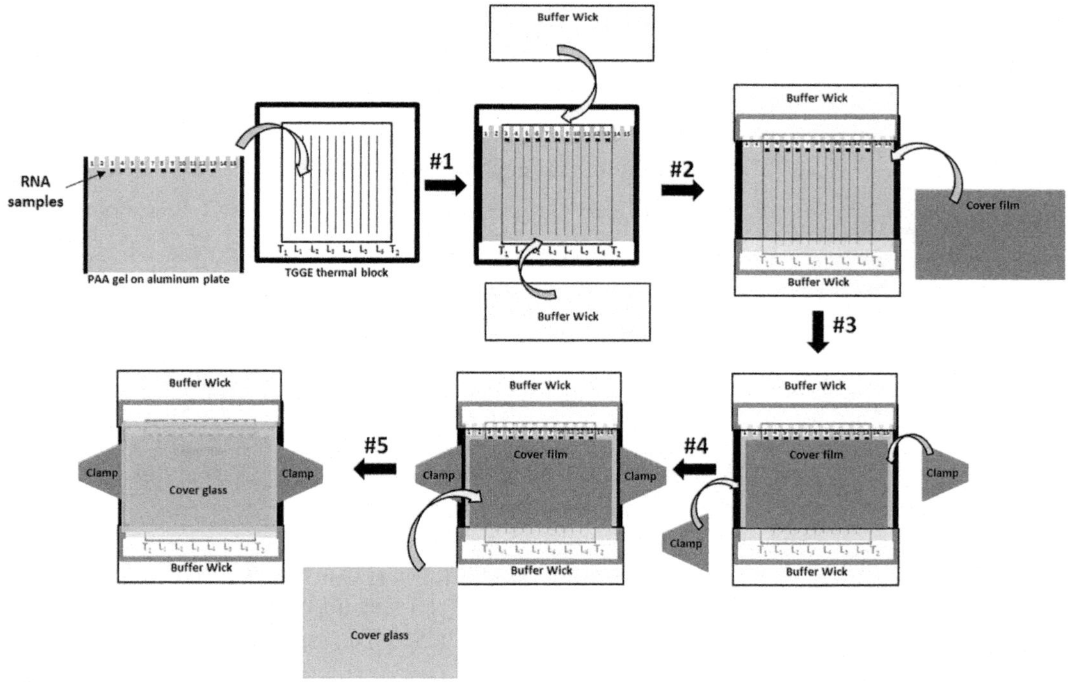

Fig. 3 A step-by-step diagram of TGGE assembly procedure. *Step 1*: transfer gel containing RNA nanoparticles into TGGE thermal block so wells 3–13 will be positioned between lower and higher temperatures T1 and T2, respectively. *Step 2*: attach presoaked buffer wicks to allow current conduction between two electrodes. *Step 3*: gently cover the gel with flexible cover film and avoid formation of any empty space or bubbles between gel and cover film. This is a critical point as bobbles will cause gel drying effect ruining experiment. *Step 4*: fix the position of the glass plate with clamps placed on the gel spacers and thermal block and finally place a cover glass on top of the overall assembly as shown in **step 5**

2. Gently wipe the surface of the thermal block with clean water. Place the glass plate containing gel on the thermal block (refer to schematic diagram of Fig. 3) (*see* **Note 7**).
3. Appropriate adjustments are required to ensure temperature ranges match the gel wells and lines L1 and L6 in the thermal block. Figure 3 **step 1** demonstrates the appropriate alignment of the gel's wells and the lines on thermal block. From the 15 wells only 11 wells undergo precise temperature changes (wells numbered 3–13, highlighted in red in Fig. 3). Wells 1 and 2 can be additionally used for a molecular weight ladder or monomer controls as needed. Noteworthy, wells 14 and 15 can be omitted since the maximum temperature is reached at T2 (corresponding to well 13).
4. Carefully attach the presoaked buffer wicks to the top and the bottom of the gel and cover the gel with a cover film available from the manufacturer (Fig. 3, **steps 2** and **3**) (*see* **Note 8**).
5. Fix the gel position on the thermal block with clamps provided by manufacturer as shown in Fig. 3 **step 4** (*see* **Note 9**).
6. Place the cover glass on the resulting assembly in such a way that it attaches to the top of the buffer wicks (*see* **Note 10**).
7. Close the safety lid which will connect the two electrodes. The TGGE system is now ready to run.
8. Program the TGGE temperature controller to the desired temperature range according to manufacturer's manual.
9. Electrophorese at constant 100 V for about 1 h (*see* **Note 11**).
10. When run is complete, carefully disassemble the whole system and place the gel into solution containing ethidium bromide for 20 min.
11. Rinse the gel twice with sufficient amount of dd H_2O to remove excess of Eth.Br. and destain the gel for another 10 min.
12. Visualize RNA bands on the gel using a gel imaging system and paper-print the actual size of the gel.

3.3 Calculate the Temperature Within Each Well and Determine Melting Temperature

1. Align the printed gel with the thermal block as it was before the experiment. Lane 3 and 13 should correspond to the starting and final temperatures T1 and T2 respectively. Since there was a linear gradient from L1 to L6, the analysis of the temperature on each well will be similar to the corresponding lines on the thermal block.
2. Calculate the estimated temperatures corresponding to each well using the following example:

As an example, let the temperature range be set to a linear gradient starting from L1 = 30 °C to L6 = 60 °C, which

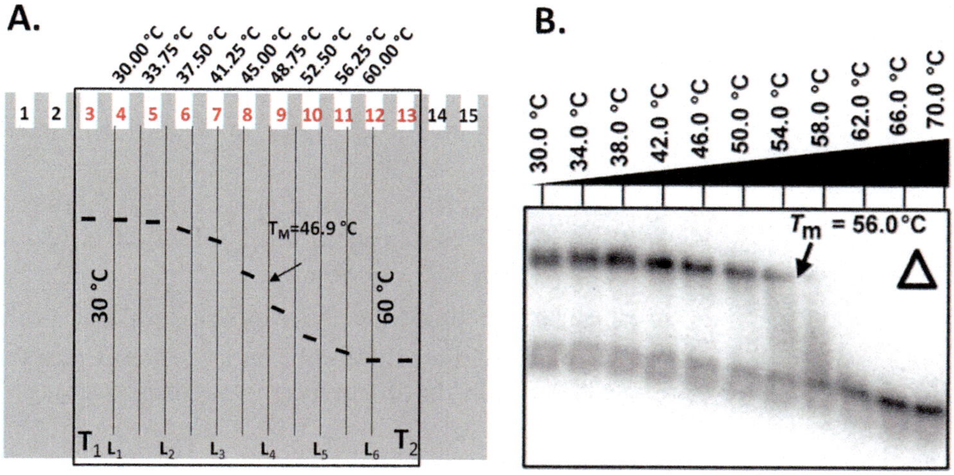

Fig. 4 Example of RNA complex melting point determination. (**a**) Calculation of temperature on each PAA well based on known the temperatures of L1–L6. (**b**) Melting temperature of P-32 labeled RNA triangle NP determined using TGGE system. (Image obtained from ref. [15] with permission from NAR, Oxford University Press)

corresponds to temperatures in the gel's wells #4 and #12, respectively (Fig. 4).

(a) The operating temperature range can be found by subtracting 60 °C (well #12)—30 °C (well #4) = 30 °C.

(b) To estimate the temperature increment corresponding to adjacent wells, divide the temperature range over the total numbers of wells excluding the first well (9–1 = 8), resulting in 30 °C /8 = 3.75 °C.

(c) Next, estimate the temperature corresponding to each well by adding the appropriate number of temperature increments. For this example, one obtains for well 4 a temperature of 30 °C (equal to T1), for well 5: 30 °C + 3.75 °C = 33.75 °C, for well 6: 33.75 °C + 3.75 °C, etc.

3. In general, the inflection point of the location of the gel bands as a function of temperature corresponds to the melting point of the analyzed nanoparticle (*see* **Note 12**). The T_M point can be determined accurately using a gel quantification software mentioned in the materials section. For these purposes, integrate the intensity of the RNA NP bands on each well and divide this value by the total intensity of the well. This will determine the relative fraction of formed nanoparticle in percent.

4. Plot the data (with temperature as x-axis and remaining fraction of NP as y-axis) using a graphing software, e.g., Origin. Fit the data points to a nonlinear fitting curve using sigmoid function to determine 50% nanoparticle formation, this corresponds to the melting temperature.

4 Notes

1. It is important to wear a mask when weighing acrylamide. Prepare aqueous solution of acrylamide powder in a fume hood only.
2. It is convenient to have 1.6 mL Eppendorf test tubes containing 1 g of APS. When needed, add 1 mL of dd H_2O and mix well to have fresh APS solution.
3. Storage at 4 °C reduces the characteristic TEMED smell.
4. It is advisable to insert the combs on its 1/5 part into the gel, this will decrease the depth of the resulting wells allowing better washing.
5. The final RNA amount should be at least 20 ng/well to detect RNA NP using total staining with ethidium bromide (Et.Br.).
6. To ensure that the PAA gel sticks to the small alumina plate but not to the large plate it is recommended to wipe the large glass plate with a siliconized reagents, e.g., Sigmacoat or a Gel Slick.
7. We found that the manufacturer's recommended thermal coupling solution can be omitted. The alumina glass plate is a very good heat conductor and it makes full contact with the surface of the thermal block.
8. Avoid dislocation of the gel at this step.
9. Small paper clips can be used as alternatives. Attach clippers on plastic T-spacers; do not clip on the aluminum plate.
10. The cover plate should gently lie on the wicks, do not squeeze them on the gel.
11. The timing depends of the size of RNA NP and the percentage of the native gel. Based on our studies, it takes 60 min to run an RNA triangle NP on 7% native TGGE [14] and 50 min for a pRNA 3WJ analyzed on 15% TGGE [13].
12. This is a coarse way of calculating TM for RNA NP, and generally useful when comparing thermal stabilities between nanoparticles where the exact melting point is not critical. The more precise way of calculating TM requires the use of a radiolabeled strand allowing the quantification of the remaining fraction of NP.

Acknowledgment

We thank Seth Abels for proofreading this work and leaving valuable comments. The research was supported by Department of Chemistry BSU start-up funds, Chemistry Research Immersion Summer Program (CRISP) at BSU and Indiana Academy of Science grant # G9000602A to Emil Khisamutdinov.

References

1. SantaLucia J Jr, Hicks D (2004) The thermodynamics of DNA structural motifs. Annu Rev Biophys Biomol Struct 33:415–440
2. Chadalavada DM, Bevilacqua PC (2009) Analyzing RNA and DNA folding using temperature gradient gel electrophoresis (TGGE) with application to in vitro selections. Methods Enzymol 468:389–408
3. Nakano M, Moody EM, Liang J, Bevilacqua PC (2002) Selection for thermodynamically stable DNA tetraloops using temperature gradient gel electrophoresis reveals four motifs: d(cGNNAg), d(cGNABg),d(cCNNGg), and d(gCNNGc). Biochemistry 41:14281–14292
4. Ellington AD, Szostak JW (1990) Invitro selection of RNA molecules that bind specific ligands. Nature 346:818–822
5. Manzano M, Cocolin L, Iacumin L, Cantoni C, Comi G (2005) A PCR-TGGE (Temperature Gradient Gel Electrophoresis) technique to assess differentiation among enological *Saccharomyces cerevisiae* strains. Int J Food Microbiol 101:333–339
6. Van den Bossche A, Van Nevel C, Herman L, Decuypere J, De Smet S, Dierick N, Heyndrickx M (2001) PCR-TGGE: a method for fingerprinting the microbial flora in the small intestine of pigs. Meded Rijksuniv Gent Fak Landbouwkd Toegep Biol Wet 66:359–363
7. Kang J, Harders J, Riesner D, Henco K (1994) TGGE in quantitative PCR of DNA and RNA. Methods Mol Biol 31:229–235
8. Myers RM, Fischer SG, Lerman LS, Maniatis T (1985) Nearly all single base substitutions in DNA fragments joined to a GC-clamp can be detected by denaturing gradient gel electrophoresis. Nucleic Acids Res 13:3131–3145
9. Danko P, Kozak A, Podhradsky D, Viglasky V (2005) Analysis of DNA intercalating drugs by TGGE. J Biochem Biophys Methods 65:89–95
10. Henco K, Harders J, Wiese U, Riesner D (1994) Temperature gradient gel electrophoresis (TGGE) for the detection of polymorphic DNA and RNA. Methods Mol Biol 31:211–228
11. Sorlie T, Johnsen H, Vu P, Lind GE, Lothe R, Borresen-Dale AL (2005) Mutation screening of the TP53 gene by temporal temperature gradient gel electrophoresis. Methods Mol Biol 291:207–216
12. Viglasky V (2013) Polyacrylamide temperature gradient gel electrophoresis. Methods Mol Biol 1054:159–171
13. Binzel DW, Khisamutdinov EF, Guo PX (2014) Entropy-driven one-step formation of Phi29 pRNA 3WJ from three RNA fragments. Biochemistry 53:2221–2231
14. Khisamutdinov EF, Jasinski DL, Guo P (2014) RNA as a boiling-resistant anionic polymer material to build robust structures with defined shape and stoichiometry. ACS Nano 8:4771–4781
15. Khisamutdinov EF, Li H, Jasinski DL, Chen J, Fu J, Guo P (2014) Enhancing immunomodulation on innate immunity by shape transition among RNA triangle, square and pentagon nanovehicles. Nucleic Acids Res 42:9996–10004
16. Severcan I, Geary C, Verzemnieks E, Chworos A, Jaeger L (2009) Square-shaped RNA particles from different RNA folds. Nano Lett 9:1270–1277
17. Grabow WW, Zakrevsky P, Afonin KA, Chworos A, Shapiro BA, Jaeger L (2011) Self-assembling RNA nanorings based on RNAI/II inverse kissing complexes. Nano Lett 11:878–887
18. Afonin KA, Bindewald E, Yaghoubian AJ, Voss N, Jacovetty E, Shapiro BA, Jaeger L (2010) In vitro assembly of cubic RNA-based scaffolds designed in silico. Nat Nanotechnol 5:676–682
19. Severcan I, Geary C, Chworos A, Voss N, Jacovetty E, Jaeger L (2010) A polyhedron made of tRNAs. Nat Chem 2:772–779
20. Binzel DW, Khisamutdinov EF, Guo PX (2014) Addition to entropy-driven one-step formation of Phi29 pRNA 3WJ from three RNA Fragments. Biochemistry 53:3709

Chapter 9

Design and Crystallography of Self-Assembling RNA Nanostructures

Mark A. Boerneke and Thomas Hermann

Abstract

Biological RNA architectures are composed of autonomously folding modules which can be tailored as building blocks for the construction of RNA nanostructures. Designed base pair interactions allow complex nano-objects to self-assemble from simple RNA motifs. X-ray crystallography plays an important role in both the design and analysis of such RNA nanostructures. Here, we describe methods for the design and X-ray crystallographic structure analysis of an RNA square and two different triangles, which self-assemble from short oligonucleotides and serve as a platform for building functional nano-sized nucleic acid architectures.

Key words IRES, Hepatitis C virus, Nanostructure design, RNA nanotechnology, RNA structure, RNA conformational switch, Self-assembly, Senecavirus A, X-ray crystallography

1 Introduction

Nucleic acid is used as a soft material in nanotechnology to build structures which form by self-assembly through designed base pair interactions between building blocks. The diversity of RNA structure and function, and the ease of chemical or enzymatic nucleic acid synthesis, render RNA an attractive material to construct nano-sized objects. Complex three-dimensional RNA architectures with a wide variety of natural functions have been incorporated into structures of distinct size and shape. Various self-assembling RNA nanostructures have been designed from motifs observed in crystal structures [1–16], but X-ray crystallography has rarely been used to analyze such nanostructures [12, 15]. Ligand-responsive RNA switches discovered in subdomain IIa of the internal ribosome entry site (IRES) in positive-strand RNA viruses regulate viral protein synthesis by adopting distinct and stable bent or elongated conformations [17, 18]. RNA switches from hepatitis C virus (HCV) and Senecavirus A (SVA, synonymous with Seneca Valley Virus [SVV]) in their ligand-free bent conformation have been used

Eckart Bindewald and Bruce A. Shapiro (eds.), *RNA Nanostructures: Methods and Protocols*, Methods in Molecular Biology, vol. 1632, DOI 10.1007/978-1-4939-7138-1_9, © Springer Science+Business Media LLC 2017

as corner building blocks to construct small, circularly closed and completely double stranded RNA nanostructures (Fig. 1) [12, 15].

Here, we describe the design, assembly, and crystallography of an RNA square and two different RNA triangles. The 100 nucleotide RNA square was designed to self-assemble from four symmetrical corner units of two single strands, each of which were derived from the HCV IRES IIa RNA switch motif (Fig. 1a) [12]. RNA triangles of 81 and 138 nucleotides, respectively, were designed to self-assemble from three symmetrical corner units of two single strands each which were derived from the SVA IRES IIa RNA switch motif (Fig. 1b, c) [15].

As the understanding of the roles that RNAs play in biological processes continues to expand, structural studies will reveal new intricate three-dimensional architectures and autonomously folding modules. Nanostructure design and analysis strategies discussed here provide useful tools to incorporate this growing repertoire of RNA building blocks into nanostructures of increasing complexity and function. Detailed characterization of nanostructures validates design principles and allows for further optimization and functionalization. X-ray crystallography provides the most detailed structural information of available techniques to structurally characterize RNA nanostructures. This protocol provides detailed steps to design, characterize, and crystallize RNA nanostructures.

2 Materials

All solutions described below are prepared using ultrapure water and analytical grade reagents. Ultrapure water was prepared by purifying deionized water to attain a conductivity of 18 MΩ cm at 25 °C.

2.1 RNA Preparation

1. RNA reconstitution buffer: 10 mM sodium cacodylate buffer, pH 6.5. Mix 10 μL of 1 M sodium cacodylate buffer, pH 6.5, with 990 μL water in a 1.6 mL microcentrifuge tube.
2. Chemically synthesized single stranded RNA oligonucleotides (ssRNA) for solution and crystallography studies with the following sequences: 5′ CCG GAG GAA CUA CUG 3′ (Square Outer), 5′ CCG GCA GCC U 3′ (Square Inner), 5′ GUC UAC CCA CCU CGC C 3′ (Small Triangle Outer), 5’AGA GGA CGG CG 3′ (Small Triangle Inner), 5′ UGC UAC CCA CCU CGG AUC ACU GCG CG 3′ (Large Triangle Outer), 5′ CAG GAA CCG AGA GGC ACG CG 3′ (Large Triangle Inner) (*see* **Note 1**).
3. 3× RNA self-assembly buffer: 15 mM $MgCl_2$, 10 mM sodium cacodylate buffer, pH 6.5. Mix 15 μL 1 M $MgCl_2$, 10 μL 1 M sodium cacodylate buffer, pH 6.5, and 975 μL H_2O. Store at 4 °C.

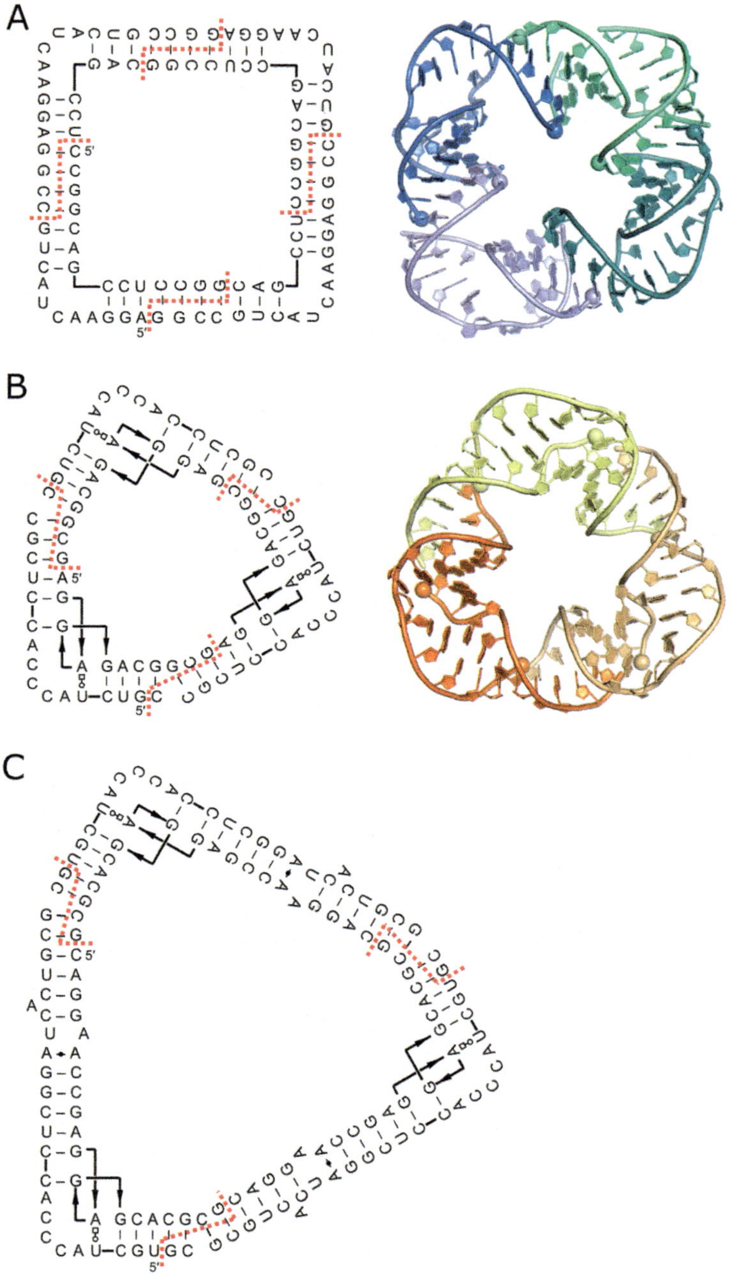

Fig. 1 Self-assembling RNA nanostructures. (**a**) Secondary structure and crystal structure (PDB ID: 3P59) of a 100-nucleotide square containing four identical corner units which were derived from the HCV IRES IIa RNA by adding overhanging complementary single strands for self-assembly by base pairing. (**b**) Secondary structure and crystal structure (PDB ID: 5CNR) of an 81-nucleotide triangle comprising three identical corner units which were derived from the SVA IRES IIa RNA by adding overhanging complementary single strands. (**c**) Secondary structure of a 138 nucleotide triangle also based on the SVA IRES IIa RNA

2.2 Native Polyacrylamide Gel Electrophoresis (PAGE)

1. Native polyacrylamide gel solution: 13% polyacrylamide, 2× MOPS buffer, pH 7.0 (40 mM 3-morpholinopropane-1-sulfonic acid, 10 mM sodium acetate), and 2.5 mM $MgCl_2$ (*see* **Note 2**). Mix 4.875 mL 40% acrylamide–bis-acrylamide (29:1), 3.0 mL 10× MOPS buffer, pH 7.0, 37.5 μL 1 M $MgCl_2$, and 7.088 mL H_2O. Store at 4 °C.
2. Ammonium persulfate: 10% solution in water (*see* **Note 3**). Store at 4 °C.
3. N,N,N,N′-tetramethylethylenediamine (TEMED): Store at 4 °C.
4. 2× gel loading buffer: 50% glycerol, 5 mM $MgCl_2$. Mix 3.571 mL 70% glycerol, 25 μL 1 M $MgCl_2$, and 1.404 mL H_2O. Store at 4 °C.
5. Gel running buffer: 1× MOPS buffer, 7.0, 2.5 mM $MgCl_2$. Mix 45 mL 10× MOPS buffer, pH 7.0, 1.125 mL 1 M $MgCl_2$, and 403.875 mL H_2O. Store at 4 °C.
6. Low range single-stranded RNA ladder: Store at −20 °C.
7. Ethidium bromide gel staining solution: 0.5 μg/mL ethidium bromide solution. Mix 10 μL of 10 mg/mL ethidium bromide solution with 200 mL H_2O, being careful to avoid contact with skin.

2.3 Crystallography

1. Square crystallizing solution: 150 mM $MgCl_2$, 50 mM KCl, 8–10 mM $[Co(NH_3)_6]Cl_3$, 50 mM sodium cacodylate, pH 6.5, and 15–20% PEG400.
2. Triangle crystallizing solution: 55 mM NaCl, 12 mM spermine, 40 mM Tris hydrochloride buffer, pH 7.5, and 20% methyl-2,4-pentanediol (MPD).
3. Pregreased 24-well plates for crystallization.
4. 22 mm diameter siliconized glass slides.
5. Cryo-loops for crystal mounting.
6. X-ray data processing program (HKL2000 [19] or equivalent).
7. Software suites for automated macromolecular structure determination by X-ray crystallography (for example, PHENIX [20] and CCP4 [21],or equivalent).
8. Macromolecular visualization software (Coot [22] and PYMOL [23] or equivalent).

3 Methods

3.1 Design of RNA Square

1. Download the file 2NOK.pdb (HCV IIa RNA switch structure) from the RSCB Protein Data Bank (http://www.rcsb.org).

2. View RNA structure in molecular graphics software (such as PyMOL).
3. Manually model four individual identical RNA corners in a square orientation
4. Overlap the upper helix of one corner with the lower helix of the neighboring corner.
5. Adjust the number of overlaying helix base pairs so that each RNA corner is in the same plane facing the same direction.
6. Repeat for additional two corners.
7. Trim down the length of each corner so that the ends touch. Treat these touching ends as continuous strands.
8. Count the ten helix base pairs in each side connecting neighboring RNA corners (*see* **Note 4**).
9. In the center of each side place one nick in both the outer and inner strands four nucleotides apart, creating four nucleotide overhanging single strands (Fig. 1). These sticky ends allow for the assembly of four corners by complementary base pairing.
10. Change the sequence of overhanging nucleotides to be identical for each side (5′-CCGG) confirming no self-complementarity issues (*see* **Note 5**).
11. Identify the sequence of the outer strand: 5′ CCG GAG GAA CUA CUG 3′.
12. Identify the sequence of the inner strand: 5′ CCG GCA GCC U 3′.

3.2 Design of RNA Triangles

1. For the small triangle, download 4P97.pdb file (short SVA IIa RNA switch structure) from the RSCB Protein Data Bank. For the large triangle, download 4PHY.pdb file (long SVA IIa RNA switch structure).
2. View RNA structure in molecular graphics software.
3. Reconstruct neighboring asymmetric units (*symexp* command in PyMOL).
4. Identify pseudocontinuously closed triangles formed in crystal lattice packing using single overhanging nucleotides for inter-corner base pairing.
5. Adjust nicks in these triangles to be four nucleotides apart, creating four nucleotide overhanging single strands for complementary base pairing during self-assembly (*see* **Note 6**).
6. Change the sequence of overhanging nucleotides to be identical for each side. Small triangle: 3′-CCGC, 3′-GCGG; large triangle: 3′-GCGC, confirming no self-complementarity issues (*see* **Note 5**).

7. Identify the sequence of the outer strand: 5′ GUC UAC CCA CCU CGC C 3′ (Small Triangle Outer), 5′ UGC UAC CCA CCU CGG AUC ACU GCG CG 3′ (Large Triangle Outer).
8. Identify the sequence of the inner strand: 5′AGA GGA CGG CG 3′ (Small Triangle Inner), 5′ CAG GAA CCG AGA GGC ACG CG 3′ (Large Triangle Inner).

3.3 Preparation of RNA Stocks

1. Prepare 0.4–0.5 mM RNA stock solutions by dissolving lyophilized ssRNA oligonucleotides in reconstitution buffer for 10 min at room temperature (*see* **Note 7**).
2. Vortex for several seconds.
3. Centrifuge for 10 s at 3300 × *g* to allow contents to settle at the bottom of the tube.
4. Confirm the concentration of each ssRNA oligonucleotide by UV spectrometry, using the extinction coefficients for each corresponding oligonucleotide (*see* **Note 8**).
5. Aliquot each ssRNA stock into separate tubes with appropriate volumes for specific solution or crystallography studies to avoid repeated freeze–thaw cycles.

3.4 Preparation of Self-Assembling RNA Nanostructures for Solution Studies

1. Mix 1 μL of both 0.4 mM ssRNA Inner and Outer stock solutions (total 2 μL) with 1 μL 3× RNA self-assembly buffer [final dsRNA concentration of 0.133 mM (*see* **Note 9**); final Mg^{2+} concentration of 5 mM]. Other Mg^{2+} and salt concentrations may be tested for their influence on structure assembly and stability (*see* **Note 10**). Lower RNA concentrations may also be used and still visualized by native PAGE (*see* **Note 11**).
2. Anneal and assemble the RNA triangles and square by heating dsRNA solutions to 65 °C for 5 min followed by snap cooling on ice and incubation at 4 °C for 5 min (*see* **Note 12**).

3.5 Native PAGE Analysis

1. Prepare gel by mixing 15 mL native polyacrylamide gel solution, 100 μL 10% ammonium persulfate, and 10 μL TEMED in 25 mL plastic beaker.
2. Cast gel in 10 × 10 cm glass plates with 1.0 mm spacers. Insert 12 well comb. Allow gel to polymerize for 30 min.
3. Remove gel comb and wash wells with ultrapure water by using a squeeze bottle or syringe.
4. Assemble gel in electrophoresis chamber in a 4 °C cold room, or constantly run 4 °C water through a compatible electrophoresis chamber (*see* **Note 13**).
5. Fill the upper and lower buffer reservoirs with 4 °C gel running buffer.
6. Thoroughly rinse wells with gel running buffer from a syringe.

7. At 4 °C, mix 3 μL assembled RNA nanostructure solution with 3 μL 2× gel loading buffer.
8. Load 5 μL assembled RNA nanostructure–loading buffer mixture on the gel.
9. Load 5 μL ssRNA ladder on the gel.
10. Run gel at 4 °C for 1.5 h at ~220 V, but not exceeding 22 mA.
11. Remove gel from electrophoresis chamber and submerge in gel staining solution for 20 min (*see* **Note 14**).
12. Rinse gel thoroughly with ultrapure water three times.
13. Visualize under UV light.
14. Analyze results by visual inspection to confirm formation of species consistent with the size of designed structures (Fig. 2) (*see* **Notes 14** and **15**).

3.6 Preparation of Self-Assembling RNA Nanostructures for Crystallography Studies

1. For the small triangle, mix stoichiometric amounts of ssRNA Inner and Outer stock solutions to obtain 0.2 mM dsRNA, diluted with reconstitution buffer (*see* **Note 16**). For the square, mix stoichiometric amounts of ssRNA Inner and Outer stock solutions to obtain 0.2 mM dsRNA in 5 mM $MgCl_2$, diluted with reconstitution buffer (*see* **Note 17**).
2. Anneal and assemble the RNA triangle and square by heating dsRNA solutions to 65 °C for 5 min, followed by slow cooling to room temperature on benchtop over 45 min (*see* **Note 18**).

3.7 Nanostructure Crystallization

1. Add 700 μL crystallization solution to a well in a 24-well plate (*see* **Notes 19** and **20**).
2. Ensure a siliconized glass slide is clean and dust free by blowing clean compressed air over its surface.
3. Add 1 μL of assembled RNA nanostructure solution to the top of this glass slide.
4. Add 1 μL of crystallizing solution to this 1 μL RNA-containing drop.
5. Invert and seal this 2 μL drop on the glass slide over the pregreased, solution-containing well (hanging drop method).
6. Incubate this 24-well plate at 16 °C for the square or at 6 °C for the small triangle.
7. Rapidly but carefully inspect the plates once per week, to prevent plates from warming to room temperature (*see* **Note 21**).

3.8 Crystal Mounting

1. Mount a crystal from the crystallization drop on a cryo-loop and flash-freeze in liquid nitrogen (*see* **Note 22**).

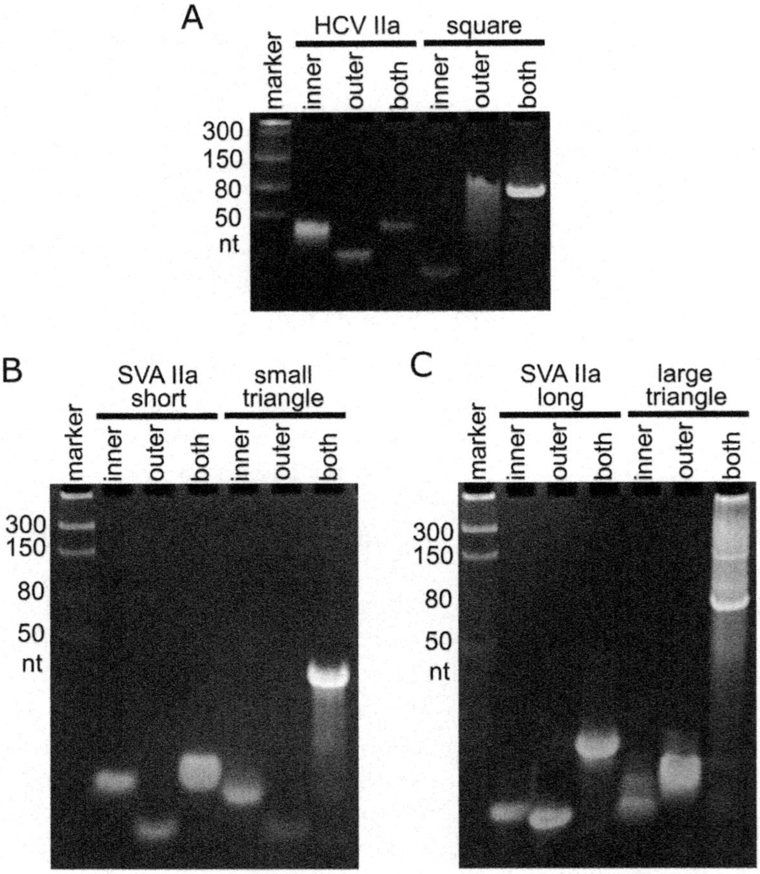

Fig. 2 Assembly of RNA nanostructures by native PAGE analysis. All gels were run in the presence of 2.5 mM $MgCl_2$ at 4 °C, stained with ethidium bromide, and visualized under UV light. (**a**) Assembly and PAGE migration of the RNA square (Square—Both) compared to its individual single-stranded components (Square—Inner, Outer), a non-self-assembling single corner unit (HCV IIa—Both, and its single-stranded components [HCV IIa—Inner, Outer]), and an ssRNA size marker. (**b**) Assembly and PAGE migration of the small RNA triangle (Small Triangle—Both) compared to its individual single-stranded components (Small Triangle—Inner, Outer), a non-self-assembling single corner unit (SVA IIa Short—Both, and its single-stranded components [SVA IIa Short—Inner, Outer]), and an ssRNA size marker. (**c**) Assembly and PAGE migration of the large RNA triangle (Large Triangle—Both) compared to its individual single-stranded components (Large Triangle—Inner, Outer), a non-self-assembling single corner unit (SVA IIa Long—Both, and its single-stranded components [SVA IIa Long—Inner, Outer]), and an ssRNA size marker

2. Transfer the mounted crystal to the goniometer head of the X-ray diffractometer under a steady stream of ultracold nitrogen gas.
3. Collect diffraction data. Collect greater than 180 images with 1° oscillations between images. Choose exposure time to

obtain strong diffraction pattern but avoid overexposure (typically 5–30 min exposure, depending on crystal size and quality, on a rotating anode X-ray generator).

3.9 Square Data Collection and Processing

1. Square RNA X-ray diffraction frames were collected on beamline 17-ID-B at the Advanced Photon Source, Argonne National Laboratory.
2. Identify crystal space group as $P3_121$ in HKL2000 software.
3. Integrate dataset in HKL2000.
4. Scale dataset in HKL2000.

3.10 Square Structure Solution

1. Calculate initial phases in PHENIX by the single wavelength anomalous diffraction method using the anomalous scattering from a Br atom incorporated at the 5-position of the U59 nucleobase [12].

3.11 Square Structure Refinement

1. Use the automatically built initial model from PHENIX.
2. Refine model using the program Refmac [24] within the CCP4 package following software manual's recommended refinement strategy [24].
3. Manually rebuild model in Coot by adjusting bond torsion angles and conformations based on $2F_o–F_c$ and $F_o–F_c$ density maps.
4. Iteratively refine in Refmac and rebuild structure in Coot until a satisfactory model is obtained with no violations of bond angles, torsions, or clashes and no significant changes in R_{work}/R_{free} are observed.
5. Place metal ions (such as magnesium cations) at appropriate binding sites around the RNA based on electron density, B-factor refinement, and distance and geometry of coordinating ligands, and taking into account established parameters of metal coordination environment [25].
6. Run final refinement steps in PHENIX with combined translation–libration–screw motion, individual isotropic atomic displacement parameters, and water picking.

3.12 Small Triangle Data Collection and Processing

1. Small triangle RNA X-ray diffraction frames were collected on a RIGAKU rotating anode X-ray generator (λ = 1.54 Å) equipped with a MAR345 imaging plate detector.
2. Identify crystal space group as R32 in HKL2000.
3. Integrate dataset in HKL2000.
4. Scale dataset in HKL2000.

3.13 Small Triangle Structure Solution

1. Build an initial search model from the short SVV IIa RNA crystal structure (RCSB Protein Data Bank accession codes 4P97) and the designed small triangle (*see* Subheading 3.2).
2. Use the program Phaser [26] within the CCP4 package to solve the RNA nanotriangle three-dimensional structure by molecular replacement using this initial search model.

3.14 Small Triangle Structure Refinement

1. Refine model using the program Refmac [24] within the CCP4 package following software manual's recommended refinement strategy [24].
2. Manually rebuild model in Coot by adjusting bond torsion angles and conformations based on $2F_o–F_c$ and $F_o–F_c$ density maps.
3. Iteratively refine in Refmac and rebuild structure in Coot until a satisfactory model is obtained with no violations of bond angles, torsions, or clashes and no significant changes in R_{work}/R_{free} are observed.
4. Place metal ions (such as magnesium cations) at appropriate binding sites around the RNA based on electron density, B-factor refinement, and distance and geometry of coordinating ligands taking into account established parameters of coordination environment [25] (*see* **Note 23**).
5. Run final refinement steps in PHENIX with individual isotropic atomic displacement parameters and water picking.

4 Notes

1. Many custom RNA oligonucleotide manufacturers offer various purification methods for chemically synthesized oligonucleotides. In these studies, oligonucleotide components for the square were obtained as HPLC-purified RNA, while small and large triangle components were used after standard desalting and without further purification. For RNA nanostructures designed from small self-assembling oligonucleotides, it is more cost-effective to screen constructs using oligonucleotides purified by desalting only. Oligonucleotides may also be further purified by denaturing PAGE. While HPLC-purified oligonucleotides may be generally beneficial for crystallization trials, our laboratory has also had success in crystallizing small RNA constructs such as the small triangle composed of oligonucleotides purified by standard desalting only [15, 18].
2. In addition to including Mg^{2+} in RNA self-assembly buffer, PAGE gels must also contain Mg^{2+} to maintain assembled RNA square and triangle structure.

3. Prepare 10% ammonium persulfate solution and separate into 1 mL aliquots in 1.6 mL microcentrifuge tubes. To prevent thermal decomposition, keep extra aliquots stored at −20 °C, and the working aliquot stored at 4 °C.
4. Manual modeling and design of RNA nano-objects from crystal structures provides a useful starting point in nano-object construction. The number of base pairs per nano-object side often must be adjusted (increased or decreased) to achieve an experimentally observed optimal stable size.
5. The sequence of theses overhangs may be varied to produce other self-assembling (same overhang sequence on each side) or programmable (different overhang sequences on each side) nano-objects [12, 15]. Sequences rich in G and C residues provide the best structure stability when utilizing four base pair overhangs.
6. Overhang length may need to be optimized for each structure depending on its size and stability [15, 27]. Overhang length may affect all or nothing assembly by changing the stability of an individual corner outside the context of a fully assembled structure.
7. A minimum ssRNA stock concentration of 0.4 mM is desirable as 0.2 mM dsRNA is a useful starting point to obtain crystals for RNA nanostructures. 0.1–0.3 mM dsRNA concentrations have also been successful in obtaining crystals, but concentrations above 0.2 mM typically yield a greater number of crystals per drop, but not larger crystals. Concentrations below 0.2 mM typically yield fewer crystals per drop and occasionally smaller or no crystals.
8. Data sheets including extinction coefficients for chemically synthesized oligonucleotides are commonly provided by manufacturers (such as IDT). Online programs such as OligoAnalyzer 3.1 (https://www.idtdna.com/calc/analyzer) also provide oligonucleotides properties such as extinction coefficients.
9. The prepared RNA nanostructures described here (small and large triangles and square) are self-assembled from two short oligonucleotides which form symmetrical corner units in a triangle or square arrangement (square: four identical inner strands, four identical outer strands; small and large triangles: three identical inner strands, three identical outer strands). The described dsRNA concentration of 0.133 mM corresponds to a single double stranded corner unit. When successfully assembled, this corresponds to 0.033 mM square or 0.044 mM triangle. In a 3 μL volume, this corresponds to 0.099 nMol square and 0.132 nMol triangle to be loaded on the gel.

10. The square and small and large triangles all require a minimum of ~1 mM Mg^{2+} to efficiently self-assemble. They also assemble equally efficient in the presence of greater Mg^{2+} concentrations (tested up to 10 mM Mg^{2+}). Monovalent cation concentration has little influence on the assembly of these three structures in solution but may affect crystallization.
11. Lower nanostructure concentrations may be analyzed by native PAGE to save materials, depending on the nanostructure size, stability, and assembly efficiency (formation of single discrete structure, no extensive band smearing, and all single strands are incorporated into the assembled structure). 0.010 mM square or small and large triangle can still be accurately analyzed by native PAGE.
12. Slow cooling annealing protocols also result in efficient assembly of the triangles and square.
13. Native PAGE gels with high Mg^{2+} concentration generate high electrical current and heat. Gels should be run in a cold environment to prevent gel warming and RNA structure denaturing. If no cold room access is available, ice-cooled water may be run through a compatible electrophoresis chamber. A common aquarium pump may be placed in a bucket of ice water to cycle a flow of cold water through the gel chamber and back into the ice water bucket.
14. Ethidium bromide in the gel staining solution more effectively intercalates double-stranded RNA than single-stranded RNA. Hence, double-stranded nanostructure bands are stained with greater intensity than unassembled single-stranded RNA components (Fig. 2). Higher band staining intensities indicate successful nanostructure assembly. Band intensities are usually compared by visual inspection, but may also be compared quantitatively by using image processing software such as ImageJ [28].
15. It is useful to run several controls in wells next to an RNA nanostructure of interest, to compare general size and band visualization intensity. Single-stranded components of nanostructures may be run in control wells and can provide clues about assembly efficiency and optimal stoichiometry (Fig. 2). Non-self-assembling single corner units (no overhanging sticky ends) may be run in control wells and can provide clues about all-or-nothing assembly of nanostructure from single strands, or stepwise assembly of single corners first from single strands before nanostructure assembly from double-stranded corners (Fig. 2). Single- or double-stranded RNA (or DNA) markers are useful in determining general nanostructure mass, though circular double-stranded nanostructures often migrate faster through a gel due to their reduced hydrodynamic radius relative to linear markers of comparable mass.

16. Attempts to crystallize the large triangle were performed in a similar fashion to the small triangle but did not result in well-diffracting crystals.
17. The described dsRNA concentration of 0.2 mM corresponds to a single double stranded corner unit. When successfully assembled, this corresponds to 0.050 mM square or 0.067 mM triangle.
18. Slow or snap cooling of thermally denatured RNA squares or small triangles did not have an effect on their crystallization, as they self-assemble in diverse conditions. Other structures may have more complex annealing protocols.
19. A similar methodology for crystallographic procedures described in Subheadings 3.7–3.10 was previously published by Dibrov and Hermann [29].
20. The well volume may be changed as a crystallization optimization variable. Precipitant and overall salt concentration in the well solution are the most important components of the well solution, as specific metal ions in the well solution do not make contact with the RNA in the drop but rather help dictate the speed of crystal drop desiccation. Here, all well crystallization solutions are identical to drop crystallization solutions to minimize the number of different solutions to prepare (as all components are relatively low-cost reagents or salts). Adjusting well and drop crystallization solutions independently plays a more important role when screening minimally different drop conditions or when more expensive heavy metals are being used in crystallization conditions.
21. Small triangle crystals should appear within 1 month. Square crystals should appear within 1–2 months.
22. No additional cryoprotectant is needed crystallization conditions contain greater that 15% PEG400 or MPD.
23. These parameters provide clear evidence for a symmetrical ion in each of the three triangle corners. No magnesium cations were present in the nanotriangle preparation or crystallization. Chlorine anions were present and their assignment confirmed by electron density, B-factor refinement, and distance as well as geometry of coordinating ligands.

Acknowledgment

The research outlined here was supported in part by the National Institutes of Health, grant R01 AI72012, and the UCSD Academic Senate, grant No. RM069B. M. A. B. was supported by a GAANN fellowship from the US Department of Education. Instrumentation at the UCSD Biomolecule Crystallography Facility was acquired with funding from the National Institutes of Health, grant OD011957. Use of the Advanced Photon Source for X-ray diffraction data collection was supported by the US Department of Energy, Office of Science, Office of Basic Energy Sciences, under Contract No. W-31-109-Eng-38. Use of the IMCA-CAT beamline 17-ID at the Advanced Photon was supported by the companies of the Industrial Macromolecular Crystallography Association through a contract with the Center for Advanced Radiation Sources at the University of Chicago.

References

1. Jaeger L, Leontis N (2000) Tecto-RNA: one-dimensional self-assembly through tertiary interactions. Angew Chem Int Ed Engl 39:2521–2524
2. Shu D, Huang LP, Hoeprich S et al (2003) Construction of phi29 DNA-packaging RNA monomers, dimers, and trimers with variable sizes and shapes as potential parts for nanodevices. J Nanosci Nanotechnol 3:295–302
3. Chworos A, Severcan I, Koyfman AY et al (2004) Building programmable jigsaw puzzles with RNA. Science 306:2068–2072
4. Nasalean L, Baudrey S, Leontis NB et al (2006) Controlling RNA self-assembly to form filaments. Nucleic Acids Res 34:1381–1392
5. Bindewald E, Grunewald C, Boyle B et al (2008) Computational strategies for the automated design of RNA nanoscale structures from building blocks using NanoTiler. J Mol Graph Model 27:299–308
6. Severcan I, Geary C, Verzemnieks E et al (2009) Square-shaped RNA particles from different RNA folds. Nano Lett 9:1270–1277
7. Severcan I, Geary C, Chworos A et al (2010) A polyhedron made of tRNAs. Nat Chem 2 (9):772
8. Afonin KA, Bindewald E, Yaghoubian AJ et al (2010) In vitro assembly of cubic RNA-based scaffolds designed in silico. Nat Nanotechnol 5:676–682
9. Geary C, Chworos A, Jaeger L (2011) Promoting RNA helical stacking via A-minor junctions. Nucleic Acids Res 39:1066–1080
10. Grabow WW, Zakrevsky P, Afonin KA et al (2011) Self-assembling RNA nanorings based on RNAI/II inverse kissing complexes. Nano Lett 11:878–887
11. Ohno H, Kobayashi T, Kabata R et al (2011) Synthetic RNA–protein complex shaped like an equilateral triangle. Nat Nanotechnol 6:116–120
12. Dibrov SM, McLean J, Parsons J et al (2011) Self-assembling RNA square. Proc Natl Acad Sci U S A 108:6405–6408
13. Bindewald E, Afonin K, Jaeger L et al (2011) Multistrand RNA secondary structure prediction and nanostructure design including pseudoknots. ACS Nano 5:9542–9551
14. Geary C, Rothemund PWK, Andersen ES (2014) A single-stranded architecture for cotranscriptional folding of RNA nanostructures. Science 345:799–804
15. Boerneke MA, Dibrov SM, Hermann T (2016) Crystal-structure-guided design of self-assembling RNA nanotriangles. Angew Chem Int Ed 55:4097–4100
16. Parlea L, Bindewald E, Sharan R et al (2016) Ring catalog: a resource for designing self-assembling RNA nanostructures. Methods 103:128–137
17. Boerneke MA, Hermann T (2015) Ligand-responsive RNA mechanical switches. RNA Biol 12:780–786
18. Boerneke MA, Dibrov SM, Gu J et al (2014) Functional conservation despite structural divergence in ligand-responsive RNA switches. Proc Natl Acad Sci U S A 111:15952–15957
19. Otwinowski Z, Minor W (1997) Processing of X-ray diffraction data collected in oscillation mode. Methods Enzymol 276:307–326
20. Adams PD, Grosse-Kunstleve RW, Hung LW et al (2002) PHENIX: building new software for automated crystallographic structure determination. Acta Crystallogr D Biol Crystallogr 58:1948–1954
21. Collaborative Computational Project, Number 4 (1994) The CCP4 suite: programs for protein crystallography. Acta Crystallogr D Biol Crystallogr 50:760–763
22. Emsley P, Cowtan K (2004) Coot: model-building tools for molecular graphics. Acta Crystallogr D Biol Crystallogr 60:2126–2132

23. WL DeLano (2002) The PyMOL molecular graphics system, version 11, Schrödinger LLC http://wwwpymolorg
24. Murshudov GN, Vagin AA, Dodson EJ (1997) Refinement of macromolecular structures by the maximum-likelihood method. Acta Crystallogr D Biol Crystallogr 53:240–255
25. Auffinger P, Grover N, Westhof E (2011) Metal ion binding to RNA. Met Ions Life Sci 9:1–35
26. McCoy AJ, Grosse-Kunstleve RW, Adams PD et al (2007) Phaser crystallographic software. J Appl Cryst 40:658–674
27. MA Boerneke (2016) Ligand-responsive RNA switches: viral translation regulators, therapeutic targets and tunable building blocks for nanotechnology. Dissertation, University of California, San Diego, 2016
28. Abràmoff MD, Magalhães PJ, Ram SJ (2004) Image processing with imageJ. Biophotonics International 11:36–42
29. Dibrov SM, Hermann T (2015) Structure of the HCV internal ribosome entry site subdomain IIa RNA in complex with a viral translation inhibitor. In: Nucleic acid crystallography: methods and protocols, pp 1–356

Chapter 10

X-Aptamer Selection and Validation

Ganesh L. Lokesh, Hongyu Wang, Curtis H. Lam, Varatharasa Thiviyanathan, Nancy Ward, David G. Gorenstein, and David E. Volk

Abstract

Aptamers and second generation analogs, such as X-Aptamers (XAs), SOMAmers, locked nucleic acids (LNAs), and others are increasingly being used for molecular pathway targeting, biomarker discovery, or disease diagnosis by interacting with protein targets on the surface of cells or in solution. Such targeting is being used for imaging, diagnostic evaluation, interference of protein function, or delivery of therapeutic agents. Selection of aptamers using the original SELEX method is cumbersome and time-consuming, often requiring 10–15 rounds of selection, and provides aptamers with a limited number of functional groups, namely four bases of DNA or RNA, although newer SELEX methods have increased this diversity. In contrast, X-Aptamers provide an unlimited number of functional groups and thus are superior targeting agents. Here, we discuss the X-Aptamer selection process.

Key words Aptamer, X-Aptamer, Molecular targeting, Biotin labeling, Dye labeling, Bead-based selection, Split-pool synthesis, Drug aptamer conjugation, Aptamer nanoparticle conjugation, Aptamer siRNA chimeras, Chemical cross-linking, Proteomics

1 Introduction

Oligonucleotide aptamers were first derived in the labs of Gold [1] and Szostak [2] nearly simultaneously using extremely diverse oligonucleotide libraries with a repeated reduction of the oligonucleotide sequence space through iterative cycles of target binding, isolation, and PCR amplification steps. Aptamers have been well reviewed [3–9], and variations such as locked nucleic acids [10, 11], and aptamers containing substitutions, including 2′-fluoro [12], 2′-OMe [13], monothiophosphates and dithiophosphates [14, 15], and a variety of C5 or C7 substitutions [9] have been used. The biophysical basis of enhanced thioated aptamer binding has been studied [16, 17]. In our past work, we have developed aptamers targeting proteins from the HIV [18], VEE [19], and Dengue 4 [20] viruses, and cancer-associated proteins such as E-selectin [21],

Eckart Bindewald and Bruce A. Shapiro (eds.), *RNA Nanostructures: Methods and Protocols*, Methods in Molecular Biology, vol. 1632, DOI 10.1007/978-1-4939-7138-1_10, © Springer Science+Business Media LLC 2017

CD44 [22, 23], and TGF-β1 [24], and have found that aptamer conjugated liposomes [25–27] and multistage vehicles [28] are efficacious therapeutic agents. Here, we describe the selection and validation process for X-Aptamers [23] containing protein-like or drug-like side chains, which are built upon our earlier one-bead, one-sequence aptamer technology [29]. This allows the creation of affinity reagents that simultaneously contain the best features of antibodies, small molecules and aptamers. The X-Aptamer Selection Kit has been used successfully for the creation of electronic X-Aptamer-based biosensors [30], nano-surface plasmon resonance imaging sensors [31], and by industrial clients (confidential communications).

2 Materials

Prepare all solutions in either RNAase-free water (when indicated) or ultrapure deionized water (18 MΩ cm at 25 °C), and use analytical grade reagents. Conduct all experiments at room temperature unless otherwise noted, and dispose of wastes according to your local regulations. It is assumed that the reader will purchase the X-Aptamer selection kit from AM Biotechnologies and target proteins and antibodies from commercial vendors, and that the reader is familiar with basic biochemical techniques, including PCR methods.

2.1 Preparing the Protein for Coupling with Biotin

1. 0.5 mL 3000 MWCO centrifugal filter.
2. 50 mM sodium borate buffer pH 8.5: 3.84 mg $Na_2B_4O_7 \cdot 10H_2O$, 124 mg H_3BO_3, 100 mL H_2O.

2.2 Preparing Biotinylated Protein

1. Target protein prepared in Subheading 3.1.
2. Biotinylation reagent: Preweighed sulfosuccinimidyl-6-[biotin-amido]hexanoate (EZ-Link Sulfo-NHS-LC-Biotin No Weigh format, 1 mg microtubes).
3. Preparation of biotinylation reagent: Add 90 μL of water to the microtube containing 1 mg of Sulfo-NHS-LC-Biotin to make a 20 mM solution. This reagent needs to be prepared fresh immediately before use.
4. 2 M (20×) sodium borate buffer, pH 8.5: 3.84 g $Na_2B_4O_7.10H_2O$, 1.24 g H_3BO_3, 50 mL H_2O.
5. Microcentrifuge.
6. Spin desalting column—0.5 mL 7K molecular weight cutoff (MWCO), 25 columns (Zeba).
7. Nuclease-free water (10 × 50 mL).
8. 0.5 mL 3K centrifugal filters 3000 MWCO (Millipore).
9. Nuclease-free phosphate buffered saline (10× PBS, pH 7.4).

10. Coupling buffer: 50 mM borate, pH 8.5. Combine 50 μL of 20× Borate buffer and 950 μL nuclease-free water in a 1.6 mL vial.
11. Quench buffer: 1 M Tris pH 8.0.
12. Phosphate buffered saline (1× PBS): To 5 mL of 10× PBS, pH 7.4, add 45 mL of nuclease-free water.
13. Microvolume UV spectrometer (such as a NanoDrop instrument).

2.3 Preparation of Prewashed Magnetic Beads

1. Streptavidin-functionalized magnet beads (DynaBeads M-280 Streptavidin).
2. Magnet or magnetic separator. The magnetic separator used in our work was purchased from Pure Biotech.
3. Buffer PBSTM (1× PBS pH 7.4 with 1 mM $MgCl_2$ and 0.05% Tween 20): Mix 5 mL of 10× PBS, pH 7.4 with 50 μL of 1 M $MgCl_2$ and 25 μL Tween 20 in a 50 mL conical tube and dilute to 50 mL with nuclease-free water.

2.4 Preparing a Protein–Magnetic Bead Complex

1. Prewashed streptavidin-functionalized magnet beads (from Subheading 3.3).
2. (Method 1 only) biotinylated target protein from Subheading 3.2.
3. (Method 2 only) biotinylated antibody specific for the target protein(s).
4. Nuclease-free phosphate buffered saline (10× PBS, pH 7.4).
5. Magnetic separator. The magnetic separator used in our work was purchased from Pure Biotech.
6. Rotisserie shaker.
7. Buffer PBSTM (1× PBS pH 7.4 with 1 mM $MgCl_2$ and 0.05% Tween 20): Mix 5 mL of 10× PBS, pH 7.4 with 50 μL of 1 M $MgCl_2$ and 25 μL Tween 20 in a 50 mL conical tube and dilute to 50 mL with nuclease-free water.
8. Buffer PBSTMB (1× PBS pH 7.4 with 1 mM $MgCl_2$ and 0.05% Tween 20): Mix 5 mL of 10× PBS, pH 7.4 with 50 μL of 1 M $MgCl_2$, 100 mg BSA and 25 μL Tween 20 in a 50 mL conical tube and dilute to 50 mL with nuclease-free water.

2.5 X-Aptamer Library Preparation

1. X-Aptamer Selection Kit (AM Biotechnologies, LLC, Houston, TX, USA). The kit contains the X-Aptamer bead library, a set of eight primers that include a forward primer and seven barcoded reverse primers for PCR amplification of DNA from selected beads.
2. Buffer PBSTM (1× PBS pH 7.4 with 1 mM $MgCl_2$ and 0.05% Tween 20): Mix 5 mL of 10× PBS, pH 7.4 with 50 μL of 1 M

$MgCl_2$ and 25 μL Tween 20 in a 50 mL conical tube and dilute to 50 mL with nuclease-free water.

3. Vortex mixer (optional).
4. Centrifuge with swinging buckets.
5. Bovine serum albumin (BSA).
6. Tween 20 (synonyms: polysorbate 20, Aktest TW 20, Scattics, among others).
7. 1 M $MgCl_2$.
8. Buffer PBSTMB (1× PBS pH 7.4 with 1 mM $MgCl_2$ and 0.05% Tween 20): Mix 5 mL of 10× PBS, pH 7.4 with 50 μL of 1 M $MgCl_2$, 100 mg BSA and 25 μL Tween 20 in a 50 mL conical tube and dilute to 50 mL with nuclease-free water.

2.6 Negative Selection Against Naked Magnetic Beads or Antibody-Bound Magnetic Beads

1. (Method 1 only) prewashed (naked) magnetic beads prepared in Subheading 3.3.
2. (Method 2 only) magnetic beads conjugated to protein target antibody.
3. Prepared XA-library from Subheading 3.5.
4. Rottiserie shaker (Lab Quake).
5. Nuclease-free water.
6. PBSTM buffer (1× PBS pH 7.4 with 1 mM $MgCl_2$ and 0.05% Tween 20): Mix 5 mL of 10× PBS, pH 7.4 with 50 μL of 1 M $MgCl_2$ and 25 μL Tween 20 in a 50 mL conical tube and dilute to 50 mL with nuclease-free water.

2.7 Bead-Based Positive Selection

1. Protein–magnetic bead complex created in Subheading 3.4.
2. Depleted XA bead library created in Subheading 3.6.
3. Microcentrifuge.
4. PBSTMB buffer (1× PBS pH 7.4 with 1 mM $MgCl_2$ and 0.05% Tween 20): Mix 5 mL of 10× PBS, pH 7.4 with 50 μL of 1 M $MgCl_2$, 100 mg BSA and 25 μL Tween 20 in a 50 mL conical tube and dilute to 50 mL with nuclease-free water.
5. Rotisserie shaker.
6. Magnet or magnetic separator.
7. PBSTM buffer (1× PBS pH 7.4 with 1 mM $MgCl_2$ and 0.05% Tween 20): Mix 5 mL of 10× PBS, pH 7.4 with 50 μL of 1 M $MgCl_2$ and 25 μL Tween 20 in a 50 mL conical tube and dilute to 50 mL with nuclease-free water.

2.8 Preparing a Solution Phase Library from the Positively Selected XA Beads

1. Positive-selection beads from Subheading 3.7.
2. Sodium hydroxide solution: 0.2 M NaOH. 200 mg NaOH in 25 mL nuclease-free water.
3. Water bath set to 65 °C.
4. Water bath set to 95 °C.
5. Microcentrifuge.
6. PBSTM buffer (1× PBS pH 7.4 with 1 mM $MgCl_2$ and 0.05% Tween 20): Mix 5 mL of 10× PBS, pH 7.4 with 50 μL of 1 M $MgCl_2$ and 25 μL Tween 20 in a 50 mL conical tube and dilute to 50 mL with nuclease-free water.
7. Centrifugal concentrator, 0.5 mL capacity 3 kD MWCO.

2.9 Solution-Based Positive Selection

1. Protein–magnetic bead complex created in Subheading 3.4.
2. Solution-phase XA library created in Subheading 3.8.
3. Magnet or magnetic separator.
4. PBSTM buffer described in Subheading 2.3.
5. Nuclease-free water.

2.10 PCR Amplification of Selected Sequences

1. The selected DNA from each round of selection, including any negative selection steps.
2. PCR tubes.
3. PCR primers supplied as part of the X-Aptamer Selection Kit (AM Biotechnologies).
4. 10× Stock primer stock solution (100 μM) for each primer, including bar coded primers, provided in the X-Aptamer Selection Kit.
5. 1× Working Primer solutions (10 μM) for each primer, including bar coded primers, provided in the X-Aptamer Selection Kit.
6. Choice Taq 10× PCR buffer: 100 mM Tris–HCl, pH 9.0, 100 mM KCl, 80 mM $(NH_4)_2SO_4$, 15 mM $MgCl_2$ and 0.5% NP40.
7. Choice Taq: Taq DNA polymerase (5 U/μL).
8. Nuclease-free water.
9. Thermal cycler for PCR.
10. 10 mM dNTP solution.

2.11 Electrophoresis for Validation of PCR Products

1. PCR products from Subheading 3.10.
2. 6× gel loading buffer (50% glycerol, 0.05% w/v bromophenol blue).
3. DNA ladder solution (with bottom band 50 bp or smaller).

4. A minigel electrophoresis apparatus (such as XCELL SureLock) with power supply.
5. Precast minigel (10% polyacrylamide) sized for the electrophoresis unit.
6. Staining solution (2.5 ng/mL ethidium bromide).
7. Destaining solution—Deionized water.
8. Gel imaging apparatus capable of detecting ethidium bromide-stained nucleic acids.

2.12 Preparations for Acquisition of Next-Generation Sequencing (NGS)

1. PCR reaction products from Subheading 3.10.
2. Centrifugal concentrator, 0.5 mL capacity, 10 kD MWCO.
3. Microcentrifuge.
4. Microspectrometer capable of determining concentration from 1 μL of sample.
5. Access to a next-generation sequencing service.

2.13 Analysis of NGS Sequencing Data

1. NGS analysis software (*see* **Note 1**).
2. X-Aptamer library design file.

2.14 Synthesis of X-Aptamers

This step requires an oligosynthesizer, specialized reagents, and chemistry experience not typically possessed in most labs. Therefore, the X-Aptamer synthesis should be outsourced to a commercial service.

2.15 Validation of X-Aptamers

1. 1 μM solution of biotinylated X-Aptamer to be validated.
2. 10 μM solution of the target protein.
3. Chemiluminescent Nucleic Acid Detection Module (Thermo Scientific).
4. Streptavidin–horseradish peroxidase solution (a proprietary component in Thermo Nucleic acid detection chemiluminescent module).
5. Luminol and enhancer solution (a proprietary component in Thermo nucleic acid detection chemiluminescent module).
6. Blocking buffer (a proprietary component in Thermo nucleic acid detection chemiluminescent module).
7. 4× Wash Buffer (a proprietary component in Thermo nucleic acid detection chemiluminescent module).
8. Substrate equilibration buffer (a proprietary component in Thermo nucleic acid detection chemiluminescent module).
9. Nylon 66 membrane, 0.45 μm thick, 8 × 12 cm.
10. Mini-Gel electrophoresis apparatus (Xcell Sure Lock).

11. Wet trans-blot transfer apparatus (Xcell II blot module, Thermo Scientific).
12. Gel imaging apparatus capable of measuring chemiluminescence.
13. Pipette tip box lid or similarly sized container.
14. Gel shaker.
15. Lint-free wipes, such as Kimwipes.
16. Software for graphing binding data, such as GraphPad Prism or Sigma Plot.

3 Methods

This section assumes that the readers are familiar with typical molecular biology procedures such as PCR and the creation of buffers, and further that the reader will purchase DNA primers, aptamers, X-Aptamers, and their labeled forms as needed, from commercial entities. Two variations of X-Aptamer selection are described: Method 1 requires biotinylation of the target protein, while Method 2 relies on the availability of an antibody specific to the target protein. These procedures can be modified for proteins labeled with other tags, such as his-tag or GST-tag.

3.1 Preparing the Protein for Coupling with Biotin

1. Protein Preparation: Target proteins supplied in Tris/amine containing buffers need to be exchanged to an amine-free buffer, such as a borate or phosphate buffer. We prefer 50 mM borate buffer (1× coupling buffer) for performing labeling reactions.
2. Load 50–100 μg of target protein in 100 μL of liquid into a 0.5 mL 3000 MWCO centrifugal filter.
3. Centrifuge at 14,000 × *g* for 10 min.
4. Remove the flow-through liquid.
5. Add 400 μL of coupling buffer to the residual protein on the top of the concentrator.
6. Centrifuge for 14,000 × *g* for 10 min.
7. Remove the flow-through liquid.
8. Repeat **steps 3–5**.
9. Repeat **steps 3–5**.
10. Repeat **steps 3–5** (*see* **Note 2**).
11. Collect the protein from the top of the centrifugal filter and add coupling buffer sufficient to make the protein concentration 3–5 mg/mL for optimal biotinylation in the next step.

3.2 Preparing Biotinylated Protein (Method 1 Only)

1. To 100 μg of protein in 30 μL of coupling buffer at room temperature, add 4 μL of biotinylation reagent (80 nmol) to yield a 20- to 40-fold molar excess of biotinylation reagent.
2. Shake the reaction for 2 h.
3. Add 10 μL of quench buffer to halt the reaction.
4. Snap the bottom plug of a 0.5 mL spin desalting column.
5. Centrifuge the (protein-free) spin desalting column at 1000 × *g* for 1 min in a collection tube to remove the spin column's storage buffer.
6. Add 100 μL of PBS to the top of the slurry.
7. Centrifuge at 1000 × *g* for 1 min.
8. Discard the flow-through.
9. Add 50 μL of PBS to the top of the slurry.
10. Centrifuge the spin column at 1000 × *g* for 1 min.
11. Discard the flow-through.
12. Add 50 μL of PBS to the top of the slurry.
13. Centrifuge the spin column at 1000 × *g* for 1 min.
14. Discard the flow-through.
15. Place the spin desalting column in a new collection tube.
16. Add up to 50 μL of the quenched biotinylation reaction to the slurry/column.
17. Centrifuge the spin desalting column at 1000 × *g* for 1 min.
18. Collect the flow-through that contains purified biotinylated target protein for downstream use.
19. Place 1 μL of PSB buffer onto a microvolume UV-Vis spectrometer (such as a Thermo Scientific NanoDrop).
20. Blank the spectrometer against the PBS solution.
21. Mix 1 μL of purified protein sample and 4 μL of PBS.
22. Measure the optical density (OD) of this sample at 280 nm.
23. Calculate the protein concentration as [P] = (OD280)(5)(1 mg/mL).

3.3 Preparation of Prewashed Magnetic Beads

1. Aliquot 500 μL (5 mg) of streptavidin functionalized magnetic bead solution (Dynabead M-280) into a microcentrifuge tube.
2. Apply the magnet to settle the beads.
3. Use a fine-tipped pipette to remove the storage solution from the bead pellet.
4. Add 500 μL of PBSTM to the beads.
5. Apply the magnet to settle the beads.

6. Use a fine-tipped pipette to remove the PBSTM wash solution from the bead pellet.
7. Remove the magnet.
8. Repeat **steps 4**–7.
9. Repeat **steps 4**–7.
10. Add 250 μL of PBSTM to the beads. These "prewashed" beads will be used for both negative and positive selection experiments in subsequent steps.

3.4 Preparing a Protein–Magnetic Bead Complex

Method 1 (Magnetic Bead–Streptavidin–Biotinylated Protein Complex)

1. Place 100 μL (~2 mg beads) of the prewashed beads from Subheading 3.3 into a 1.6 mL microcentrifuge tube.
2. Add 5 μg of biotinylated target protein to the same microcentrifuge tube.
3. Add 195 μL of PBS to the microcentrifuge tube.
4. Place the biotinylated protein–bead mixture on a rotisserie shaker set to ~30 rpm.
5. Incubate the biotinylated target protein and bead solution for 1 h while rotating on the rotisserie shaker at ~30 rpm.
6. Apply the magnet to the biotinylated protein–bead mixture to pellet the beads.
7. Use a fine-tipped pipette to remove the biotinylated protein solution from the bead pellet.
8. Remove the magnet.
9. Add 500 μL of PBSTM to wash the bead protein complex.
10. Apply the magnet to the biotinylated protein containing beads to pellet the beads.
11. Repeat **steps 6–8**.
12. Repeat **steps 6–8**.
13. Add 200 μL of PBSTMB (Fig. 1a) to the biotinylated protein containing beads.
14. Store the biotinylated protein–bead complex at 4 °C for future selection steps, which are outlined in Fig. 1.

Method 2 (Magnetic Bead–Steptavidin–Antibody Protein Complex)

1. Place 250 μL (~5 mg beads) of the prewashed beads from Subheading 3.3 into a 1.6 mL microcentrifuge tube.
2. Add 10 μg (10 μL) of biotinylated antibody specific to your target protein into the same microcentrifuge tube.
3. Add 190 μL of PBS to the microcentrifuge.

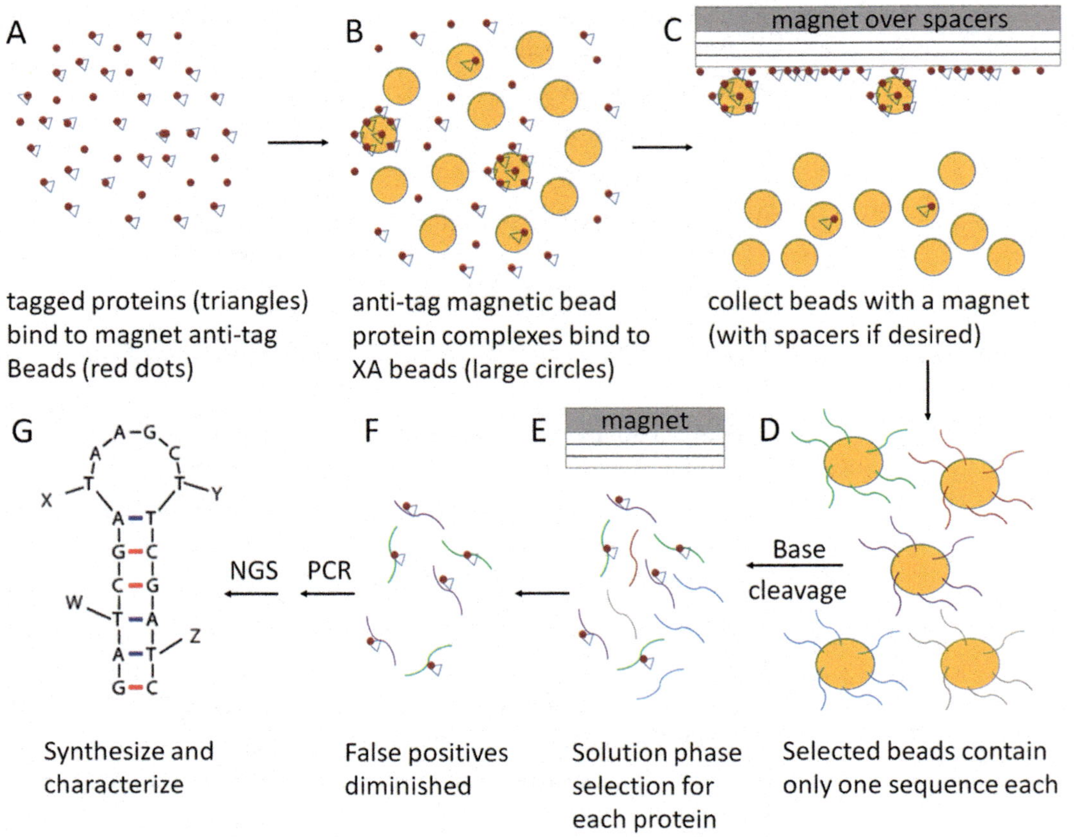

Fig. 1 Work flow illustrating the two-step X-Aptamer selection process. (**a**) The biotin-tagged protein targets (*triangles*) bind strongly with magnetic anti-tag beads (*red dots*). (**b**) These complexes are then mixed with the X-Aptamer bead-based library (*orange circles*). Some XA beads bind very strongly to the protein, while others bind weakly or not at all. (**c**) XA beads with many magnetic beads on them can be easily pulled from solution with a magnet. (**d**) The selected X-Aptamer beads contain many copies of a single unique X-Aptamer sequence, which can be cleaved from the X-Aptamer bead by treatment with base. (**e**) After base-mediated cleavage of the X-Aptamers from the beads, the solution phase XAs are subjected to a second round of positive selection for each protein. (**f**) After removal of the sequences in solution that are not bound to the magnet, a library depleted in false positives remains. Subsequently, PCR amplification, next-generation sequencing and analysis, and (**g**) X-Aptamer synthesis are performed for validation experiments.

4. Place the biotinylated antibody–bead mixture on a rotisserie shaker set to ~30 rpm.
5. Incubate the biotinylated antibody and bead solution for 1 h while rotating the sample at ~30 rpm on the rotisserie shaker.
6. Apply the magnet to the biotinylated antibody–bead mixture to pellet the beads.
7. Use a fine-tipped pipette to remove the biotinylated antibody solution from the bead pellet.
8. Remove the magnet.
9. Add 500 μL of PBSTM to wash the bead antibody complex.

10. Apply the magnet to the biotinylated antibody containing beads to pellet the beads.
11. Repeat **steps 6–8**.
12. Repeat **steps 6–8**.
13. Add 250 μL of PBSTMB to the biotinylated protein containing beads.
14. Store the biotinylated antibody–bead complex at 4 °C for future positive and negative selection steps.

3.5 X-Aptamer Library Preparation

The XA bead library is supplied as a dry powder in a small glass vial that can be screened against up to five target proteins simultaneously.

1. Add 1 mL of PBSTM to the glass vial containing the XA bead library.
2. Gently shake or vortex the XA bead solution.
3. Use a pipette to quickly transfer the beads in solution into a sterile 15 mL conical tube before most of the beads settle out of solution.
4. Add 1 mL of PSBSTM to the original XA glass vial containing some residual XA beads.
5. Gently shake or vortex the XA bead solution.
6. Use a pipette to quickly transfer the residual XA beads in solution into the same sterile 15 mL conical tube.
7. Repeat **steps 4–6**.
8. Repeat **steps 4–6**.
9. Add 6 mL of PBSTM to the 15 mL conical tube containing the XA beads, bringing the total volume to 10 mL.
10. Centrifuge the 15 mL conical tube at 3000 × *g* for 10 min at room temperature in a swing bucket rotor to pellet the beads.
11. Use a 1 mL pipette to carefully remove all but 100–200 μL of the supernatant liquid above the beads while not disturbing the beads.
12. Add 3 mL of PBSTM (no BSA) to the XA beads.
13. Heat the XA bead tube in a beaker placed in a 95 °C water bath for 5 min to anneal the XAs.
14. Turn off the water bath, or removed the beaker from the bath, to allow the XAs to cool slowly over 30 min and form thermodynamically stable structures.
15. Add 7 mL of PBSTMB (which contains BSA) to the XA beads.
16. Gently shake or vortex the XA bead solution.

17. Centrifuge the 15 mL conical tube at 3000 × *g* for 10 min at room temperature in a swing bucket rotor to pellet the beads.
18. Use a 1 mL pipette to carefully remove the supernatant liquid above the beads while not disturbing the beads.
19. Add 3 mL of PBSTMB to the bead pellet. The XA library is now ready for use for screening.
20. (Optional) The library may be screened against a mixture of proteins, with specification determined in later steps. However, if one wishes to screen multiple proteins individually, the XA beads should be fractionated, one fraction for each protein target.

3.6 Negative Selection Against Naked Magnetic Beads or Antibody-Bound Magnetic Beads

1. (Method 1 only) Add 3 mg of prewashed naked magnetic beads (from Subheading 3.3) to a 1.6 mL microcentrifuge tube.
2. (Method 2 only) For each protein target, add 150 μL (~3 mg) of magnet beads conjugated to the target antibody to individual 1.6 mL microcentrifuge tubes. For Method 2, negative selection must be performed against each magnet bead–antibody complex, leading to increased negative selection rounds (*see* **Note 3**).
3. Add 1 mL of the prepared XA library from Subheading 3.5 into the microcentrifuge tube(s) from **step 1** above.
4. Place the microcentrifuge onto a rotisserie shaker set to ~30 rpm.
5. Incubate for 1 h at room temperature while rotating at ~30 rpm.
6. Remove centrifuge tube from the rotisserie shaker.
7. Place the tube in the magnetic separator to attract magnetic bead–X-Aptamer bead complexes and free magnetic beads.
8. Gently tilt the separator about 45°.
9. Using a pipette, transfer the unbound X-Aptamer beads (not the pellet) into a new 1.6 mL microcentrifuge tube. Do not throw away the pellet.
10. Place this new second 1.6 mL microcentrifuge tube containing unbound X-Aptamers beads into the magnetic separator.
11. Gently tilt the separator about 45°.
12. Using a pipette, transfer the unbound X-Aptamer beads (not the pellet at the magnet) into a third 1.6 mL microcentrifuge tube. Do not throw away the pellet. This third tube contains a diminished XA library (fewer false-positives) that will be used in the bead-based positive selection in Subheading 3.8.

13. Add 0.5 mL of PBSTM to the second 1.6 mL microcentrifuge tube.
14. Transfer the contents of this second microcentrifuge tube to the initial microcentrifuge tube containing most of the magnet particles and magnetic bead–X-Aptamer complexes.
15. Add 1.0 mL of PBSTM to this initial microcentrifuge tube.
16. Place the tube into the magnetic separator.
17. Gently tilt the separator about 45°.
18. Using a pipette, gently remove the PBSTM wash liquid.
19. Repeat **steps 14–18**.
20. Repeat **steps 14–18**.
21. Add 100 μL of nuclease-free water to the beads in this first microcentrifuge tube.
22. With gentle agitation to keep beads in solution, partition these magnetic particles equally into two 200 μL PCR tubes.
23. Add 50 μL of nuclease-free water to this first microcentrifuge tube.
24. With gentle agitation to keep beads in solution, partition these magnetic particles equally into same two 200 μL PCR tubes.
25. Store these two PCR tubes at −20 °C for PCR amplification later with a unique bar coded primer that will serve as a negative selection control.

3.7 Bead-Based Positive Selection

1. Spin the microcentrifuge tube containing the depleted XA bead library from Subheading 3.6 after negative selection (or the one appropriate for each protein if using Method 2) in a microcentrifuge at 3000 × *g* for 5 min at room temperature.
2. Carefully remove and discard the solution with a pipette while retaining the depleted XA beads.
3. Gently agitate your DynaBead–Protein complex that has been created in Subheading 3.4.
4. Quickly transfer 100 μL of the DynaBead–Protein complex into the depleted XA bead library.
5. Add 400 μL of PBSTMB to the depleted XA bead library (Fig. 1b).
6. Incubate for 1 h at room temperature on a rotisserie shaker.
7. Apply a magnet to attract the XA beads that are coated by many much smaller magnetic beads due to strong affinity between the XAs and the protein (Fig. 1c).
8. Using a pipette and gentle agitation, remove the XAs beads that are not attracted to the magnet.
9. Remove the magnet.

10. Add 1 mL of PBSTM to the positively selected beads.
11. Apply the magnet.
12. Using a pipette and gentle agitation, remove the weakly bound XA beads that are not attracted to the magnet.
13. Repeat **steps 9–12**.
14. Repeat **steps 9–12**. Each of the beads contains many copies of a single XA sequence (Fig. 1d).
15. Proceed to Subheading 3.8 or store the beads at −20 °C for later use.

3.8 Preparing a Solution Phase Library from the Positively Selected XA Beads

1. Suspend the selected beads from the end of Subheading 3.7 in 100 μL of 0.2 M NaOH by gentle agitation via pipette.
2. Incubate these selected XA- beads at 65 °C for 30 min to cleave the single base-sensitive RNA linker to release the XA candidates into solution.
3. Centrifuge at 3000 × *g* for 10 min at room temperature to settle the beads.
4. Transfer the solution phase library to a new 1.6 mL microcentrifuge tube using a pipette.
5. Add 125 μL of PBSTM to the beads.
6. Centrifuge the beads at 3000 × *g* for 10 min at room temperature to settle the beads.
7. Transfer this residual solution phase library to the same microcentrifuge tube (from **step 4** above) using a pipette.
8. Repeat **steps 5–7**.
9. Transfer the collected solution phase library from the microcentrifuge tube into a 0.5 mL capacity 3 kD MWCO centrifugal filter.
10. Centrifuge the filter at 14,000 × *g* for 10 min at room temperature.
11. Empty the contents of the centrifugal filter's collection tube.
12. Add sufficient PBSTM to bring the volume in the top of the filter to 450 μL.
13. Centrifuge at 14,000 × *g* for 10 min at room temperature.
14. Repeat **steps 11–13** to reduce residual sodium hydroxide.
15. Repeat **steps 11–13** to reduce residual sodium hydroxide.
16. Using a pipette, transfer the retentate from the top of the centrifugal concentrator to a new 1.6 mL microcentrifuge tube (this is the major fraction).
17. Transfer 2 μL of this retentate into another microcentrifuge tube (this is the minor fraction).

18. Dilute the major fraction of the retentate with PBSTM up to a total volume of 200 μL.
19. Heat the major fraction of the retentate in a 95 °C water bath for 5 min.
20. Allow the major fraction to cool down on the bench top. This is the solution phase library to be used later in the second, solution-phase aptamer screen.
21. Add 148 μL of water to the 2 μL aliquot of retentate (minor fraction).
22. Split the dilute minor fraction equally into two 200 μL capacity PCR tubes.
23. Store these two minor fractions at −20 °C for later PCR amplification.

3.9 Solution-Based Positive Selection

1. For each protein target, place 100 μL of Dynabead–protein complexes into separate 1.6 mL microcentrifuge tubes (*see* **Note 3**).
2. Add 30 μL of the solution phase X-Aptamers from Subheading 3.8, **step 20** to each of these microcentrifuge tubes.
3. Incubate the mixture for 1 h at room temperature.
4. Place the tubes into the magnetic holder to attract the magnetic particles (Fig. 1e).
5. Remove the supernatant solution (containing nonbinding X-Aptamers) with a pipette.
6. Remove the magnet.
7. Add 500 μL of PBSTM.
8. After 1 min, apply the magnet to pellet the positive X-Aptamers once again.
9. Remove the supernatant solution (containing any residual nonbinding X-Aptamers) with a pipette.
10. Remove the magnet.
11. Repeat **steps 7–10**.
12. Repeat **steps 7–10**.
13. Add 100 μL of nuclease-free water to each microfuge tube.
14. Gently agitate the beads into solution via pipette.
15. Partition the beads in each microcentrifuge tube equally (50 μL each) into two 200 μL PCR tubes.
16. Add an additional 50 μL of nuclease-free water to each (nominally empty) microcentrifuge tube.
17. Gently agitate any remaining beads into solution via pipette.

18. Partition these remaining beads in each microcentrifuge tube equally (25 μL each) into the same two 200 μL PCR tubes from **step 15**.
19. Store these two PCR samples (per protein target) at −20 °C for later PCR amplification.

3.10 PCR Amplification of Selected Sequences

To identify X-Aptamer sequences from the control (negative) and positive rounds of selection, one must perform barcoded PCR reactions. A unique barcode must be used for each fraction of selected material. Encoded reverse primers (up to 7) necessary for PCR preparation are supplied with the X-Aptamer selection kit. If selection experiments have been performed on multiple proteins, so that more than five selection rounds exist (including negative, bead-based, and solution-phase selection for each protein), *see* **Note 3**.

1. For each round of selection that has been performed, transfer 75 μL of selected sequences (template mix) into a 200 μL PCR tube.
2. Add 10 μL 10× PCR buffer to each PCR tube.
3. Add 3 μL 10 mM dNTPs to each PCR tube.
4. Add 3.0 μL 10 μM forward primer to each PCR tube.
5. Add 5 μL of nuclease-free water to each PCR tube.
6. Add 1 μL of Taq DNA polymerase to each PCR tube.
7. Select and document a unique barcoded reverse primer to be used for each round of selection. It is critical to keep track of which barcoded primer corresponds to which fraction of your selection procedures.
8. Referring to **step** 7 above, add 3.0 μL of the appropriate unique barcoded reverse primer to each PCR tube.
9. Use a thermal cycler to PCR to amplify the X-Aptamers in each of the PCR tubes (*see* **Note 4**).

3.11 Electrophoresis for Validation of PCR Products

1. Place 5 μL of each PCR product into individual microcentrifuge tubes.
2. Add 1 μL of 6× gel loading buffer. Final 1× buffer contains 5% glycerol and 0.05% w/v bromophenol blue.
3. Load the 6 μL from each tube into individual lanes of a precast 10% polyacrylamide minigel.
4. In another lane of the gel, add 1 μL of a DNA ladder solution containing a bottom band of 50 bp or smaller.
5. Electrophorese the gel at 180 V for 60 min in a minigel apparatus appropriate for the gel used (Similar to Xcell Sure-Lock, Invitrogen).
6. Remove the gel from the cassette.

7. Stain the gel for 100 min in 15 mL of gel staining solution.
8. Destain the gel for 20 min in 25 mL of gel destaining solution.
9. Image the gel on an AlphaImager or similar instrument under UV light, using the appropriate filter for detection of ethidium bromide, to verify the presence PCR product with an apparent length of about 80 bases. The presence of residual primers can be ignored because such short sequences are easily culled from NGS sequencing data.

3.12 Preparations for Acquisition of Next-Generation Sequencing (NGS)

In order to produce a sample for NGS with approximately equal contributions from each selection round, excess dNTPs should be removed in order to accurately quantify each PCR product. Additional considerations are required if multiple protein targets are used (*see* **Note 3**).

1. Load each PCR reaction into separate 0.5 mL capacity, 10 kD MWCO centrifugal concentrators.
2. Dilute each sample in the concentrators up to 0.5 mL total volume with water.
3. Centrifuge each sample at 14,000 × *g* for 5 min in a microcentrifuge.
4. Transfer the flow-through in the receptacle tube of each concentrator into separate 15 mL conical tubes and retain in case one or more of the filters should break during centifugation.
5. Dilute the retentate on top of each concentrator up to 0.5 mL total with water.
6. Repeat **steps 3–5**.
7. Repeat **steps 3–5**.
8. Collect the retentate on top of each concentrator into individual 1.6 mL microcentrifuge tubes.
9. Measure the absorbance of each PCR product at 260 nm by loading 1 μL of PCR product onto a microvolume UV–Vis spectrometer (such as a NanoDrop).
10. For each PCR product, estimate the DNA concentration by assuming that 1 absorbance unit (through a 1 cm path length) equals 40 μg/mL of DNA.
11. Using the concentrations determined above, calculate the volume of each PCR product that is needed to provide 30 ng of material.
12. Using the calculations above, mix 30 ng of each PCR product together into a new PCR tube. The total volume is typically 20–50 μL.
13. Submit this NGS sample, without providing any primer or barcode information, to an NGS sequencing service (*see* **Note 5**)

to obtain a FASTQ formatted sequencing data file or return the materials to AM Biotechnologies for NGS sequencing service and post-sequencing analysis.

3.13 Analysis of NGS Sequencing Data

The NGS data must be subjected to bioinformatics analysis for several reasons. First and foremost, the sequencing data must be appropriately aligned in order to decode the specialized bases (5′-X-dU), which are all interpreted as thymidines in the sequencing data. Knowledge of the XA library design is required for this step. Secondly, one must determine the number of times each sequence occurs in each selection round. The analysis should also include clustering analysis, because true XA candidates typically have "relatives" which also bind to a lesser degree.

1. Use an NGS analysis software program (*see* **Note 1**) to align the sequences against a library design file.
2. Use an NGS analysis software program to decode the position and identity of the unusual (X) DNA bases within the X-Aptamer (*see* **Note 1**).
3. Use an NGS analysis software program to determine sequence frequencies for all selection rounds based on the barcodes (*see* **Note 1**).
4. Use an NGS analysis software program to perform clustering analysis of the sequences to find families of DNA sequences (*see* **Note 1**).
5. Select candidate X-Aptamers to be synthesized for validation experiments (*see* **Note 6**).

3.14 Synthesis of X-Aptamers

1. For validation, X-Aptamers with 5′-biotin labels will need to be synthesized by a commercial vendor (*see* **Note 7**).

3.15 Validation of X-Aptamers

Validation of X-Aptamers is usually performed using biotinylated XAs in electrophoretic mobility shift assays (EMSA) to verify binding, but a variety of other methods are available (*see* **Note 8**). A series of samples with a constant DNA concentration and varying protein concentrations (~0.25 nM to 1 μM) are used in the EMSA described here.

1. Transfer 5 μL of the 1 μM X-Aptamer solution into a microcentrifuge tube.
2. Add 95 μL of water to the microcentrifuge tube to create a 50 nM X-Aptamer solution. Set this aside.
3. Add 1 μL of the 10 μM target protein solution to a new microcentrifuge tube (call it P1).
4. Add 4 μL of water to tube P1 to create a 2000 nM solution (*see* **Note 9**).

5. Transfer 2.5 μL from tube P1 into a new microcentrifuge tube (call it P2).
6. Add 2.5 μL of water to tube P2 to create a 1000 nM solution.
7. Transfer 2.5 μL from tube P2 into a new microcentrifuge tube (call it P3).
8. Add 2.5 μL of water to tube P3 to create a 500 nM protein solution.
9. Transfer 2.5 μL from tube P3 into a new microcentrifuge tube (call it P4).
10. Add 2.5 μL of water to tube P4 to create a 250 nM protein solution.
11. Transfer 2.5 μL from tube P4 into a new microcentrifuge tube (call it P5).
12. Add 2.5 μL of water to tube P5 to create a 125 nM protein solution.
13. Transfer 2.5 μL from tube P5 into a new microcentrifuge tube (call it P6).
14. Add 2.5 μL of water to tube P6 to create a 62.5 nM protein solution.
15. Transfer 2.5 μL from tube P6 into a new microcentrifuge tube (call it P7).
16. Add 2.5 μL of water to tube P7 to create a 31.25 nM protein solution.
17. Transfer 2.5 μL from tube P7 into a new microcentrifuge tube (call it P8).
18. Add 2.5 μL of water to tube P8 to create a 15.625 nM protein solution.
19. Transfer 2.5 μL from tube P8 into a new microcentrifuge tube (call it P9).
20. Add 2.5 μL of water to tube P9 to create a 7.8125 nM protein solution.
21. Transfer 2.5 μL from tube P9 into a new microcentrifuge tube (call it P10).
22. Add 2.5 μL of water to tube P10 to create a 3.906 nM protein solution.
23. Transfer 2.5 μL from tube P10 into a new microcentrifuge tube (call it P11).
24. Add 2.5 μL of water to tube P11 to create a 1.953 nM protein solution.
25. Add 2.5 μL of water into a new microcentrifuge tube (call it P12).

26. To each tube P1–P12, add 1 μL (50 fmol) of the 50 nM X-Aptamer solution.
27. Incubate all the samples for 30 min at room temperature.
28. To each tube P1–P12, add 3.5 μL of 2× DNA loading dye (*see* **Note 9**).
29. Load the contents of tubes P1–P12 into separate wells of a precast 6% native TBE polyacrylamide gel with 12 loading wells.
30. Electrophorese the gel for 60 min at 100 V.
31. Electrophoretically transfer the biotinylated DNA from the gel to the nylon membrane using a wet trans-blot apparatus (Xcell II blot module, Invitrogen) device for 90 min at 25 V.
32. After transfer, place the nylon membrane in a pipette tip box lid and add 15 of blocking buffer from the Nucleic Acid Detection kit (Thermo Scientific, *see* **Note 10**).
33. Place the box on the gel rocker and rock it on a medium setting for 20 min.
34. Add 65 μL of streptavidin–HRP conjugate to 15 mL of blocking buffer in a conical tube and mix.
35. Gently decant and discard the blocking buffer from the box containing the nylon membrane.
36. Add 65 μL of fresh blocking buffer with streptavidin–HRP conjugate to the box containing the nylon membrane.
37. Repeat **step 33**.
38. Make 50 mL 1× wash buffer by adding 37.5 mL of deionized water to 12.5 mL of 4× wash buffer (from the kit) in a 50 mL conical tube.
39. Decant and discard blocking buffer in the box with streptavidin–HRP conjugate.
40. Add 15 mL of 1× wash buffer to the box containing the nylon membrane.
41. Place the box on the Gel rocker and rock it on a medium setting for 5 min.
42. Gently decant and discard the 1× wash buffer from the box containing the nylon membrane.
43. Repeat **step 40–42**.
44. Repeat **step 40–42**.
45. Add 15 mL of substrate buffer from the kit to the box containing the nylon membrane and rock it on a medium setting for 5 min.

46. Mix 5 mL of luminol solution (Brown bottle) with 5 mL of Enhancer solution (White bottle) in a 15 mL conical tube (Call it LE mix).
47. With clean forceps hold and lift the nylon membrane from box containing substrate buffer.
48. Place the nylon membrane on a wad of 5–6 Kimwipes for 1 min.
49. Flip the membrane and place it on a fresh wad of Kimwipes for 1 more minute.
50. Transfer the membrane to a dry tip box lid and add the LE mix.
51. Spread the LE mix evenly on the membrane by gentle tilting of the box and leave for 5 min.
52. Repeat **steps 47–49**.
53. Image the nylon membrane for chemiluminescense on an AlphaImager or similar instrument under different exposure times such as 1, 2, and 5 min to find the optimal exposure.
54. Perform densitometry analysis on the acquired images to determine the fraction of DNA bound to protein.
55. Determine the dissociation constant by plotting a graph of fraction bound vs. protein concentration using any graphing software program.

4 Notes

1. We use the Aptaligner software program [32], which requires a sequencing data file in FASTQ format and a library design file to perform Markov-model based alignments. The library design file must be obtained from AM Biotechnologies. It automatically decodes the specialized bases within X-Aptamers, enumerates the frequencies of each sequence by bar code (selection round), performs clustering of the top sequences, and employs both noise reduction and length filters. Aptaligner can also automatically analyze and separate multiple aptamer projects from a single FASTQ file. Aptaligner requires a Linux operating system, several commonly used programs, and familiarity with software installation.
2. Although three buffer exchanges may remove most of the original buffer components, an additional three rounds of buffer exchange are recommended to optimize the yield in subsequent coupling reactions.
3. The method described here, for a single protein target, is just one of many possible variations. For example, rather than creating magnetic bead protein complexes first, one could

instead create protein target X-Aptamer complexes first, and then add magnetic beads, in the solution phase positive selection. Also, if selection experiments will be performed on multiple proteins, so that more than seven total selection rounds are expected (including negative, bead-based, and solution-phase selection for each protein), one must choose among the following three options: (a) purchase additional X-Aptamer Selection Kits, (b) perform next-generation sequencing individually for each protein's selection rounds, or (c) purchase a set of unique forward primers such that selection rounds for each protein use a unique forward primer.

4. The final (1×) PCR buffer contains 1.5 mM $MgCl_2$, but a concentration as high as 2.5 mM can be used. Use the following protocol for the PCR amplification of selected X-Aptamers: 94 °C for 2 min (the initial denaturation), 20 PCR cycles (94 °C for 30 s, 50 °C for 30 s, and 72 °C for 45 s), and a final extension step of 72 °C for 5 min. Twenty PCR cycles exhaust the supply of free primers and provides a homogenous PCR product of defined length so that purification steps to remove free primers are not needed.
5. In our experience, it is best to not provide primer and bar code information to the NGS provider because they invariably trim these critical data from the sequence files returned to the customer. We typically use Ion Torrent for sequencing using either an Ion 316™ or 318™ chip, which nominally provide up to 3 or 5.5 million reads, respectively.
6. X-Aptamers with good selectivity tend to have high frequency in the solution phase data, are present in the bead-based selection data, and no/low frequency in the negative-selection data. In addition, they tend to have closely related sequences that also fit the above criteria while not possessing too many hydrophobic X-modifications.
7. At present, AM Biotechnologies, LLC, in Houston, TX, is the only commercial vendor providing synthesis of X-Aptamers derived from their X-Aptamer Selection Kit. Additional information on intellectual property rights is available from AM Biotechnologies, LLC.
8. Any method capable of measuring binding constants can be used in place of the EMSA method outlined here. For example, devices that detect binding by interferometry or by plasmon resonance methods could be used instead of EMSA.
9. The concentrations of protein noted during Subheading 3.15 **steps 4** through **28** are temporary. After the biotinylated DNA is added (Subheading 3.15, **step 29**) and the solutions are diluted (Subheading 3.15, **step 30**), the final protein concentrations of tubes P1–P13 are 1000, 500, 250, 125, 62.5,

31.25, 15.63, 7.81, 3.91, 1.95, 0.98, 0.49, and 0.24, respectively.

10. All of the reagents required for chemiluminescence are typically purchased as a kit. For our work, we typically use the Chemiluminescent Nucleic Acid Detection Module Kit from Thermo Scientific.

References

1. Tuerk C, Gold L (1990) Systematic evolution of ligands by exponential enrichment: RNA ligands to bacteriophage T4 DNA polymerase. Science 249:505–510
2. Ellington A, Szostak J (1990) In vitro selection of RNA molecules that bind specific ligands. Nature 346:818–822
3. Yang X, Wang H, Beasley DWC, Volk DE, Zhao X, Luxon BA, Lomas LO, Herzog NK, Aronson JF, Barrett ADT, Leary JF, David G, Gorenstein DG (2006) Selection of thioaptamers for diagnostics and therapeutics. Ann N Y Acad Sci 1082(1):116–119
4. Cho EJ, Joo-Woon Lee J-W, Ellington AD (2009) Applications of aptamers as sensors. Annu Rev Anal Chem 2:241–264
5. Keefe AD, Pai P, Ellington A (2010) Aptamers as therapeutics. Nat Rev Drug Discov 9:537–550
6. Kang KN, Lee YS (2013) RNA aptamers: a review of recent trends and applications. Adv Biochem Eng Biotechnol 131:153–169
7. Sun H, Zhu X, Lu PY, Rosato RR, Tan W, Zu Y (2014) Oligonucleotide aptamers: new tools for targeted cancer therapy. Mol Ther Nucleic Acids 3:e182
8. Blind M, Blank M (2015) Aptamer selection technology and recent advances. Mol Ther Nucleic Acids 4:e223
9. Lapa SA, Chudinov AV, Timofeev EN (2016) The toolbox for modified aptamers. Mol Biotechnol 58(2):79–92
10. Obika S, Nanbu D, Hari Y, Morio K-I, In Y, Ishida T, Imanishi T (1997) Synthesis of 2′-O,4′-C-methyleneuridine and -cytidine. Novel bicyclic nucleosides having a fixed C3′-endo sugar puckering. Tetrahedron Lett 38 (50):8735–8738
11. Koshkin AA, Singh SK, Nielsen P, Rajwanshi VK, Kumar R, Meldgaard M, Olsen CE, Wengel J (1998) LNA (locked nucleic acids): synthesis of the adenine, cytosine, guanine, 5-methylcytosine, thymine and uracil bicyclonucleoside monomers, oligomerisation, and unprecedented nucleic acid recognition. Tetrahedron 54(14):3607–3011
12. Rhie A, Kirby L, Sayer N, Wellesley R, Disterer P, Sylvester I, Gill A, Hope J, William James W, Tahiri-Alaoui A (2003) Characterization of 2′-fluoro-RNA aptamers that bind preferentially to disease-associated conformations of prion protein and inhibit conversion. J Biol Chem 278:39697–39705
13. Biesecker G, Dihel L, Enney K, Bendele RA (1999) Derivation of RNA aptamer inhibitors of human complement C5. Immunopharmacology 42:219–230
14. Yang X, Fennewald S, Luxon BA, Aronson J, Herzog NK, Gorenstein DG (1999) Aptamers containing thymidine 3′-O-phosphorodithioates: synthesis and binding to nuclear factor-κ B. Bioorg Med Chem Lett 9:3357–3362
15. Volk DE, Yang X, Fennewald SM, King DJ, Bassett SE, Venkitachalam S, Herzog N, Luxon BA, David G, Gorenstein DG (2002) Solution structure and design of dithiophosphate backbone aptamers targeting transcription factor NF-κB. Bioorg Chem 30 (6):396–419
16. Volk DE, Power TD, Gorenstein DG, Luxon B (2002) An ab initio study of phosphorothioate and phosphorodithioate interactions with sodium cation. Tetrahedron Lett 43 (25):4443–4447
17. Anderson KM, Nguyen D, Esadze A, Zandrashvili L, Gorenstein DG, Iwahara J (2015) A chemical approach for site-specific identification of NMR signals from protein side-chain NH^{3+} groups forming intermolecular ion pairs in protein-nucleic acid complexes. J Biomed NMR 62:1–5
18. Ferguson MR, Rojo DR, Somasunderam A, Thiviyanathan V, Ridley B, Yang X, Gorenstein DG (2006) Delivery of double-stranded DNA thioaptamers into HIV-1 infected cells for antiviral activity. Biochem Biophys Res Commun 344(3):792–797
19. Kang J, Lee MS, Watowich SJ, Gorenstein DG (2007) Combinatorial selection of a RNA thioaptamer that binds to Venezuelan Equine Encephalitis virus capsid protein. FEBS Lett 581(13):2497–2502

20. Gandham SHA, Volk DE, Lokesh GLR, Neerathilingam M, Gorenstein DG (2014) Thioaptamers targeting dengue virus type-2 envelope protein domain III. Biochem Biophys Res Commun 453(3):309–315
21. Mann AP, Somasunderam A, Nieves-Alicea R, Li X, Hu A, Sood AK, Ferrari M, Gorenstein DG, Tanaka T (2010) Identification of thioaptamer ligand against E-selectin: potential application for inflamed vasculature imaging. PLoS One 5(9):e13050
22. Somasunderam A, Thiviyanathan V, Tanaka T, Li X, Neerathilingam M, Lokesh GL, Mann A, Peng Y, Ferrari M, Klostergaard J, Gorenstein DG (2010) Combinatorial selection of DNA thioaptamers targeted to the HA binding domain of CD44. Biochemistry 49:9106–9112
23. He W, Elizondo-Riojas M-A, Li X, Lokesh GLR, Somasunderam A, Thiviyanathan V, Volk DE, Durland RH, Englehardt J, Cavasotto CN, Gorenstein DG (2012) X-Aptamers: a bead-based selection method for random incorporation of drug-like moieties onto next-generation aptamers for enhanced binding. Biochemistry 51(42):8321–8323
24. Kang J, Lee MS, Copland JA, Luxon BA, Gorenstein DG (2008) Combinatorial selection of a single stranded DNA thioaptamer targeting TGF-beta-1 protein. Bioorg Med Chem Lett 18(6):1835–1839
25. Mann AP, Bhavane RC, Somasunderam A, Montalvo-Ortiz BL, Ghaghada KB, Volk D, Nieves-Alicea R, Suh KS, Ferrari M, Annapragada A, Gorenstein DG, Tanaka T (2011) Thioaptamer conjugated liposomes for tumor vasculature targeting. Oncotarget 2(4):298–304
26. Mann AP, Somasunderam A, Liu X, Gorenstein DG, Ferrari M (2011) E-selectin-targeted porous silicon particle for nanoparticle delivery to the bone marrow. Adv Mater 23(36): H278–H282
27. Tanaka T, Kang S-A, Hasan N, Mann AP, Zheng W, Zhao L, Zhao D, Suh KS, Volk D, Gorenstein DG, Cristofanilli M, Rui H (2015) Blocking the adhesion cascade at the premetastatic niche for prevention of breast cancer metastasis. Mol Ther 23(6):1044–1054
28. Mai J, Huang Y, Mu C, Zhang G, Xu R, Guo X, Xia X, Volk DE, Lokesh GL, Thiviyanathan V, Gorenstein DG, Liu X, Ferrari M, H Shen H (2014) Bone marrow endothelium-targeted therapeutics for metastatic breast cancer. J Control Release 187:22–29
29. Yang X, Bassett SE, Li X, Luxon BA, Herzog NK, Shope RE, Aronson J, Prow TJ, Leary JF, Kirby R, Ellington AD, Gorenstein DG (2002) Construction and selection of bead-bound combinatorial oligonucleotide phosphorothioate and phosphorodithioate aptamer libraries designed for rapid PCR-based sequencing. Nucleic Acids Res 30:e132
30. Fetter L, Richards J, Daniel J, Roon L, Rowland TJ, Bonham AJ (2015) Electrochemical aptamer scaffold biosensors for detection of botulism and ricin toxins. Chem Commun 51 (82):15137–15140
31. Zeidan E, Shivaji R, Henrich VC, Sandros MG (2016) Nano-SPRi aptasensor for the detection of progesterone in buffer. Sci Rep. doi:10.1038/srep26714
32. Lu E, Elizondo-Riojas M-A, Chang JT, Volk DE (2014) Aptaligner: automated software for aligning pseudo-random DNA X-Aptamers from next-generation sequencing data. Biochemistry 53(22):3523–3525

Chapter 11

Design and Preparation of Aptamer–siRNA Chimeras (AsiCs) for Targeted Cancer Therapy

Sven Kruspe and Paloma H. Giangrande

Abstract

Conjugation of cell-internalizing DNA or RNA aptamers to tumor-suppressing siRNAs represents a novel promising approach for cancer therapy. Here we describe how to employ RNA aptamers that bind to cell surface receptors as carriers for cell-targeted siRNA (or miRNA) delivery. This protocol was optimized to improve the efficiency of the aptamer–siRNA/miRNA conjugates and facilitate its implementation in most molecular biology labs. The single working steps include (1) outlining the optimal sequences of the RNA strands bearing the aptamer and siRNA sequence, for which further (2) a dsDNA template is synthesized and then (3) transcribed into an RNA that will be (4) folded and annealed to a chemically synthesized siRNA complementary strand. Moreover, we reference recent examples and advances in the aptamer delivery field.

Key words Aptamers, siRNA, miRNA, RNAi, Targeted delivery, Primer extension, In vitro transcription, Targeted cancer therapy

1 Introduction

Aptamers, a class of artificial single-stranded oligonucleotides with high affinity for their target, have been explored for several therapeutic purposes, such as the detection, blocking/inhibition, or activation of accessible cellular components [1]. One recent and promising application, includes the use of aptamers for targeted drug delivery. In particular, the combination of aptamers with RNA interference (RNAi) inducing RNAs, such as siRNAs or miRNAs, has been highly sought out [2, 3]. The ability to deliver RNAi-based drugs to organs beyond the liver remains a major bottleneck for the broad clinical translation of this relatively new class of biodrugs. A general approach used nowadays is based on lipoplex formulations, but these particles lack cell specificity and provoke severe side effects, such as a nonspecific immune response, due to their nanoscopic size. Another drawback is the complexity of these multicomponent assemblies, leading to laborious development

Eckart Bindewald and Bruce A. Shapiro (eds.), *RNA Nanostructures: Methods and Protocols*, Methods in Molecular Biology, vol. 1632, DOI 10.1007/978-1-4939-7138-1_11,

steps, high production cost, and requiring extensive assessment of biological tolerance.

Various approaches of combining cell-targeting aptamers with therapeutically active siRNAs have been described [4–6]. The protocol herein describes an aptamer–siRNA chimera (AsiC) consisting of a 2′ fluoro (2′F)-pyrimidine modified aptamer and a partially modified (only the passenger strand is modified with 2′F pyrimidines) double-stranded siRNA. The AsiC is made by annealing two strands: a *long strand*, which includes the aptamer sequence and the sequence of the siRNA passenger strand and a *short strand*, which is the sequence of the siRNA guide strand. The required two-nucleotide 3′ overhang (e.g., 3′UU) for proper Dicer processing is positioned on the 3′ end of the long strand. This forces Dicer processing to be asymmetric, enhancing the Dicer cleavage to occur in the desired position 19 nt downstream and preventing steric hindrance by the aptamer [7] (*see* **Note 1**). Our protocol outlines the working steps for a first evaluation of such an AsiC by tissue culture experiments. For cost efficiency reasons, we outline the enzymatic synthesis of the major component, a long 2′F-modified RNA strand, by means of T7 RNA polymerase (T7 RNAP). Alternatively, the long strand can be made using chemical synthesis. Although costlier, chemical synthesis may be important for clinical translation. To increase yields and to reduce costs of chemical synthesis it is advised to use an aptamer sequence below 40 nucleotides in length.

This protocol further allows for the preparation of different AsiC candidates in parallel (using different siRNA sequences) and appropriate negative control analogs with minimal expenses. The negative controls will either have to bear a nonfunctional aptamer variant or the tumor suppressor siRNA being replaced with a negative control siRNA sequence.

Before evaluating the AsiCs for their ability to mediate an RNAi effect in cells in culture, AsiCs should be tested for retention of aptamer activity (e.g., binding to the target or target inhibition or activation). This can be accomplished using the AsiC in an in vitro assay established for the respective aptamer or by examination of the AsiC as a competitor to the original aptamer (e.g., by flow cytometry, *see* ref. 4). *See* **Note 2** for further considerations on enzymatically synthesized RNA.

2 Materials

Prepare all solutions using RNase-free ultrapure water (to assure absence of any RNase activity use DEPC treated water or prepare it in house by adding 1/1000 volume of DEPC, stirring for 2 h at room temperature and subsequent autoclaving to break down DEPC leftovers). Furthermore, use analytical grade reagents.

Prepare and store all reagents at room temperature (unless indicated otherwise). Increased caution should be used by working with unmodified RNA samples (such as the counter strand of the siRNA part, also called *guide strand* or *antisense strand*).

2.1 Primer Extension

1. Annealing buffer 10×:100 mM Tris–HCl, pH 8.0. Add about 50 mL water to a 250 mL graduated cylinder or glass beaker. Weigh 3.03 g Tris base (*see* **Note 3**) and transfer to the cylinder. Add water to a volume of 200 mL. Mix and adjust pH with HCl (*see* **Note 4**). Fill up to a volume of 250 mL. Store at 4 °C.
2. 50 mM $MgCl_2$: Use either ultrapure solid $MgCl_2$ or a commercially available 1 M $MgCl_2$ solution. If prepared from solid $MgCl_2$ weigh 1.02 g $MgCl_2{\cdot}6H_2O$ or 0.48 g of anhydrous $MgCl_2$, transfer to a 100 mL graduated cylinder and fill up to 100 mL and mix until completely dissolved. If prepared from a 1 M stock solution, pipet 5 mL of this solution into a graduated cylinder and fill up to 100 mL and mix.
3. Taq Polymerase and optimized buffer for this enzyme as a 10× concentrated stock solution.
4. Deoxynucleoside triphosphates (dNTPs) as a mix with a concentration of 10 mM of each dNTP (*see* **Note 5**).
5. Two kind of oligodeoxynucleotides (ODNs): One carrying the aptamer sequence part or a nonbinding variant of the aptamer. The other overlapping reverse ODNs bear selected siRNA sequences. Both ODNs are described in detail in Subheading 3.

2.2 In Vitro Transcription

1. 5× T7 RNAP reaction buffer (optimized for 2′-F-transcription): 200 mM Tris–HCl (pH 8.0), 60 mM $MgCl_2$, 5 mM spermidine–HCl, 20% (w/v) PEG 8000, 0.01% (w/v) Triton X-100, and 25 mM DTT. Prepare 1 M Tris–HCl, pH 8.0 analogously as the 100 mM annealing buffer in Subheading 2.1 (use 30.3 g Tris base for a final volume of 250 mL and adjust pH using 10 M HCl). Weigh and add 20 g PEG 8000 to a 100 mL graduated cylinder. Add 20 mL of 1 M Tris–HCl to it. Weigh 1.22 g of $MgCl_2{\cdot}6H_2O$ and add it to the graduated cylinder. Weigh and add 127 mg of spermidine-HCl and add it to the graduated cylinder. Fill up to 80 mL with DEPC-treated water, stir until all compounds are dissolved. Fill up to 100 mL with DEPC-treated water. For final use add 25 μL of 1 M DTT to 975 μL of the above buffer immediately before using for a final DTT concentration of 25 mM.
2. Mix of NTPs (*see* **Note 5**): 30 mM 2′-F-CTP, 30 mM 2′-F-UTP, 10 mM 2′-OH-ATP, and 10 mM 2′-OH-GTP.

3. For the transcription use T7 RNAP which can be prepared according to the protocol in [8] which follows the preparation described in [9].
4. Pyrophosphatase for T7 transcription reaction (final concentration: 1 unit/100 μL).

2.3 RNA Gel Purification

1. Electrophoresis chamber (1 × 180 × 250 mm gel size, or similar) and respective power supply.
2. Tris–borate–EDTA buffer (TBE): 10× solution can be prepared by adding 108 g Tris base and 55 g boronic acid together with 40 mL 0.5 M EDTA (pH 8.0) to 600 mL deionized water. After dissolving fill up to a volume of 1 L.
3. 250 mL of 10% acrylamide gel solution: In a chemical bottle combine 62.5 mL 40% acrylamide–bis (29:1), 50 mL water, and 120 g urea and place it on a magnetic stirrer until the urea has dissolved. Add 12.5 mL 10× TBE and fill up to 250 mL with water. Filter with bottle-top filter and store protected from light at 4 °C.
4. Tetramethylethylenediamine (TEMED).
5. 10% ammonium persulfate (APS): Dissolve 1 g anhydrous APS in 6 mL deionized water and fill up to a volume of 10 mL.
6. 2× formamide gel loading buffer: Mix 0.01 g xylene cyanol, 0.01 g bromophenol blue and 500 μL 10× TBE with 10 mL formamide. Tabletop centrifuge for 1.5 mL reaction tubes.
7. Spin columns (1.5–2 mL reaction tube size) for regular PCR product workup.
8. Low EDTA elution buffer: 10 mM Tris–HCl, pH 7.5 and 0.1 mM EDTA.

2.4 Folding and Annealing

1. Heat block unit for 1.5 mL reaction tubes.

3 Methods

3.1 AsiC Design

1. General considerations. For ideal chimera efficacy we recommend a design similar to formerly described examples [5, 10]. The final construct is depicted in Fig. 1. Enzymatic synthesis of the upper longer strand is described in the following paragraphs (*see* **Note 6**). The counterstrand, the final RNAi inducing antisense strand, has to be purchased by a vendor for RNA oligonucleotides without any chemical modifications.
2. Choice of an appropriate siRNA. This can be achieved either by using experimentally established sequences or by using a siRNA-finder online tool as it is provided by most of the siRNA vendors (*see* **Note 8**). In both cases such an siRNA

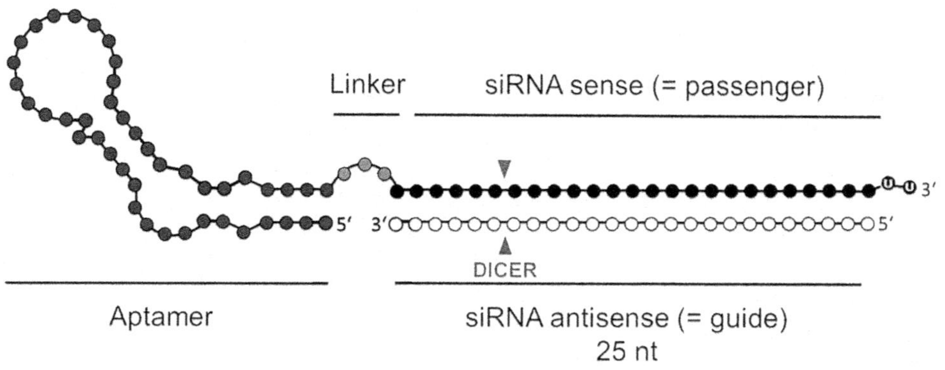

Fig. 1 AsiC design. The final construct consists of two strands of which the shorter strand (comprising the siRNA guide strand sequence—depicted as *open circles*) is reverse complementary to the 3′ terminal region of the longer strand (*gray/black*). The longer strand includes the aptamer sequence as well as the sense (passenger) part of the siRNA, both separated by a 3 nt linker (UUU, *see* **Note 7**). This strand comprises 2′-F-pyrimidines and is transcribed using T7-RNAP variant Y639F [11]. The shorter strand is often left unmodified, unless modifications have been previously validated and confirmed not to disrupt RNAi processing, and may be chemically synthesized/purchased

sequence will have a length of 19–21 nt. As the intended AsiC here displays a 25 nt Dicer substrate additional nucleotides will be required. We recommend for this partial sequence to use the sequence matching the mRNA of the target gene, adding 4–6 nt upstream of the sequence targeted by the siRNA. The mRNA sequence can be obtained using bioinformatics resources available (for example) at http://www.ncbi.nlm.nih.gov/. A quick guide how to use the website for this purpose can be found at: http://www.ncbi.nlm.nih.gov/guide/howto/find-transcript-gene/.

We also recommend consulting two studies on the conjugation of siRNAs and aptamers. The first focusses on the general effect of different aptamers attached to each terminus and strand of the siRNA [7]. The second study addresses optimization of Dicer substrate part of AsiCs by means of bridging "sticky" sequences [12].

3. Design of template ODNs. As stated above, one of two RNA strands of the final AsiC will be transcribed using T7 RNAP Y639F. For this transcription a double-stranded DNA template is generated from two overlapping ODNs. The aptamer sequence will be incorporated into the first of these two ODNs This should be used in the most truncated version and must be functional with 2′F-pyrimidines. The final ODNs to be ordered will look like this:

   (a) *aptamer_template ODN*:

   5′-GCCGGGAATT-**TAATACGACTCACTATA-**<u>**G**</u>**GG**-N_x-<u>*NNNNNNNNNNNNN*</u>-3′.

|< T7 promoter with optimized upstream seq.>|<Aptamer sequence >|
Also design an analogous ODN bearing a nonbinding variant of the aptamer.

This ODN has a T7 promoter plus a 10 nt sequence upstream of the promoter for optimized binding of T7 RNAP to the transcription template (*see* **Note 9**). A prerequisite for the aptamer sequence is a 5′-terminal GGG in order for it to be suitable for in vitro transcription with T7 RNAP (*see* **Note 10**). The first nucleoside that will be transcribed by the T7 RNAP is the underscored bold G.

The underscored italic 3′-terminus displays the overlap that will be annealed to the 5′-terminus of the second ODN (*see* **Note 11**). The part "Nx" stands for the aptamer sequence excluding the terminal primer annealing site. The second ODN for the DNA template will therefore look like this:

(b) *3′-siRNA_template ODN*:
5′-AATATT-**TAAGGAGGGTGATCTTC**-AAA-*yyyyyyyyyyyyyy*-3′.
|< siRNA seq. (*here*: anti-Luciferase) >| * |< overlap to aptamer 3′-terminal seq. >|
*: linker part (coding a "UUU" in the final RNA).

Choose any siRNA antisense sequences for this ODN. Thus, one has to use the reverse complement of the mRNA target site. We depicted a sequence targeting Luciferase mRNA, which displays an appropriate all negative siRNA in Mammalia. The central AAA-part will be the template for the 3 nt-Linker (UUU) of the chimera, that separates aptamer and siRNA sequence parts (*see* **Note** 7).

These two ODNs can be used in a primer extension reaction to obtain a completely double-stranded DNA. This is described in Subheading 3.2.

4. Design of RNA oligonucleotides functioning as siRNA antisense strands. Since RNAi might be hindered by chemical modifications in the RNAi mediating antisense strand of the chimera, we advise to order this strand as an unmodified RNA, prepared from chemical synthesis. Its sequence is the analogous RNA sequence of the grey outlined sequence part in the 3′-siRNA_template ODN. Hence, for the example given above one will have to order this RNA:

   5′-r(AAUAUUUAAGGAGGGUGAUCUUC)-3′.

3.2 Synthesis of a DNA Template by Primer Extension

1. Prepare the annealing mix in a 1.5 mL reaction vessel by combining 9 μL of 100 mM Tris–HCl, pH 8.0, 15 μL of 50 mM $MgCl_2$, 20 μL of 100 μM 3′-siRNA_template_ODN (2 nmol), and 10 μL of 100 μM 5′ aptamer_template_ODN (1 nmol). Add 36 μL water to a final volume of 90 μL.

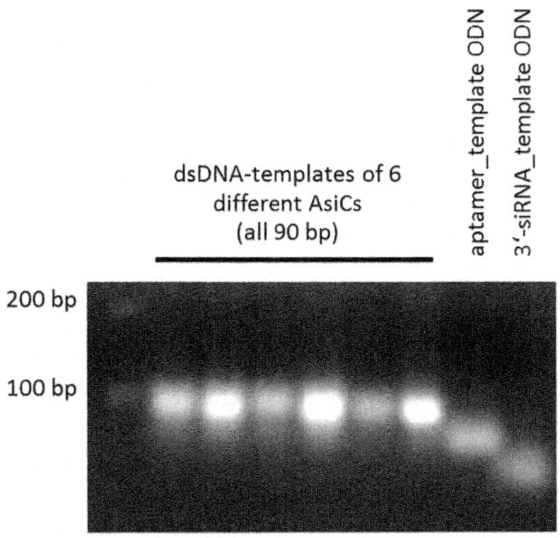

Fig. 2 Analyses of dsDNA templates on a 3% agarose gel. The gel shows six different 90-bp dsDNA templates after successful primer extension (*first six sample lanes*) and examples of ODNs (aptamer_template ODN and 3′-siRNA_template ODN) used for the primer extension reaction with a length of 60 nt and 42 nt, respectively

2. In another 1.5 mL reaction vessel prepare the primer extension reaction mix. Combine 50 μL of 10× PCR buffer (without Mg^{2+}), 10 μL of 10 mM dNTP mix, 4 μL of Taq Polymerase (5 U/μL) and 348 μL water.
3. Place both reaction vessel in a thermal block and heat to 95 °C for 5 min (*see* **Note 12**).
4. Allow both tubes to cool to room temperature for 20 min (*see* **Note 13**).
5. Mix together both reactions and distribute 100 μL aliquots into five PCR tubes.
6. Heat all five tubes to 72 °C for 10 min. Allow to cool to room temperature.
7. Use PCR Purification columns or miniprep spin columns in accordance with the manufacturer's protocol to purify the DNA duplexes (*see* **Note 14**).
8. Determine template concentration on a NanoDrop photometer and also check the reaction product on a 3% agarose gel, use 5 μL of the PCR product and 10 pmol of each AsiC strand as a reference (Fig. 2).

3.3 In Vitro Transcription

1. For a decent amount of chimera RNA for initial experiments set up a 250 μL transcription reaction. In a 1.5 mL reaction vessel combine in the following order 50 μL 5× T7 RNAP buffer, 125 pmol template DNA (*see* **Note 15**) from the previous work

step (Subheading 3.2) in 171 μL DEPC-treated water, 25 μL 10 NTP mix, 2 μL pyrophosphatase (~2.5 units) and 2 μL T7 RNAP (Y639F; ~6000 units).

2. Incubate for 2–6 h at 37 °C. (*see* **Note 16**).
3. Add 1 μL DNase I (RNase free, 10 units/μL).
4. Incubate for 10 min at 37 °C.
5. Add 200 μL of chloroform for extraction and vortex the reaction tube.
6. Centrifuge for 10 min at 10,000 × *g*.
7. Collect the top layer and proceed with gel purification.

3.4 Gel Purification of RNA AsiC

1. Cast a polyacrylamide gel (1 × 180 × 250 mm, or similar) using 10% acrylamide (ratio of monomeric–dimeric acrylamide may range from 19:1 to 29:1) and 8 M urea in TBE. To start polymerization, add first 25 μL tetramethylethylenediamine (TEMED) then 150 μL 10% ammonium persulfate (APS) to 25 mL of the gel solution.
2. Set up the gel in the electrophoresis chamber, pour TBE buffer into the chamber and rinse the wells with a syringe.
3. Prerun gel at 20 W for at least 20 min (*see* **Note 17**).
4. Add 1 volume of 2× formamide gel loading buffer (containing bromophenol blue) to transcription samples and heat to 65 °C for 5 min:
   (a) Rinse wells prior to loading samples.
   (b) Load directly from heat block.
5. Run at 20 W until first bromophenol blue front reaches the bottom (~20–25 min).
6. Excise RNA bands after visualizing with UV-shadowing (*see* **Note 18**).
7. Extract RNA from gel by incubating in 4 mL 10 mM Tris–HCl, pH 7.5, 0.1 mM EDTA (TE buffer low EDTA) at 37 °C for 1 h.
8. Filter buffer from gel/buffer slurry with a spin filter according to the manufacturer's protocol in a tabletop centrifuge.
9. Re-elute RNA from left over gel by repeating the extraction step (add 4 mL TE buffer low EDTA at 37 °C for 1 h).
10. Concentrate RNA with Centricon YM-10 spin filters (1000 × *g* for 15 min at RT).
11. Wash the RNA 2× (3× if excessive gel was cut) by adding 4 mL 10 mM Tris–HCl, pH 7.5, 0.1 mM EDTA (TE low EDTA) and repeating the spin.
12. Transfer the concentrated RNA to a clean reaction tube.
13. Quantify the RNA (1:50 dilution in TE low EDTA) by measuring the OD260.

3.5 Folding and Annealing

1. For final formation of a functional AsiC two processes have to be accomplished successfully: Folding of the aptamer part and annealing of the antisense siRNA part. These two reactions might interfere with each other leading to nonfunctional products. To minimize probability of such misfolding events we advise to use the following working steps.
2. In a second step the antisense siRNA strand can be annealed. However, for proper annealing temperatures above the Tm of the antisense strand should be reached. For some aptamers this will cause partial or complete aptamer unfolding. It is advisable to use the following protocol, which was optimized for PMSA aptamer A9g. If you discover loss of affinity in the AsiC, reduce the temperature in the second step from 65 °C sequentially by 2 °C until you regain aptamer affinity.
3. Dilute the long aptamer containing RNA strand up to a concentration of 10 μM in aptamer binding buffer (*see* **Note 19**).
4. In a second tube heat-dilute 1.5 equivalents of the shorter antisense siRNA strand in aptamer binding buffer.
5. Place both samples at 98 °C (heat block) for 10 min.
6. Transfer to 65 °C (heat block) for 10 min.
7. Add the solution of the short complementary strand and keep the combined solution sample at 65 °C for additional 10 min.
8. Transfer to 37 °C (heat block) for 10 min.
9. The resulting solution is ready to use (*see* **Note 20**).

4 Notes

1. The guide and passenger strands can also be swapped (so the guide strand would be part of the long strand with a 3′UU). This design is thought to favor loading of the guide strand into RISC leading to an enhanced RNAi effect. We have observed this for some of our chimeras [5] and advise the reader to test both versions.
2. As RNAs prepared by bacteriophage RNA polymerases (e.g., T7 RNAP) will contain 5′-triphosphates, one has to take into account activation of immune responses. We therefore recommend to use this cost-efficient compound only for initial cell culture experiments. For cell viability examination cells treated with the negative controls have to be compared to the actual AsiC candidates, and incubation times should be kept as short as possible (<12 h) and RNAi effect should be determined on the mRNA level by methods such as quantitative RT-PCR on RNA extracted from treated cells. For final experiments,

especially in vivo studies, all RNAs should be purchased as chemically synthesized oligonucleotides.

Another issue that has to be considered is T7 RNAPs tendency to append one or even a few more nontemplated nucleotides to the 3′-terminus [13, 14], which will lead to a small fraction of RNA molecules not bearing the required 3′-UU overhang that facilitates Dicer binding.

3. Make sure not to use Tris–HCl instead of Tris base.
4. For adjusting pH use 1 N HCl. As you will need about 13 mL of 1 N HCl you can start with 6 N or 10 N HCl.
5. Avoid too many freeze–thaw cycles for dNTPs and NTPs as this will reduce the active number of triphosphates.
6. The AsiC can also be prepared as a non-Dicer substrate, such as the examples given in [5], which worked as well in our lab.
7. The 3 nt linker (UUU) can be extended up to 7–9 nt which will prevent interference of aptamer folding with the tethered sense strand. However, in this case it is not recommended to use a linker only consisting of uridines since longer A/U-strings (>8 nt) can cause abrogation of T7 transcription [13]. A reasonable linker sequence would be 5′-UUGUUGUGU-3′.
8. The online tool provided at www.Dharmacon.com is preferred by the authors.
9. One way to increase yield in T7 in vitro transcription is to use an extended T7 promoter instead of the minimal 17 nt sequence part. The T7 enzyme will grab any potential template DNA and slide along the double strand until it finds the promoter. Having a 5′-addition of 6–10 nt simply prevents the enzyme from sliding off from the template. The present 10 nt sequence displays the canonical phage genome sequence, a truncation lacking the 5′-most 4 nt works almost as well [15].
10. For in vitro transcription using T7-RNAP one requirement is to have the final RNA starting with GGG (a sequence only starting with two consecutive Gs might work as well, but on the cost of lower yield). If the desired aptamer does not start with GGG on its 5′-terminus, try if adding three guanines at the 5′-terminus does not interfere with aptamer target affinity.
11. For the overlap a sequence length of 10–14 nt is recommended with a Tm ~50 °C. A high CG content is preferred as it allows shorter overlap length with such a Tm, which will result in less chance of off-target site binding. However, one should avoid any self-complementarity in both ODNs. Check your sequences with: http://unafold.rna.albany.edu/?q=mfold/DNA-Folding-Form.
12. Spin reaction tubes shortly in a centrifuge after heating step to prevent loss of condensate at tube lid.

13. This heating and cooling step is necessary to both melt and anneal the oligonucleotides and to activate the Taq enzyme.
14. Carefully chose the right number of columns to avoid loss of material or risking impurities after workup by knowing the recommended sample amount per column from the manufacturer's protocol.
15. For example, the amount of 125 pmol for a 100 bp template, a typical template size, is approximately 8.3 ng.
16. In general, 2 h should already yield >90% of a possible final yield and as NTPs are consumed reaction velocity is normally predominantly restricted by this parameter. However, different aspects can lead to reduced yield, such as disadvantageous transcript sequence with a high A/U content in position 1–8. If observing inappropriate yield for a AsiC transcript, extending transcription to 6 h is recommended. Also testing higher template or $MgCl_2$ concentrations (up to 2×) in small scale reactions (20 μL reaction volume) is advisable.
17. Carefully raise the voltage and check for heat generation when running the gel for the first time as overheating might result in gel burst. A power of 20 W can be applied on a gel with the denoted dimensions. However, a gel with a smaller cross-section (width × thickness) will have a higher electrical resistance and thus produce more heat so the power will have to be limited at a lower value.
18. In UV shadowing the absorption of nucleobases in the UV range is used to visualize nucleic acids band in a gel. After electrophoresis is complete, remove your gel from the glass plates and transfer it on a transparent foil and also cover it with another piece of transparent foil. Place your foil protected gel on a fluorescent thin layer chromatography plate and visualize the RNA bands in the gel by irradiating the gel with a handheld UV lamp (254 nm). Wear goggles to protect your eyes from UV light! For best visualization conduct this step in a dark place. Mark the RNA band areas on the upper foil with a marker pen and excise these gel parts turning the gel upside down (using the marks as a pattern).
19. It may be possible to go higher than 10 μM, but you will have to test this for your aptamer. Only fold up to 1 mL total volume in a 1.5 mL flip-flop tube at a time. If a total volume of more than 1 mL is required, utilize as many 1.5 mL tubes as necessary.
20. Prepare a nondenaturing gel (3% agarose or 15% polyacrylamide) and run it with a sample of the AsiC annealed product (>5 pmol) and separate aliquots of the two separate AsiC strands.

Acknowledgment

This work was supported by grants to P.H.G. from the National Institutes of Health (R01CA138503 and R21DE019953), Mary Kay Foundation (9033-12 and 001-09), Elsa U Pardee Foundation (E2766), and the Roy J Carver Charitable Trust (RJCCT 01-224); and to S.K. from the German Research Fundation (Deutsche Forschungsgemeinschaft, DFG).

References

1. Keefe AD, Pai S, Ellington A (2010) Aptamers as therapeutics. Nat Rev Drug Discov 9 (7):537–550. doi:10.1038/nrd3141
2. Zhou J, Rossi JJ (2014) Cell-type-specific, aptamer-functionalized agents for targeted disease therapy. Mol Ther Nucleic Acids 3:e169. doi:10.1038/mtna.2014.21
3. Kruspe S, Mittelberger F, Szameit K, Hahn U (2014) Aptamers as drug delivery vehicles. ChemMedChem 9(9):1998–2011. doi:10.1002/cmdc.201402163
4. McNamara JO II, Andrechek ER, Wang Y, Viles KD, Rempel RE, Gilboa E, Sullenger BA, Giangrande PH (2006) Cell type-specific delivery of siRNAs with aptamer-siRNA chimeras. Nat Biotechnol 24(8):1005–1015. doi:10.1038/nbt1223
5. Dassie JP, Liu XY, Thomas GS, Whitaker RM, Thiel KW, Stockdale KR, Meyerholz DK, McCaffrey AP, McNamara JO II, Giangrande PH (2009) Systemic administration of optimized aptamer-siRNA chimeras promotes regression of PSMA-expressing tumors. Nat Biotechnol 27 (9):839–849. doi:10.1038/nbt.1560
6. Dassie JP, Giangrande PH (2013) Current progress on aptamer-targeted oligonucleotide therapeutics. Ther Deliv 4(12):1527–1546. doi:10.4155/tde.13.118
7. Berezhnoy A, Brenneman R, Bajgelman M, Seales D, Gilboa E (2012) Thermal stability of siRNA modulates aptamer-conjugated siRNA inhibition. Mol Ther Nucleic Acids 1: e51. doi:10.1038/mtna.2012.41
8. Bonner G, Patra D, Lafer EM, Sousa R (1992) Mutations in T7 RNA polymerase that support the proposal for a common polymerase active site structure. EMBO J 11(10):3767–3775
9. Sousa R, Rose JP, Chung YJ, Lafer EM, Wang BC (1989) Single crystals of bacteriophage T7 RNA polymerase. Proteins 5(4):266–270. doi:10.1002/prot.340050403
10. Gilboa-Geffen A, Hamar P, Le MT, Wheeler LA, Trifonova R, Petrocca F, Wittrup A, Lieberman J (2015) Gene knockdown by EpCAM aptamer-siRNA chimeras suppresses epithelial breast cancers and their tumor-initiating cells. Mol Cancer Ther 14(10):2279–2291. doi:10.1158/1535-7163.MCT-15-0201-T
11. Padilla R, Sousa R (1999) Efficient synthesis of nucleic acids heavily modified with non-canonical ribose 2′-groups using a mutant T7 RNA polymerase (RNAP). Nucleic Acids Res 27(6):1561–1563
12. Zhou J, Swiderski P, Li H, Zhang J, Neff CP, Akkina R, Rossi JJ (2009) Selection, characterization and application of new RNA HIV gp 120 aptamers for facile delivery of Dicer substrate siRNAs into HIV infected cells. Nucleic Acids Res 37(9):3094–3109. doi:10.1093/nar/gkp185
13. Milligan JF, Uhlenbeck OC (1989) Synthesis of small RNAs using T7 RNA polymerase. Methods Enzymol 180:51–62
14. Milligan JF, Groebe DR, Witherell GW, Uhlenbeck OC (1987) Oligoribonucleotide synthesis using T7 RNA polymerase and synthetic DNA templates. Nucleic Acids Res 15 (21):8783–8798
15. Tang GQ, Bandwar RP, Patel SS (2005) Extended upstream A-T sequence increases T7 promoter strength. J Biol Chem 280 (49):40707–40713. doi:10.1074/jbc.M508013200

Chapter 12

Cellular Delivery of siRNAs Using Bolaamphiphiles

Kshitij Gupta

Abstract

Discovery of RNA interference (RNAi) has opened up a new arena of therapeutic intervention for the treatment of cancerous as well as noncancerous diseases. The RNAi pathway utilizes RNAi inducers such as small interfering RNAs (siRNAs) to target and silence disease causing genes. However, efficient delivery of siRNAs for eliciting efficacious RNAi has remained a daunting challenge. Nonviral vectors such as lipids have shown great promise in delivering siRNAs. Recently, a novel class of cationic lipid molecules "bolaamphiphile lipids" or "bola lipids" has been shown to deliver siRNAs to cause effective gene silencing in cells. The present chapter showcases the ability of bola lipids to form micelles, bind with nucleic acids and protect nucleic acids against nucleases. Also, high in vitro transfection efficiency for silencing green fluorescent protein (GFP) using Dicer substrate siRNAs (dsiRNAs) designed against GFP at nontoxic dose in a human breast cancer model is demonstrated. Our results showed that these cationic bola lipids are promising siRNA delivery agents.

Key words Bolaamphiphiles, RNAi, RNA–DNA hybrids, dsiRNAs, Green fluorescent protein, Human breast cancer cells

1 Introduction

RNA interference (RNAi) became an attractive tool against diseases due to the specific and robust suppression of the targeted gene [1]. Typically, RNAi involves the cleavage of a long double-stranded RNAs by the enzyme Dicer to generate 19–21 nt siRNA duplexes comprising sense and antisense strands. Both the strands have 2 nt 3′ overhangs and a phosphorylated 5′ terminus [2]. These duplexes along with the argonaute protein are part of the RNA induced silencing complex (RISC). Argonaute catalyzes the degradation of the sense strand, whereas the remaining antisense strand together with RISC complementarily binds to the target mRNA to effect the silencing of the gene [3–6].

siRNAs have many potential therapeutic applications for the treatment of diseases, however naked siRNAs encounter several impediments before they reach the target location. These include high anionic charge and hydrophilicity that make them unable to

Eckart Bindewald and Bruce A. Shapiro (eds.), *RNA Nanostructures: Methods and Protocols*, Methods in Molecular Biology, vol. 1632, DOI 10.1007/978-1-4939-7138-1_12, © Springer Science+Business Media LLC 2017

cross cell membranes. Also, they suffer degradation from nucleases circulating in the blood stream, immune responses due to nonspecific effects, and off-target effects due to sequence homology between target and nontarget mRNAs [7–9].

Viral and nonviral vectors both have been employed to deliver siRNAs. Viral vectors are effective, but they pose safety concerns and the high cost of their development makes them unfavorable for further use. Nonviral vectors endure the advantage of being noninfectious, less immunogenic, and economically viable [10].

In past years, lipid-based nonviral vectors [11] have shown success in causing efficacious RNA interference using siRNAs and a few of them are under clinical trials [12]. Recently, Shapiro and colleagues [9, 13] demonstrated that a new class of lipids known as cationic "bolaamphiphiles" or "bola lipids" could efficiently deliver siRNAs [9, 13], RNA–DNA hybrids [14], and RNA Nanostructures [15]. Typically, the bola lipids that we investigate are amphiphilic molecules composed of a hydrophobic alkyl chain connected at both ends with positively charged hydrophilic acetyl choline head groups (ACh-HG). Bola lipids have the tendency to form micelles [9, 13] or vesicles [16, 17], by presenting their positively charged head groups toward the exterior, thus providing opportunity to condense negatively charged nucleic acid molecules [9, 13, 16, 17].

In the present chapter, we describe bolaamphiphile mediated delivery of siRNAs against the green fluorescent protein (GFP) in human breast cancer cells (MDA-MB-231) as a model example [9]. In the presented study, Dicer substrate siRNAs (dsiRNAs) are used as a precursor to generate siRNAs via Dicer processing because dsiRNAs have been shown to be more potent than classical synthetic 21-mer siRNAs for efficient assembly of the silencing complex, RISC [18, 19]. Bola lipids form stable micelles, interact to bind nucleic acids, protect nucleic acids against nuclease degradation, and promote intracellular delivery of nucleic acids. GFP-siRNAs delivered via bola lipids showed significant silencing of GFP in vitro with no toxicity at the effective silencing dose [9, 13]. Bolas potential for in vivo delivery of siRNAs was examined in MDA-MB-231 tumor-bearing athymic nude mice. Fluorescently labeled dsiRNAs complexed with bolas upon systemic administration demonstrated efficiently higher uptake of siRNAs in tumor in comparison to the other major organs [13]. In this chapter, we do not discuss the in vivo experiment since it showed only uptake of siRNAs and did not demonstrate silencing of genes via siRNAs. The methods described below for evaluating the biophysical and biological characteristics of bola micelles and bola micelle–nucleic acid complexes can be extended to evaluate the characteristics of bola vesicles and bola vesicles–nucleic acid complexes as well. Readers are requested to *see* the references for preparing bola vesicles [16, 17, 20]. Our results suggest that the bolaamphiphiles are effective and safe nonviral delivery vectors for siRNAs.

2 Materials

1. Synthesized bola lipids GLH-19, GLH-20 [13], GLH-58, and GLH-60 [9] are used (*see* Fig. 1) (*see* **Note 1**). Store the bolas at −20 °C.
2. Sense and antisense strands for Dicer substrates siRNAs (dsiRNAs) designed against GFP, fluorescently labeled DNA sequences (*see* **Notes 2–5**) for RNA–DNA hybrid duplexes and quenched duplexes (Integrated DNA Technologies) that have resemblance to the GFP-dsiRNAs are listed in Table 1. Store all the nucleic acids at −20 °C.
3. Nuclease/protease-free water (Quality Biological Inc.); store desired amounts of aliquots at −20 °C.

2.1 Preparation of Bola Films

1. Synthesized bola lipids GLH-19, GLH-20, GLH-58 and GLH-60.
2. Chloroform ($CHCl_3$) bottle.
3. Nitrogen cylinder or rotavap (Heidolph).
4. Desiccator.

2.2 Preparation of Bola Micelles

1. 1 mg of aliquoted bolas films.
2. 1 mL of nuclease/protease-free water.
3. Vortexer or vortex mixer.

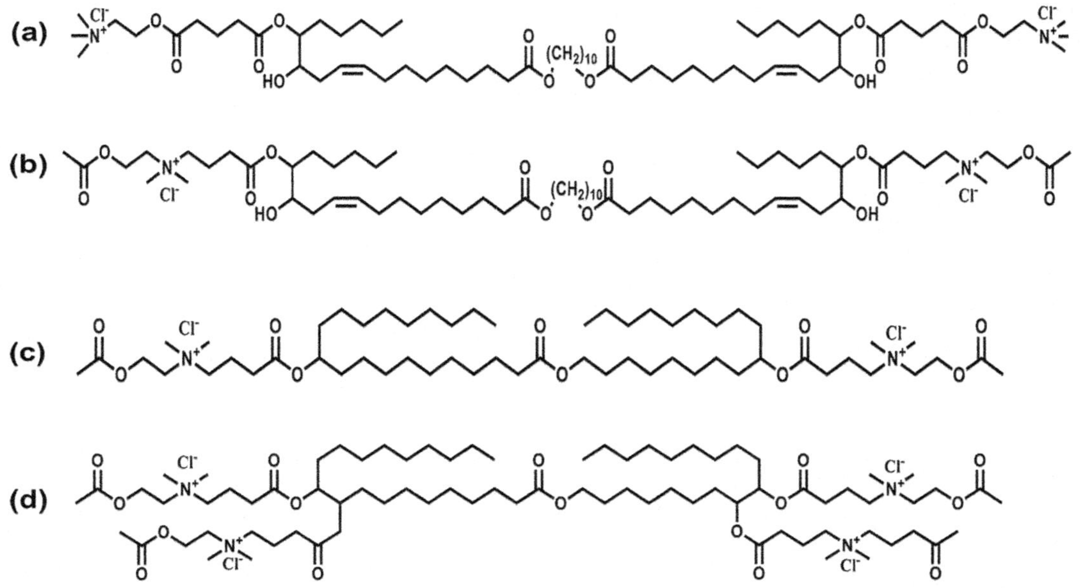

Fig. 1 Chemical structures of bolas with acetyl choline head groups (Ach-HG), (**a**) GLH-19, (**b**) GLH-20, (**c**) GLH-58, and (**d**) GLH-60

Table 1
Nucleic acids used in the study

Type of nucleic acids	Sequence
Dicer Substrate siRNA	Sense: 5′-pACCCUGAAGUUCAUCUGCACCACCG* Antisense: 5′-CGGUGGUGCAGAUGAACUUCAGGGUCA
Fluorescently labeled DNA	DNA antisense_Alexa488: 5′-CGGTGGTGCAGATGAACTTCAGGGTCAtt/3AlexF488N
Fluorescently labeled DNA	DNA sense Iowa Black FQ: 5′–/5IAbFQ/ ACCCTGAAGTTCATCTGCACCACCG

2.3 Assembly of Nucleic Acid Duplexes

1. Manufacturer supplied sense and antisense strands of dsiRNAs, fluorescent RNA–DNA hybrids, and DNA duplexes.
2. Assembly buffer: 50 mM NaCl and 2 mM $MgCl_2$. Store at room temperature.
3. Water bath.

2.4 Preparation of Bola–Nucleic Acid Complexes

1. Freshly resuspended bola micelles in nuclease/protease-free water.
2. Assembled nucleic acids (dsiRNAs or fluorescent RNA–DNA hybrids and DNA duplexes) as per requirement of the experiment.

2.5 Cryo-Electron Microscopy of Bola–DsiRNA Complexes

1. Freshly resuspended bola micelles in nuclease/protease-free water.
2. Assembled dsiRNAs that target the mRNA of GFP or other target genes.
3. Quantifoil Copper 200 mesh R 3.5/1 grids.
4. Liquid ethane using Vitrobot III (FEI).
5. Transmission electron cryo-microscope JEM-2010F (JEOL) with a field emission gun is used in this work. The JEM-2010F capable of imaging at ultrahigh resolution of 0.17 nm with a $\pm 25°$ tilt stage. The instrument operates at 200 kV with a Gatan cryo-holder (model 626) (Gatan). Images are recorded on DE-12, a 12.6 megapixel (3072×4096) Direct Detection Device sensor (Direct Electron LP).

2.6 Size Measurements of Bola–DsiRNA Complexes

1. Freshly resuspended bola micelles in nuclease/protease-free water.
2. Assembled dsiRNAs against GFP.
3. Disposable cuvettes (Malvern Instruments) for measuring size of micelles.

4. Nuclease/protease-free water.
5. Zetasizer Nano ZS (Malvern Instruments) a particle and molecular size/zeta potential analyzer allowing size measurements ranging from 0.3 nm (diameter) to 10 μm is used. This analyzer uses the principle of dynamic light scattering with noninvasive backscatter optics for size measurements and equipped with a He-Ne ion laser of 633 nm.

2.7 Fluorescent Anisotropy Measurement for Binding of Nucleic Acids to Bolas

1. Freshly resuspended bola micelles in nuclease/protease-free water.
2. Assembled fluorescent RNA–DNA hybrid duplexes where DNA antisense is labeled with Alexa 488.
3. Nuclease/protease-free water.
4. 96-well plate (Corning).
5. Microplate reader capable of fluorescence intensity reading. In this work, Tecan Infinite M1000 is used for fluorescence anisotropy measurements. The equipment has provision for flexible wavelength selection, top and bottom fluorescence read, and fast fluorescence polarization.

2.8 Protection of Bound DNA Duplexes to Bolas Against Nucleases

1. Freshly resuspended bola micelles in nuclease/protease-free water.
2. Assembled quenched DNA duplexes. DNA sense is labeled with Iowa black fluorescence quencher and DNA antisense is labeled with Alexa 488.
3. Nuclease/protease-free water.
4. RQ1 DNase buffer (Promega). Store in 100 μL aliquots at −20 °C.
5. RQ1 RNase-free DNase (Promega). Store in 10 μL aliquots at −20 °C.
6. Quartz cuvettes of 3 mL (Hellma).
7. Scanning Thermocouple Thermometer (Cole-Parmer).
8. Spectrofluorometer. In this work, Fluoromax-3 fluorimeter (Jobin-Yon, Horiba Scientific) with excitation and emission monochromators, data acquisition DataMax software and real time display application is used. The instrument is equipped with a Peltier sample temperature controller, a magnetic stirrer, and a water bath.

2.9 In Vitro Transfection of Bola–Nucleic Acids in Cells

1. Human breast cancer cells MDA-MB-231 (ATCC) and the corresponding stably GFP-expressing strain MDA-MB-231/GFP (Cell Biolabs Inc.)
2. Dulbecco's Modified Eagle's Medium (DMEM)-high-glucose. It is a variant of the original formulation provided by Dulbecco, R. and Freeman, G [21]. Store at 4 °C.

3. Fetal bovine serum (FBS). Store at −20 °C.
4. Penicillin–streptomycin 10,000 U/mL. Store at −20 °C.
5. Gibco Opti-MEM reduced serum transfection medium. Opti-MEM is a modification of Eagle's Minimum Essential Medium, buffered with HEPES and sodium bicarbonate. It is supplemented with hypoxanthine, thymidine, sodium pyruvate, L-glutamine, trace elements, and growth factors. Store at 4 °C.
6. Cell dissociation buffer (CDB) from Gibco. CDB is a membrane-filtered, calcium and magnesium-free phosphate-buffered saline (PBS) based enzyme-free isotonic salt solution with chelating, and cell-conditioning agents for dissociating the adherent cells from cell culture flask or plates. Store at room temperature.
7. Incubator (Thermo Scientific) under humidification and 5% CO_2 at 37 °C.
8. Sorvall Legend T Plus/RT Plus Centrifuge with swinging bucket rotor (Thermo Scientific).

2.9.1 Uptake of Bola–RNA-DNA Hybrid Complexes

1. MDA-MB-231 cells.
2. 12 well plates (Corning).
3. DMEM plus 10% FBS and 1% penicillin–streptomycin (complete media). Store at −4 °C.
4. Opti-MEM.
5. Freshly resuspended bola micelles in nuclease/protease-free water.
6. Assembled fluorescent RNA–DNA hybrids where DNA antisense is labeled with Alexa 488.
7. CDB.
8. Flow cytometer or fluorescence assisted cell sorter (FACS) with data acquisition software. This work is based on an instrument from BD Biosciences FACSCalibur™ using the Cell Quest software. The instrument is equipped with air-cooled argon ion laser of 488 nm and red diode laser of 635 nm.

2.9.2 Silencing of GFP by Bola–DsiRNA Complexes

1. MDA-MB-231/GFP cells.
2. 12-well plates (Corning).
3. Complete media.
4. Opti-MEM.
5. Freshly resuspended bola micelles in nuclease/protease-free water.
6. Assembled dsiRNAs against GFP.
7. CDB.
8. Flow cytometer or FACS.

2.10 Cell Viability on Exposure to Bola–DsiRNA Complexes

1. MDA-MB-231 cells.
2. 96-well plates (Corning).
3. Complete media.
4. CellTiter-Blue reagent (Promega). Store at −20 °C.
5. Freshly resuspended bola micelles in nuclease/protease-free water.
6. Assembled dsiRNAs against GFP.
7. UV–Vis spectrophotometer, SpectraMAX2 (Molecular Devices).

3 Methods

Use proper controls in all the experiments as required. Bola micelle solutions, nucleic acid solutions, and nuclease protease-free water are used at room temperature.

3.1 Preparation of Bola Films

1. Weigh each bola lipid, GLH-19 10 mg, GLH-20 10 mg, GLH-58 10 mg, and GLH-60 10 mg in glass vials.
2. Add 1 mL $CHCl_3$ to each vial of bolas and cap the vial tightly.
3. Vortex each bola vial containing $CHCl_3$ for 5 s to dissolve the bolas. Each bola lipid vial corresponds to 10 mg/mL.
4. Aliquot 100 μL of bola–$CHCl_3$ solution from each bola vial to make ten vials for each bola.
5. Each aliquoted bola vial corresponds to 1 mg.
6. Evaporate the $CHCl_3$ under stream of nitrogen or rotavap (Heidolph) while rotating to make thin well distributed films along the walls of the vials.
7. Desiccate the 1 mg films for each bola overnight for residual removal of $CHCl_3$. Store the prepared films at −20 °C.

3.2 Preparation of Bola Micelles

1. Add 1 mL of nuclease/protease-free water to 1 mg dried bola films at room temperature.
2. Vortex the mixture vigorously for time intervals of 1 min until you see complete dispersion of the film into the water (*see* **Note 6**).
3. Finally, make sure that all of the bola film mixed with water and well dispersed upon vortexing to form micelles. Use these micelles either fresh or within a week when stored at 4 °C. Bola micelles are shown in Fig. 2.

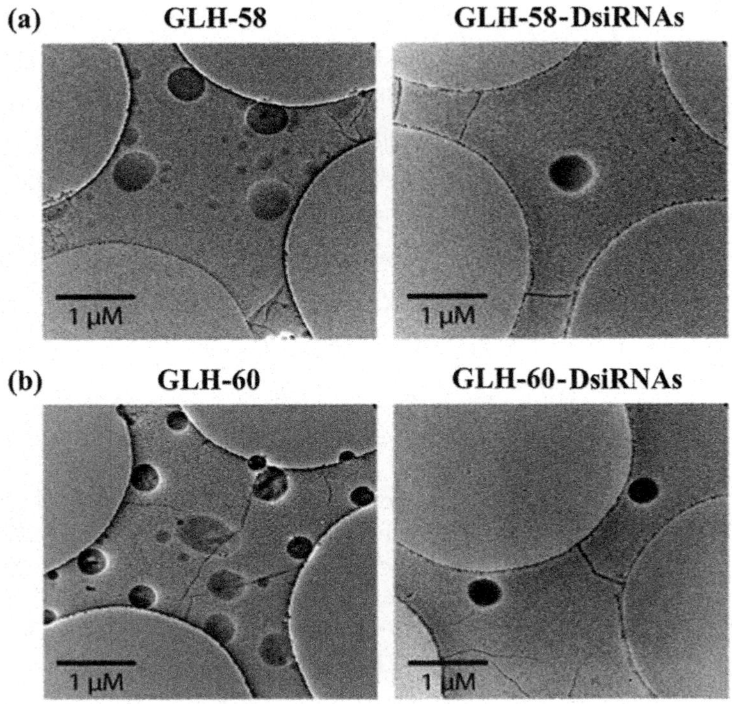

Fig. 2 Cryo-electron microscopy images to show micelles of GLH-58 and GLH-60 and GLH-58–DsiRNA and GLH-60–DsiRNAs complexes (reproduced from Gupta et al. 2015 [9] with permission from Elsevier)

3.3 Assembly of Nucleic Acid Duplexes

1. Use supplied individual nucleic acid sense and antisense strands (*see* **Note 7**). Take appropriate concentrations of these sense and antisense strands to prepare duplexes of either RNA–DNA hybrids or DNAs or dsiRNAs in assembly buffer to get 10 μM stock solutions.
2. Preheat the hot dry bath to 90 °C and heat the duplexes in assembly buffer for 1 min (*see* **Note 8**).
3. Remove the duplexes from the hot dry bath and let them anneal at room temperature (*see* **Note 9**).

3.4 Preparation of Bola–Nucleic Acid Complexes

1. Take appropriate volumes in μL from bola micelle solution (1 mg/mL) to make 1–100 μg/mL final concentration according to the requirement of the experiments.
2. According to the requirement of experiments gently mix appropriate volumes (μL) of nucleic acids to the bola micelles such that it matches the final concentration (1–100 μg/mL) for the experiments.
3. Incubate the mixture of bola and nucleic acids for 30 min at room temperature for bola–nucleic acid complex formation or micelleplexes formation.

4. Add appropriate amount of nuclease/protease-free water, Opti-MEM or DMEM plus 10% FBS, and 1% penicillin/streptomycin (complete media) to achieve final concentration of bola and nucleic acids according to the requirement of the experiments (*see* Subheadings 3.4–3.8).

3.5 Cryo-Electron Microscopy of Bola–DsiRNA complexes

1. Quantifoil Copper 200 mesh R 3.5/1 grids are washed overnight with acetone.
2. Gently mix 2.5 μL of dsiRNAs against GFP aqueous solution (10 μM) to 2.5 μL (1 mg/mL) of bola micelles. Incubate the mixture for 30 min at room temperature.
3. Add appropriate amount of nuclease/protease-free water to achieve a final volume of 500 μL. The final concentration of dsiRNAs are 50 nM and of bola micelles is 5 μg/mL.
4. Prepare a frozen, hydrated grid, by adding 2.5 μL of bolas alone or with dsiRNAs to the grid, blotted, and plunged into liquid ethane using Vitrobot III.
5. Images are collected at liquid nitrogen temperature (~100 K) on a transmission electron cryo-microscope equipped with a field emission gun.
6. Samples are imaged at 13,900× effective magnifications targeted at 3–4 μm under focus.
7. Total specimen exposure for each image of 30 e^{-}/Å2 s is used. Micelles of bolas alone and bola–dsiRNA complexes are presented in Fig. 2.

3.6 Size Measurements of Bola–DsiRNA Complexes

1. Gently mix 2.5 μL of dsiRNAs against GFP aqueous solution (10 μM) to 0.5, 2.5, 5 and 10 μL (1 mg/mL) of bola micelles. Incubate the mixture for 30 min at room temperature.
2. Add appropriate amount of nuclease/protease-free water to achieve final volume of 500 μL. The final concentrations of dsiRNAs are 50 nM and of bola micelles are 1, 5, 10, and 20 μg/mL.
3. Add bola–dsiRNA complexes of different concentrations as prepared above to the cuvettes for size measurements at room temperature. An example result is shown in Fig. 3.

3.7 Fluorescent Anisotropy Measurements for Binding Nucleic Acids to Bolas

1. Gently mix 4 μL of Alexa 488 labeled RNA–DNA hybrids duplexes aqueous solution (10 μM) to 0.4, 0.8, 1.2, 1.6, 2.0, 2.4, 2.8, 3.2, 3.6, and 4.0 μL (1 mg/mL) of bola micelles.
2. Incubate the mixture for 30 min at room temperature.
3. Add appropriate amount of nuclease/protease-free water to achieve final volume of 200 μL. The final concentrations of Alexa 488 RNA–DNA hybrids are 200 nM and of bola micelles are 2, 4, 6, 8, 10, 12, 14, 16, 18, and 20 μg/mL.

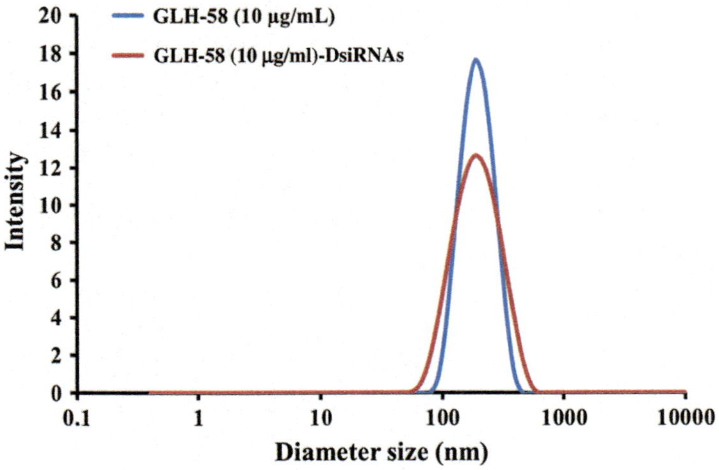

Fig. 3 Size measurements of GLH-58 alone and GLH-58–DsiRNA complexes. (reproduced from Gupta et al. 2015 [9] with permission from Elsevier)

4. Transfer the 200 µL each of the above prepared bola–hybrid complexes to 96-well plates (*see* **Note 10**).
5. Shake the plate for 1 min and read it for increment changes in fluorescent anisotropy values with respect to control hybrid duplexes for Alexa 488 (λ_{ex} 470, λ_{em} 517 nm) using Tecan Infinite M1000.
6. The obtained data is fitted using the Hill eq.

$$R = R_0 + \Delta R_{max}[\text{GLH}]^n / K_d{}^n + [\text{GLH}]^n,$$

where anisotropy is denoted as R, anisotropy of the RNA–DNA hybrids alone in solution is represented as R_0, the difference in anisotropy of the hybrids in the bola–hybrid complexes versus hybrids alone in solution is denoted as ΔR_{max}, bolaamphiphile concentration is represented as [GLH], dissociation constant is denoted as K_d and n is the representation of Hill coefficient. The fitting can be performed using a least square method available in many data analysis software packages, herein sigmaplot software is used. Fluorescent anisotropy measurements (*see* **Note 11**) of bolas with hybrid duplexes are shown in Fig. 4.

3.8 Protection of Bound DNA Duplexes to Bolas Against Nucleases

1. Gently mix 1 µL of quenched DNA duplexes labeled with Alexa 488 and Iowa fluorescence black quencher aqueous solution (10 µM) to 10 µL (1 mg/mL) of bola micelles (*see* **Note 12**).
2. Incubate the mixture for 30 min at room temperature.
3. Add 189 µL of 10× RQ1 DNase buffer to make it 200 µL final. Add appropriate amount of nuclease/protease-free water to achieve final volume of 2000 µL. The final concentrations of

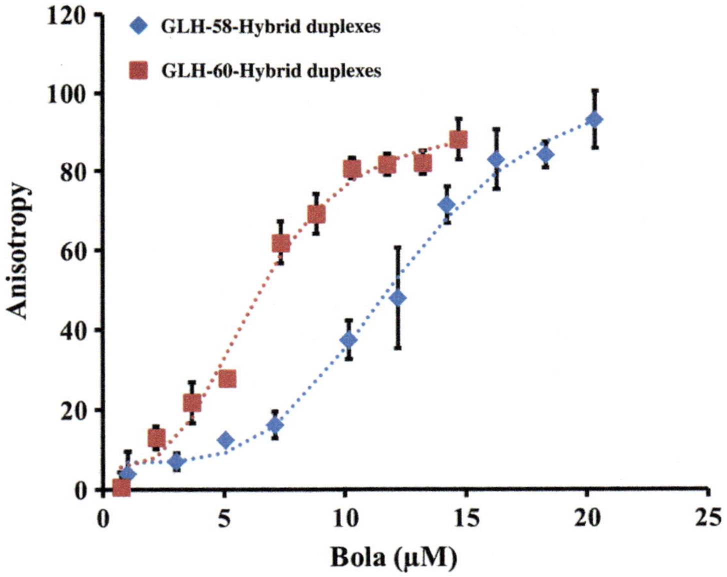

Fig. 4 Fluorescent anisotropy measurements to show binding of nucleic acids using fluorescent RNA–DNA hybrid duplexes to bolas GLH-58 and GLH-60 (reproduced from Gupta et al. 2015 [9] with permission from Elsevier)

DNA duplexes are 50 nM and of bola micelles are 5 μg/mL in bola–DNA complexes.

4. Preheat the incubation chamber of fluorimeter to 37 °C. Put 2000 μL of bola–DNA duplexes in 3000 μL of quartz cuvette under constant stirring with a magnetic bar in the fluorimeter.
5. Check the temperature of the cuvette in the incubation chamber of fluorimeter using the thermocouple thermometer (*see* **Note 13**).
6. Incubation chamber should reach 37 °C.
7. Set λ_{ex} 460 nm and λ_{em} 520 nm in the fluorimeter to monitor the fluorescence of Alexa 488.
8. Incubate the complexes at 37 °C for 5 min to a get base line.
9. Add RQ1 RNase-free DNase to the bola–DNA complexes solution according to the manufacturer's protocol.
10. The DNA degradation is followed as dequenching of Alexa 488 upon DNase digestion upon increment in the arbitrary fluorescence units keeping excitation slit width at 2 nm. The dequenching is recorded at every 30 s for more than 1000 s. An example result is shown in Fig. 5.

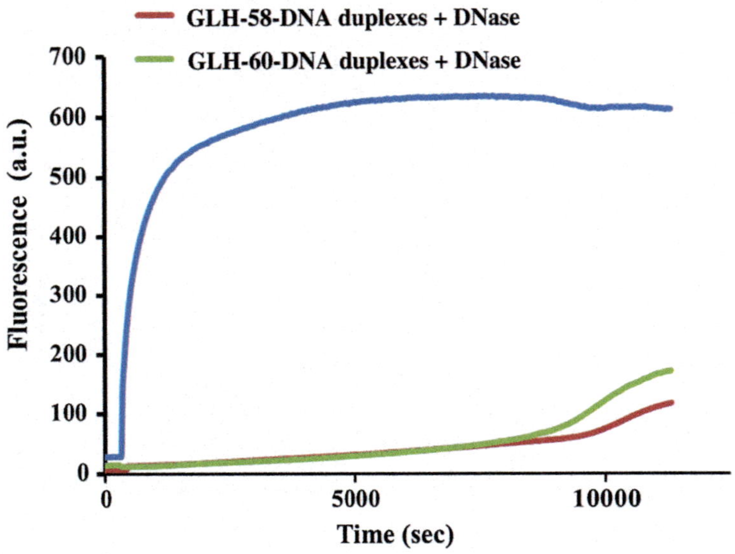

Fig. 5 Degradation of complexed quenched DNA duplexes with bolas GLH-58 and GLH-60 upon digestion with DNase (reproduced from Gupta et al. 2015 [9] with permission from Elsevier)

3.9 In Vitro Transfection of Bola–Nucleic Acid Complexes

Herein, is an example of transfection of nucleic acids for uptake in MDA-MB-231 cells and for silencing in MDA-MB-231/GFP cells mediated by bola lipids. The cells are grown under cell culture conditions of humidified 5% CO_2 incubator at 37 °C in DMEM supplemented with 10% FBS and 1% penicillin/streptomycin (complete media, store at 4 °C). 12-well plates are used for the experiments (*see* **Notes 10** and **14**).

3.9.1 Uptake of Bola–Alexa 488 RNA–DNA Hybrid Complexes

1. One day before the transfection, seed MDA-MB-231 cells at a density of 100,000 cells/well in 12-well plates in 1 mL of complete media in the incubator.
2. Mix gently 5 μL of RNA–DNA hybrid duplexes (10 μM) to 1, 5, and 10 μL of bola micelles (1 mg/mL).
3. Incubate the mix for 30 min at room temperature.
4. Before transfection add appropriate amount of Opti-MEM to achieve final volume of 1 mL. The final concentration of RNA–DNA hybrid duplexes is 50 nM. The final concentrations of bola lipids are 1, 5, and 10 μg/mL in bola–hybrid complexes.
5. Wash each well of the cells with 1 mL of prewarmed PBS at 37 °C.
6. Incubate the cells at 37 °C for 4 h with abovementioned concentrations of bola–hybrid complexes in Opti-MEM at above mentioned concentrations.

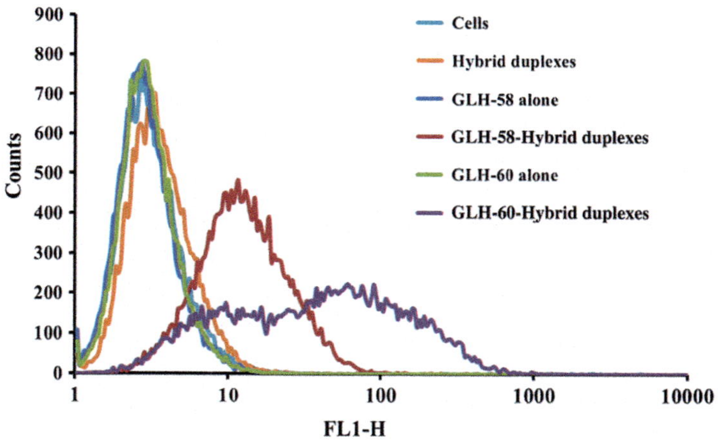

Fig. 6 Uptake of fluorescent RNA–DNA hybrid duplexes mediated by bolas GLH-58 and GLH-60 in human breast cancer (MDA-MB-231) cells (reproduced from Gupta et al. 2015 [9] with permission from Elsevier)

7. After incubation for 4 h at 37 °C, wash the cells with PBS two times.
8. Replace the Opti-MEM with 1 mL prewarmed complete media at 37 °C.
9. Incubate the cells with bola–hybrid duplex complexes for another 20 h at 37 °C.
10. After 20 h of incubation dissociate the cells with CDB.
11. Wash cells two times with CDB for 5 min each at 4 °C while centrifuging them at 484 × *g*.
12. Resuspend the cells with 500 μL of CDB and keep the samples in ice in ice bucket.
13. Read the fluorescence of the hybrid duplexes associated with cells using the flow cytometer or FACS using a 488 nm laser (*see* **Note 15**). Analyze the data using the Cell Quest data acquisition software. *See* Fig. 6 for an example result.

3.9.2 Silencing of GFP by Bola–DsiRNA Complexes

1. One day before the experiment, seed MDA-MB-231/GFP at a density of 30,000 cells/well in 12-well plates in 1 mL of complete media in the incubator.
2. Mix gently 5 μL of dsiRNAs (10 μM) to 1, 5, and 10 μL of bola micelles (1 mg/mL). Incubate the mix for 30 min at room temperature.
3. Before the transfection add appropriate amount of Opti-MEM to achieve a final volume of 1 mL. The final concentration of dsiRNAs duplexes is 50 nM. The final concentration of bola lipids are 1, 5, and 10 μg/mL in bola–hybrid duplex complexes.

4. Wash each well of the cells with 1 mL of prewarmed PBS (37 °C).
5. Incubate the cells at 37 °C for 4 h with bola–dsiRNA complexes in Opti-MEM at abovementioned concentrations.
6. After 4 h of incubation at 37 °C, wash the cells with prewarmed PBS (37 °C) two times.
7. Replace the Opti-MEM with 1 mL of prewarmed complete media (37 °C).
8. Incubate the cells with bola–dsiRNAs duplex complexes for another 68 h.
9. After 68 h of incubation dissociate the cells with CDB.
10. Then wash the cells with CDB two times for 5 min at 4 °C while centrifuging them at 484 × *g*.
11. Resuspend the cells in 500 μL of CDB and keep the samples in ice in ice bucket.
12. Analyse the samples for a decrease in fluorescence as a measure of silencing with the flow cytometer or FACS using a 488 nm laser (*see* **Note 16**). Analyze the data using the data acquisition Cell Quest software. *See* Fig. 7. for result showing silencing of GFP by bola–dsiRNA complexes or micelleplexes.

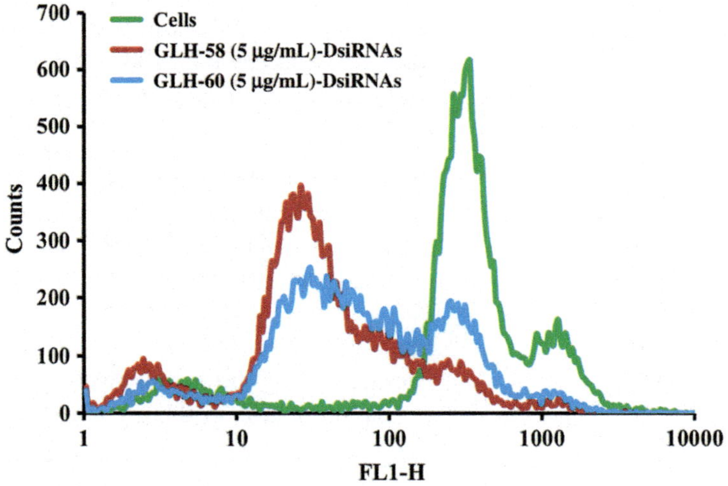

Fig. 7 Silencing of GFP by GLH-58–DsiRNA and GLH-60 complexes in GFP expressing human breast cancer (MDA-MB-231/GFP) cells (reproduced from Gupta et al. 2015 [9] with permission from Elsevier)

3.10 Cell Viability

1. One day before the viability assay, seed MDA-MB-231 cells at a density of 20,000 cells/well in 96-well plates in 200 μL complete media (37 °C) in the incubator.
2. Prepare bola–dsiRNA duplex complexes of different concentrations. Mix gently 1 μL of dsiRNAs duplexes (10 μM) to the 0.2, 1, 2, 4, 6, 10, 16, and 20 μL of bola micelles (1 mg/mL). Then add appropriate volumes of prewarmed complete media (37 °C) to obtain a final volume of 200 μL. The final concentrations of dsiRNAs are 50 nM and of bola lipids are 1, 5, 10, 20, 30, 50, 80, and 100 μg/mL.
3. Wash the cells with prewarmed PBS at 37 °C two times.
4. Incubate the cells with of bola (1–100 μg/mL)-dsiRNA (50 nM) complexes for 24 h at 37 °C.
5. After 24 h of incubation wash the cells with prewarmed PBS at 37 °C.
6. Then add CellTiter-Blue in complete media according to the manufacturer's protocol. Incubate the cells for additional 4 h for the conversion of resofurin to reazurin by viable cells (*see* **Note 17**).
7. Read the fluorescence of the reazurin after shaking the plate for 1 min at λ_{ex} 560 nm and λ_{em} 590 nm with an auto cutoff in a UV–Vis spectrophotometer equipped with the plate holder.
8. Calculate the percentage of viable cells using the following equation:

$$\% \text{ Cell Viability} = (F' - F_0) \times 100/(F_c - F_0),$$

where F' are the fluorescence values of the samples, F_c are the fluorescence values of control cells and F_0 are the fluorescence values of the CellTiter-Blue alone in DMEM. Cell viability on exposure to bolas and bola–dsiRNA complexes are shown in Fig. 8.

4 Notes

1. Bola lipids GLH-19 and GLH-20 are synthesized from veronia oil [17] and GLH-58 and GLH-60 are synthesized by using jojoba oil [9]. The products obtained from each step of synthesis are characterized by FT-IR, NMR techniques to get the desired structures of bolas. GLH-19 and GLH-20 contain identical hydrophobic domains with acetylcholine head groups (ACh-HGs). However, these bolas differ in linking the alkyl chain to ACh either via the acetyl group (GLH-19), or the nitrogen atom (GLH-20) of the choline moiety. GLH-58 and 60 has similar hydrophobic domains but differ in containing number of positively charged ACh-HGs. GLH-58 contains

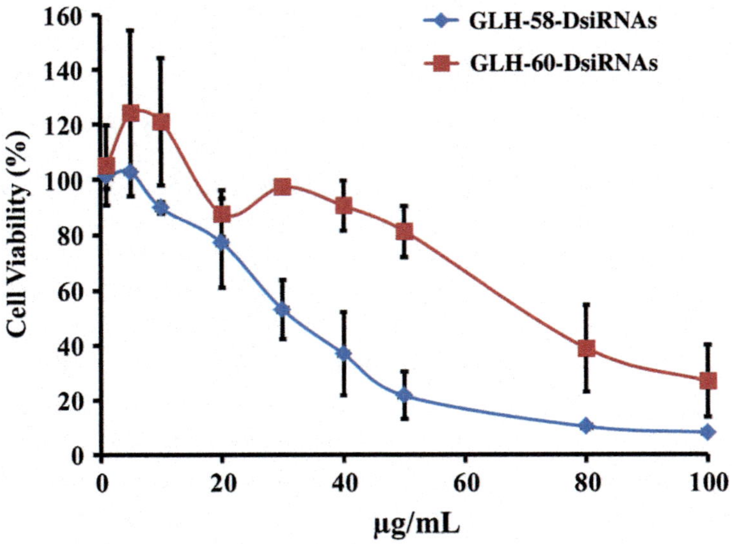

Fig. 8 Cell viability of human breast cancer (MDA-MB-231/GFP) cells on exposure to GLH-58–DsiRNA and GLH-60–DsiRNA complexes (reproduced from Gupta et al. 2015 [9] with permission from Elsevier)

one ACh-HG whereas GLH-60 contains two ACh-HGs at each end of the hydrophobic domain. Mol. Wt. of GLH-58 is 983.45 and of GLH-60 is 1359.87 (*see* Fig. 1). Letters G, L, and H in nomenclature of bolas are the last names of the inventors of the above bola lipids, i.e., Dr. Sarina Grinberg, Dr. Charles Linder, and Dr. Eliahu Heldman of Ben-Gurian University of the Negev, Beer Sheva, Israel.

2. Rossi and coworkers deduced that siRNAs of length 25–27 mers have higher efficacy than 21-mer siRNAs that are generated from the larger RNA duplexes [18]. In this work, we have used dsiRNA duplexes of 25 mer sense and 27 mer antisense strands designed against GFP [9, 22, 23]. Fluorescent DNA sense (25 mer) and antisense strands (27 mer) for RNA–DNA hybrid and DNA duplexes are designed on the basis dsiRNA sense and antisense sequences [9, 13].

3. Fluorescent DNA sequences are used for experiments. Firstly, it is more cost-effective to fluorescently label DNAs instead of dsiRNAs and secondly, DNAs and RNAs differ only in the presence of hydroxyl in the sugar moiety so presumably the fundamental interactions between the DNAs and bolas would be very similar to that of the siRNAs and the corresponding bolas. Moreover, our previous MD simulation studies confirmed that hydrogen bond interactions between the bolas and the hydroxyls in the siRNAs are minor [13].

4. Alexa 488 is tagged to 3′-end of the DNA antisense and the Iowa Black fluorescence quencher is tagged to 5′-end of the DNA sense strands to form quenched duplexes [9, 13].
5. RNA–DNA hybrids with fluorescently labeled DNAs are used. In these hybrids, 3′-end of the DNA antisense strands are fluorescently tagged with Alexa 488 and base-paired with the sense dsiRNA strands [9, 13].
6. While preparing bola micelles if some portion of the film could not mix with water during vortexing, scratch that portion of film with either glass pipette or spatula. Ensure that all the lipid film is in water otherwise there will be variation for bola concentration in experiments.
7. All the fluorescently labeled nucleic acid duplexes are sensitive to light. Therefore, avoid exposure to light while performing experiments. Keep the nucleic acids in dark (wrapped or in aluminum foil). Such nucleic acids are used in Subheadings 3.5–3.7 (Subheading 3.7, **step 1**).
8. Check the temperature of the hot dry bath always before doing the experiment with the laboratory thermometer to make sure it is at 90 °C.
9. Either use the freshly prepared nucleic acids or store at −20 °C. For all the experiments use the nucleic acids at room temperature.
10. Smaller or larger well size plates can be used for the experiments for Subheading 3.5, 3.7, and 3.8. However, experimentalist have to adjust the concentrations of the bolas and nucleic acids. The recommended well size plates are optimal for the experiments discussed.
11. Changes in the anisotropy are due to the changes of the mobility of the fluorophore upon binding with the bolas.
12. This experiment is based on Förster or fluorescence resonance energy transfer (FRET) where energy transfer between two light-sensitive molecules (chromophores) takes place. A donor chromophore, initially in its electronic excited state, may transfer energy to an acceptor chromophore within a certain distance of each other. Herein, the DNA duplexes Alexa 488 (one strand) and Iowa black quencher (another strand) are in close proximity to each other which caused quenching of the duplexes. Degradation of the duplexes by DNase relieves the quenching of the duplexes and the fluorescence of Alexa 488 is monitored over time.
13. Keep the temperature of the water bath of the fluorimeter 1–2 °C higher than 37 °C as there may be a loss of heat before reaching the incubation chamber in the fluorimeter to maintain 37 °C. Always check the temperature of the cuvette

solution with a thermocouple thermometer to make sure that it has reached to 37 °C before the experiments.

14. In experimental Subheadings 3.7 and 3.8 where cells are being used to evaluate the properties of bola–nucleic acid complexes, one has to take care of the number of cells to start with and the duration of experiment since there is possibility of losing cells during washing steps.
15. Uptake of fluorescent hybrid duplexes in cells is measured as increase in fluorescence intensity with respect to control cells.
16. GFP silencing or knockdown in GFP expressing cells is measured as decrease in the fluorescence intensity of the GFP.
17. Reazurin is a fluorescent product produced by the viable cells. Therefore, avoid light exposure while doing the experiments.

Acknowledgments

This work has been funded in whole or in part with Federal funds from the Frederick National Laboratory for Cancer Research, National Institutes of Health, under Contract No. HHSN261200800001E. This research was supported [in part] by the Intramural Research Program of the NIH, National Cancer Institute, Center for Cancer Research. The content of this publication does not necessarily reflect the views or policies of the Department of Health and Human Services, nor does the mention of trade names, commercial products, or organizations imply endorsement by the US Government. Dr. Kshitij Gupta also acknowledges "Scientist pool officer" position awarded by Council of Scientific and Industrial Research, New Delhi, India and "Young Scientist" award "YSS/2014/000937" by Science and Engineering Research Board, Department of Science & Technology, India.

References

1. Sledz CA, Williams BR (2005) RNA interference in biology and disease. Blood 106 (3):787–794
2. Elbashir SM, Lendeckel W, Tuschl T (2001) RNA interference is mediated by 21- and 22-nucleotide RNAs. Genes Dev 15(2):188–200
3. Afonin KA et al (2015) Triggering of RNA interference with RNA-RNA, RNA-DNA, and DNA-RNA nanoparticles. ACS Nano 9 (1):251–259
4. Liu J et al (2004) Argonaute2 is the catalytic engine of mammalian RNAi. Science 305 (5689):1437–1441
5. Meister G et al (2004) Human Argonaute2 mediates RNA cleavage targeted by miRNAs and siRNAs. Mol Cell 15(2):185–197
6. Valencia-Sanchez MA et al (2006) Control of translation and mRNA degradation by miRNAs and siRNAs. Genes Dev 20(5):515–524
7. Whitehead KA, Langer R, Anderson DG (2009) Knocking down barriers: advances in siRNA delivery. Nat Rev Drug Discov 8 (2):129–138
8. Parlea L et al (2016) Cellular delivery of RNA nanoparticles. ACS Comb Sci 18 (9):527–547

9. Gupta K et al (2015) Bolaamphiphiles as carriers for siRNA delivery: From chemical syntheses to practical applications. J Control Release 213:142–151
10. Zhang S, Zhao Y, Zhi D (2012) Non-viral vectors for the mediation of RNAi. Bioorg Chem 40(1):10–18
11. Zhang S, Zhi D, Huang L (2012) Lipid-based vectors for siRNA delivery. J Drug Target 20 (9):724–735
12. Kanasty R et al (2013) Delivery materials for siRNA therapeutics. Nat Mater 12 (11):967–977
13. Kim T et al (2013) In silico, in vitro, and in vivo studies indicate the potential use of bolaamphiphiles for therapeutic siRNAs delivery. Mol Ther Nucleic acids 2:e80
14. Afonin KA et al (2013) Activation of different split functionalities on re-association of RNA-DNA hybrids. Nat Nanotechnol 8 (4):296–304
15. Afonin KA et al (2014) Multifunctional RNA nanoparticles. Nano Lett 14(10):5662–5671
16. Grinberg S et al (2005) Novel cationic amphiphilic derivatives from vernonia oil: synthesis and self-aggregation into bilayer vesicles, nanoparticles, and DNA complexants. Langmuir 21(17):7638–7645
17. Popov M et al (2010) Cationic vesicles from novel bolaamphiphilic compounds. J Liposome Res 20(2):147–159
18. Kim DH et al (2005) Synthetic dsRNA Dicer substrates enhance RNAi potency and efficacy. Nat Biotechnol 23(2):222–226
19. Rose SD et al (2005) Functional polarity is introduced by Dicer processing of short substrate RNAs. Nucleic Acids Res 33 (13):4140–4156
20. Popov M et al (2013) Delivery of analgesic peptides to the brain by nano-sized bolaamphiphilic vesicles made of monolayer membranes. Eur J Pharm Biopharm 85(3 Pt A):381–389
21. Dulbecco R, Freeman G (1959) Plaque production by the polyoma virus. Virology 8 (3):396–397
22. Bindewald E et al (2016) Multistrand structure prediction of nucleic acid assemblies and design of RNA switches. Nano Lett 16(3):1726–1735
23. Afonin KA et al (2011) Design and self-assembly of siRNA-functionalized RNA nanoparticles for use in automated nanomedicine. Nat Protoc 6(12):2022–2034

Chapter 13

Preparation and Optimization of Lipid-Like Nanoparticles for mRNA Delivery

Bin Li and Yizhou Dong

Abstract

Lipid-like nanoparticles (LLNs) have shown great promise for nucleic acid delivery. Recently, we have developed N^1,N^3,N^5-tris(2-aminoethyl)benzene-1,3,5-tricarboxamide (TT) derived lipid-like compounds, formulated them into TT LLNs for mRNA delivery, and applied an orthogonal array design to facilitate formulation optimization. This chapter focuses on the following contents relevant to lipid-like nanoparticles: formulation method, particle characterization, orthogonal array design, and in vitro assays.

Key words Lipid-like nanoparticles (LLNs), Orthogonal array design, Optimization, Delivery efficiency, mRNA delivery

1 Introduction

Lipid-like compounds are termed lipidoids and have been applied to formulate lipid-like nanoparticles (LLNs) [1]. In the past decade, extensive studies have investigated a wide variety of LLNs and their broad applications including delivery of siRNA, plasmid DNA and mRNA, in various cell lines such as endothelial cells, macrophage and tumor cells [1–33]. Some of them possess enormous potential to be developed for treating diverse diseases [27, 28, 33]. Nevertheless, to maximize therapeutic benefits such as efficacy and safety in humans, new LLNs are in great demand. To this end, we have recently designed and synthesized N^1,N^3,N^5-tris (2-aminoethyl)benzene-1,3,5-tricarboxamide (TT) derived lipid-like compounds [25]. Then, we formulated them into TT LLNs for mRNA delivery based on our previous experiences [16]. LLNs were self-assembled using multiple components including lipid-like compounds, phospholipid, cholesterol, pegylated lipid, as well as the payload, mRNA. In order to maximize the tansfection efficiency of this complex system, we employed an orthogonal array design, which was widely used to optimize parameters in the field of chemical engineering (Fig. 1) [34–37]. This approach enabled us

Eckart Bindewald and Bruce A. Shapiro (eds.), *RNA Nanostructures: Methods and Protocols*, Methods in Molecular Biology, vol. 1632, DOI 10.1007/978-1-4939-7138-1_13, © Springer Science+Business Media LLC 2017

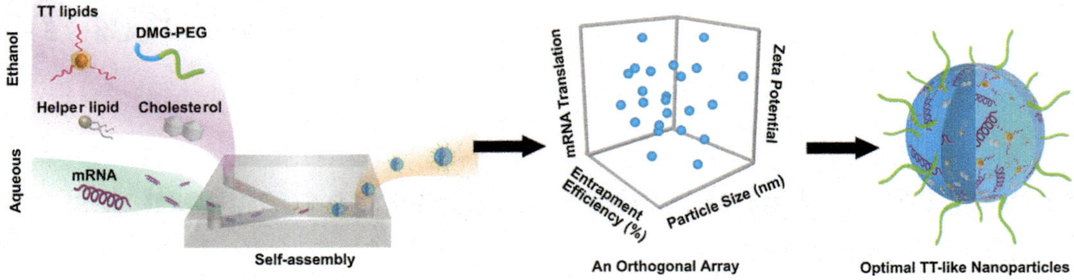

Fig. 1 A schematic illustration of preparation and optimization of lipid-like nanoparticles for mRNA delivery (reproduced from ref. 25 with permission from (2015) American Chemical Society)

to efficiently elucidate the impact of each formulation component and identify the top-performing formulation ratios with significantly reduced experimental volume. Further study indicates optimized LLNs mainly accumulate in the liver and spleen. Moreover, LLNs encapsulated with Factor IX mRNA are capable of producing a therapeutically relevant level of hFIX in both wild-type and FIX-knockout without obvious acute toxicity [25]. In the following sections, we will describe in details the experimental materials and procedures for preparing and evaluating mRNA-loaded lipid-like nanoparticles.

2 Materials

All buffers and solutions are prepared using nuclease-free water unless otherwise noted.

2.1 Preparation of Nanoparticles

1. Nuclease-free pipette tips and tubes.
2. Ethanol, 200 proof.
3. Lipid-like compounds including but not limited to TT derivates (synthesized in house according to the ref. 25). Dissolve a certain amount of lipid-like compound in ethanol to get the desired concentration (Typically, 2 mg/mL for in vitro study; 10 mg/mL for in vivo study). Keep at 4 °C.
4. Phospholipids stock solution including but not limited to 1,2-distearoyl-*sn*-glycero-3-phosphocholine (DSPC), 2-dioleoyl-*sn*-glycero-3-phosphoethanolamine (DOPE) and 1-palmitoyl-2-oleoyl-*sn*-glycero-3-phosphoethanolamine (POPE) (Avanti Polar Lipids). Dissolve a certain amount of phospholipid in ethanol to get the desired concentration (Typically, 2 mg/mL for in vitro study; 10 mg/mL for in vivo study). Keep at 4 °C.
5. Cholesterol stock solution. Dissolve a certain amount of cholesterol in ethanol to get the desired concentration (Typically, 2 mg/mL for in vitro study; 15 mg/mL for in vivo study). Keep at 4 °C.

6. DMG-PEG$_{2000}$ stock solution. Dissolve a certain amount of DMG-PEG$_{2000}$ in ethanol to get the desired concentration (Typically, 0.4 mg/mL for in vitro study; 4 mg/mL for in vivo study). Keep at 4 °C.
7. mRNAs including but not limited to firefly luciferase mRNAs (FLuc mRNA, TriLink Biotechnologies). Store at −80 °C in small aliquots.
8. Nuclease-free water.
9. 10 mM citrate buffer. Mix 41 mL of 10 mM citrate acid with 9 mL of 10 mM trisodium citrate.
10. 0.1 or 0.3 mL syringe.
11. Microfluidic mixer. LLNs used for in vivo study are prepared by fast mixing of formulation components via a microfluidic mixing device produced by Precision NanoSystems.

2.2 Evaluation of Nanoparticles

1. Ultrapure water with resistivity greater or equal 18.2 MΩ cm, obtainable from a water purification system such as the Milli-Q water purification system (Millipore). This is also referred to as Milli-Q water.
2. 96-well plates, tissue-culture treated.
3. 96-well opaque white plates.
4. Eagle's Minimum Essential Medium.
5. Heat-inactivated fetal bovine serum (FBS).
6. 0.05% trypsin–EDTA.
7. Quant-iT RiboGreen RNA reagent (Life Technologies).
8. Bright-Glo luciferase assay substrate (Promega; Store small aliquots at −20 °C in darkness).
9. Tris–EDTA solution (10 mM Tris–HCl, 1 mM EDTA).
10. 1% Triton X-100. Transfer 10 μL of Triton X-100 to 990 μL of Tris–EDTA buffer.
11. MTT (3-(4,5-dimethylthiazol-2-yl)-2,5-diphenyltetrazolium bromide) stock solution. Dissolve 0.5 g MTT in 100 mL water to obtain 5 mg/mL solution. Keep at −80 °C in small aliquots from light.
12. Dimethyl sulfoxide (DMSO).
13. Hep 3B cells.
14. Particle size and zeta potential analyzer. In this work the NanoZS Zetasizer (Malvern) has been used.
15. Microplate reader for fluorescence intensity detection (wavelength range 200–1000 nm). In this work, the SpectraMax M5 plate reader (Molecular Devices) has been used.

16. Device for automated vitrification. In this work, a Vitrobot Mark IV system (FEI) has been used.
17. Transmission electron microscope. In this work, a Tecnai F20S/TEM (FEI) has been used.
18. Cryo-transfer holder compatible with electron microscope. For this work, a Gatan 626 cryotransfer holder (Gatan) has been used.

3 Methods

All procedures are performed at room temperature unless otherwise noted.

3.1 Preparation of mRNA-Loaded Lipid-Like Nanoparticles

1. To prepare x μL (the final concentration of mRNA is y μg/μL) of LLN, dissolve each component (lipid-like compound, phospholipid, cholesterol and DMG-PEG$_{2000}$) except mRNA separately in 200 proof ethanol in individual RNase-free tubes as stock solutions (*see* **Notes 1** and **2**).
2. Combine 10xy μg of lipid-like compound with phospholipids (*see* **Note 3**), cholesterol, and DMG-PEG$_{2000}$ into a 1.5 mL RNase-free tube at molar ratio 50:10:38.5:1.5 according to the ratio from ref. 30.
3. Add $x/40$ μL of 10 mM citrate buffer.
4. Make up to the final desired volume of $x/4$ μL with ethanol. If $x < 1000$ μL and $y < 0.02$ μg/μL, proceed with **step 5a**, **b** of the protocol. Otherwise, proceed with **step 5c–f**.
5. (a) To the above solution, add an equal volume ($x/4$ μL) of mRNA (xy μg) dissolved in 10 mM citrate buffer (pH = 3) by vigorously pipetting the mixture up and down (usually 50–60 times). (b) Add $x/2$ μL of PBS with the same procedure to form homogeneous particles (*see* **Notes 4** and **5**). (c) Transfer the above solution to a 0.1 mL syringe. (d) Afterward, dissolve xy μg of mRNA in $x/4$ μL of 10 mM citrate buffer (pH = 3). (e) Combine it with twofold volume ($x/2$ μL) of PBS and transfer to a 0.1 mL syringe. (f) Particles are then assembled through a microfluidic-based mixing device following the manufacturer's instructions (*see* **Note 6**).
6. Use freshly prepared LLNs for subsequent characterization and transfection, and store the remaining formulation at 4 °C (*see* **Note 7**).

3.2 Characterization of Lipid-Like Nanoparticles

1. Measurement of particle size and zeta potential: disperse 0.3/y μL of LLNs into Milli-Q water to a final volume of 1 mL in disposable cuvette to make a 300 ng/mL working solution (*see* **Note 8**).

2. Determine the particle size and zeta potential of LLNs using a NanoZS Zetasizer after mixing well.
3. Measurement of mRNA entrapment efficiency: Prepare a 500 ng/mL working solution in TE buffer as described above.
4. Add 50 μL of working solution (use TE buffer as blank) to each well of 96-well plate in triplicate, followed by adding 50 μL of TE buffer or 1% Triton X-100 (v/v, in TE buffer) (*see* **Note 9**).
5. To this solution, add 100 μL of 200-fold diluted Quant-iT RiboGreen RNA reagent (diluted in TE buffer).
6. Incubate for 15 min at 37 °C in the dark.
7. Measure the fluorescence on a SpectraMax M5 plate reader (Molecular Devices) with an excitation wavelength of 480 nm and emission wavelength of 520 nm. The entrapment efficiency is determined as $[1 - (A_{TE} - A_{blank})/(A_{1\%\ Triton} - A_{blank})] \times 100$.
8. Cryo-TEM: Add approximate 3 μL of LLNs to a specimen grid.
9. Remove any excess solution.
10. Plunge the grid into liquid ethane using a vitrification device.
11. Transfer under liquid nitrogen to a cryotransfer holder that is loaded to an electron microscope.
12. Record Cryo-TEM images on a postcolumn 1k × 1k CCD camera at a magnification of 18,500× (Fig. 2).

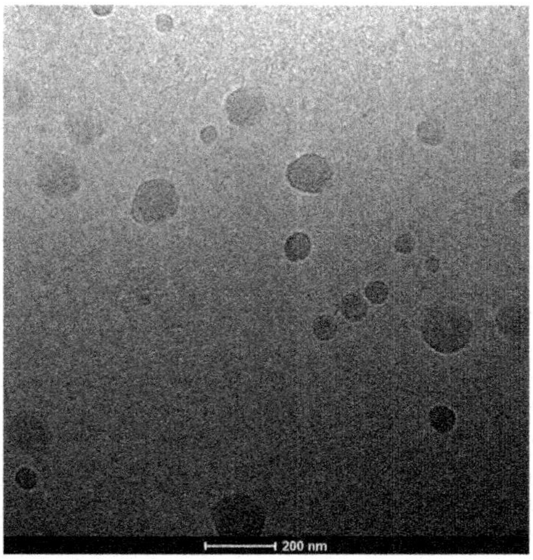

Fig. 2 A representative Cryo-TEM image of optimized LLNs. Scale bar: 200 nm (reproduced from ref. 25 with permission from (2015) American Chemical Society)

3.3 Evaluation of Cytotoxicity of Lipid-Like Nanoparticles

1. Hep 3B cells are maintained in Eagle's Minimum Essential Medium (EMEM) supplemented with 10% heat-inactivated FBS.
2. Detach cells with 0.05% trypsin when they reach approximately 70% confluence.
3. Seed cells into 96-well plates (150 μL/well) at a density of 20,000 cells per well (*see* **Note 10**).
4. Treat cells with 20 μL of freshly formulated LLNs per well for 6 h after overnight culture.
5. Add 17 μL of MTT per well for 4 h.
6. Carefully remove the whole medium.
7. Add 150 μL of DMSO.
8. Dissolve MTT by shaking the plate for about 15 min.
9. Measure the absorbance using a plate reader at 570 nm.

3.4 Evaluation of LLNs-Mediated Transfection

1. Seed and treat Hep3B cells as described above except that 96-well opaque white plates are used.
2. Six hours after treatment, carefully remove the medium containing the formulation.
3. Add 100 μL of a mixture consisting of 50 μL of serum-free EMEM and 50 μL of Bright-Glo (*see* **Note 11**).
4. Cover the plate with aluminum foil.
5. Wait at least 5 min to allow complete cell lysis.
6. Measure the luminescence values using a plate reader.

3.5 Determination of the Optimal Formulation Parameters

1. Identification of lead lipid-like compound: prepare LLNs as mentioned above using newly synthesized lipid-like compounds.
2. Conduct LLNs-mediated cell transfection assay to select the lead lipid-like compound.
3. Optimization of formulation components ratio: choose a proper orthogonal array table based on the number of formulation components (factors) and assigned molar ratios (levels) (*see* **Note 12**).
4. Prepare formulation according to the orthogonal table with assigned factors and levels.
5. Conduct LLN-mediated cell transfection assays.
6. Predict the best ratio by analyzing the average luminescence intensity (K_n) of each factor (*see* **Note 13**).
7. Readjust the intervals according to the impact trend of each component and carry out the second or even more rounds of orthogonal optimization until selecting the optimal formulation (*see* **Note 14**).

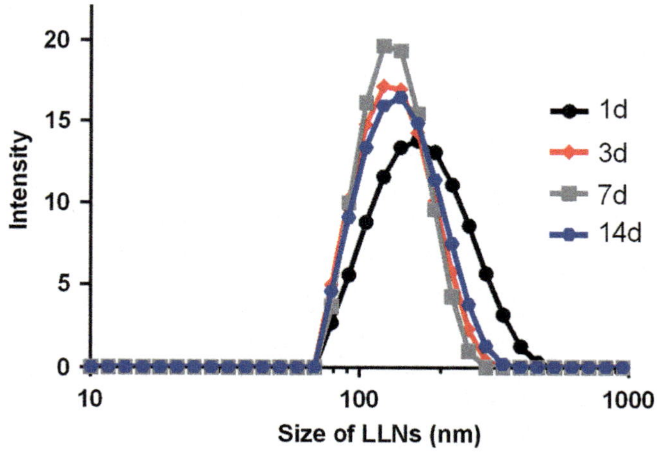

Fig. 3 The particle stability of LLNs after incorporation of DMG-PEG$_{2000}$ (reproduced from ref. 25 with permission from (2015) American Chemical Society)

8. Optimization of the ratio for pegylation: regulate the pegylation extent of the above predicted formulation while keeping the other component ratios constant.
9. Measure LLNs-mediated luciferase expression.
10. Examine the particle size of LLNs twice weekly (Fig. 3) to identify stable and potent formulation (*see* **Note 15**).

4 Notes

1. Special attention should be given when working with RNA. Assign specific benchtop and set a dedicated set of pipettes for RNA work. Spray the benchtop with RNase AWAY solution (or equivalent) to eliminate RNase contamination before preparing work solution.
2. The concentration of each component should be adjusted according to its molecular weight. For better accuracy, ensure that stock concentration is not too high. Otherwise, the required volume is too small to be accurate; conversely, if the total volume of combined components is more than $x/4$ μL, simply increase the concentration to some extent. Typically, $x = 200$, $y = 0.01$.
3. Store phospholipids stock at 4 °C and discard solutions exceeding one week.
4. To minimize ethanol evaporation, close the tube cap between intervals.

5. It is not uncommon to produce air bubbles during pipetting. Conversely, in most cases, it is an obvious feature associated with the formation of LLNs. Bubbles will disappear rapidly from solution once mixing is stopped.
6. If a large-volume and highly concentrated LLNs are required for an in vivo study, use 3 mL syringe. It is necessary to get rid of ethanol by dialyzing samples using 1 L of 1× PBS for 60–90 min at room temperature with dialysis cassettes (3500 MWCO) before injection into animals.
7. LLNs should be stable for over four weeks at 4 °C since no significant size change is observed during this period. Nevertheless, we recommend using freshly made LLNs.
8. Prior to measurement, gently invert the tube several times to obtain homogenous formulations. Do not vortex or fiercely agitate the solution, which may destroy the nanostructure of formulations.
9. Avoid creating bubbles when adding 1% Triton solution. After adding RiboGreen reagent, cover the plate with foil to avoid light since this reagent is susceptible to photo degradation.
10. Keep equivalent cell confluence for each treatment to ensure reproducibility.
11. Equilibrate the substrate Bright-Glo to room temperature to fully exert enzyme activity. The desired incubation time is 5–10 min to completely lysate cells and yield high luminescence signals.
12. For the first round of optimization, large intervals between two adjacent levels are recommended in order to identify the impact trend for each formulation component. In this study, L_{16} $(4)^4$ table (Table 1) is chosen because formulation consists of four components, and each component contains four concentrations (Table 2). For simplification, interaction effects are not considered.

Table 1
Four levels assigned to each formulation component (lipidoid, phospholipid, cholesterol, and DMG-PEG$_{2000}$)

	Formulation components (molar ratio)			
Levels	Lipidoid	Phospholipid	Cholesterol	DMG-PEG$_{2000}$
1	30	1.25	18.5	0.75
2	40	2.5	28.5	1.5*
3	50*	5	38.5*	3
4	60	10*	48.5	6

Asterisk indicates the original molar ratio 50:10:38.5:1.5

Table 2
Orthogonal array table L_{16} (4^4) used for optimization of formulation components ratio

	Formulation components (mole ratio)			
Formulation no.	**Lipidoid**	**Phospholipid**	**Cholesterol**	**DMG-PEG$_{2000}$**
1	30	2.5	38.5	3
2	40	10	18.5	1.5
3	50	10	38.5	6
4	60	2.5	18.5	0.75
5	30	5	18.5	6
6	40	1.25	38.5	0.75
7	50	1.25	18.5	3
8	60	5	38.5	1.5
9	30	1.25	48.5	1.5
10	40	5	28.5	3
11	50	5	48.5	0.75
12	60	1.25	28.5	6
13	30	10	28.5	0.75
14	40	2.5	48.5	6
15	50	2.5	28.5	1.5
16	60	10	48.5	3

13. Such a design allows rapidly predicting the top-performing formulation from 256 $(4)^4$ theoretical combinations by testing 16 combinations.
14. Theoretically, more rounds of orthogonal optimization lead to higher potency of the predicted formulation. However, the round of orthogonal optimization depends on demand, cost, statistical difference, stability of formulation and so on.
15. In general, pegylation hampers delivery efficiency of the formulation but significantly enhance particle stability. Consequently, the extent of pegylation should be regulated to balance delivery efficiency and particle stability, making it suitable for desired applications.

Acknowledgment

This work was supported by the Early Career Investigator Award from Bayer Hemophilia Awards Program, Research Reward from Trilink Biotechnologies, as well as the startup-fund from the College of Pharmacy at the Ohio State University.

References

1. Akinc A, Zumbuehl A, Goldberg M et al (2008) A combinatorial library of lipid-like materials for delivery of RNAi therapeutics. Nat Biotechnol 26:561–569
2. Whitehead KA, Langer R, Anderson DG (2009) Knocking down barriers: advances in siRNA delivery. Nat Rev Drug Discov 8:129–138
3. Cho SW, Goldberg M, Son SM et al (2009) Lipid-like nanoparticles for small interfering RNA delivery to endothelial cells. Adv Funct Mater 19:3112–3118
4. Love KT, Mahon KP, Levins CG et al (2010) Lipid-like materials for low-dose, in vivo gene silencing. Proc Natl Acad Sci U S A 107:1864–1869
5. Schroeder A, Levins CG, Cortez C et al (2010) Lipid-based nanotherapeutics for siRNA delivery. J Intern Med 267:9–21
6. Siegwart DJ, Whitehead KA, Nuhn L et al (2011) Combinatorial synthesis of chemically diverse core-shell nanoparticles for intracellular delivery. Proc Natl Acad Sci U S A 108:12996–13001
7. Chen D, Love KT, Chen Y et al (2012) Rapid discovery of potent siRNA-containing lipid nanoparticles enabled by controlled microfluidic formulation. J Am Chem Soc 134:6948–6951
8. Whitehead KA, Matthews J, Chang PH et al (2012) In vitro-in vivo translation of lipid nanoparticles for hepatocellular siRNA delivery. ACS Nano 6:6922–6929
9. Sun S, Wang M, Knupp SA et al (2012) Combinatorial library of lipidoids for in vitro DNA delivery. Bioconjug Chem 23:135–140
10. Jiang S, Eltoukhy AA, Love KT et al (2013) Lipidoid-coated iron oxide nanoparticles for efficient DNA and siRNA delivery. Nano Lett 13:1059–1064
11. Alabi CA, Love KT, Sahay G et al (2013) Multiparametric approach for the evaluation of lipid nanoparticles for siRNA delivery. Proc Natl Acad Sci U S A 110:12881–12886
12. Zhang Y, Pelet JM, Heller DA et al (2013) Lipid-modified aminoglycoside derivatives for in vivo siRNA delivery. Adv Mater 25:4641–4645
13. Sahay G, Querbes W, Alabi C et al (2013) Efficiency of siRNA delivery by lipid nanoparticles is limited by endocytic recycling. Nat Biotechnol 31:653–658
14. Kanasty R, Dorkin JR, Vegas A et al (2013) Delivery materials for siRNA therapeutics. Nat Mater 12:967–977
15. Bogorad RL, Yin H, Zeigerer A et al (2014) Nanoparticle-formulated siRNA targeting integrins inhibits hepatocellular carcinoma progression in mice. Nat Commun 5:3869
16. Dong Y, Love KT, Dorkin JR et al (2014) Lipopeptide nanoparticles for potent and selective siRNA delivery in rodents and nonhuman primates. Proc Natl Acad Sci U S A 111:3955–3960
17. Dahlman JE, Barnes C, Khan OF et al (2014) In vivo endothelial siRNA delivery using polymeric nanoparticles with low molecular weight. Nat Nanotechnol 9:648–655
18. Dong Y, Eltoukhy AA, Alabi CA et al (2014) Lipid-like nanomaterials for simultaneous gene expression and silencing in vivo. Adv Healthc Mater 3:1392–1397
19. Whitehead KA, Dorkin JR, Vegas AJ et al (2014) Degradable lipid nanoparticles with predictable in vivo siRNA delivery activity. Nat Commun 5:4277
20. Ball RL, Knapp CM, Whitehead KA (2015) Lipidoid nanoparticles for siRNA delivery to the intestinal epithelium: in vitro investigations in a Caco-2 model. PLoS One 10:e0133154
21. Eltoukhy AA, Chen D, Veiseh O et al (2014) Nucleic acid-mediated intracellular protein delivery by lipid-like nanoparticles. Biomaterials 35:6454–6461
22. Kauffman KJ, Dorkin JR, Yang JH et al (2015) Optimization of lipid nanoparticle formulations for mRNA delivery in vivo with fractional factorial and definitive screening designs. Nano Lett 15:7300–7306

23. Khan OF, Zaia EW, Jhunjhunwala S et al (2015) Dendrimer-inspired nanomaterials for the in vivo delivery of siRNA to lung vasculature. Nano Lett 15:3008–3016
24. Yin H, Song CQ, Dorkin JR et al (2016) Therapeutic genome editing by combined viral and non-viral delivery of CRISPR system components in vivo. Nat Biotechnol 34:328–333
25. Li B, Luo X, Deng B et al (2015) An orthogonal array optimization of lipid-like nanoparticles for mRNA delivery in vivo. Nano Lett 15:8099–8107
26. Li B, Luo X, Deng B et al (2016) Effects of local structural transformation of lipid-like compounds on delivery of messenger RNA. Sci Rep 6:22137
27. Leuschner F, Dutta P, Gorbatov R et al (2011) Therapeutic siRNA silencing in inflammatory monocytes in mice. Nat Biotechnol 29:1005–1010
28. Novobrantseva TI, Borodovsky A, Wong J et al (2012) Systemic RNAi-mediated gene silencing in nonhuman primate and rodent myeloid cells. Mol Ther Nucleic Acids 1:e4
29. Dong Y, Dorkin JR, Wang W et al (2016) Poly (glycoamidoamine) brushes formulated nanomaterials for systemic siRNA and mRNA delivery in vivo. Nano Lett 16:842–848
30. Jayaraman M, Ansell SM, Mui BL et al (2012) Maximizing the potency of siRNA lipid nanoparticles for hepatic gene silencing in vivo. Angew Chem Int Ed Engl 51:8529–8533
31. Wang M, Sun S, Alberti KA et al (2012) A combinatorial library of unsaturated lipidoids for efficient intracellular gene delivery. ACS Synth Biol 1:403–407
33. Fenton OS, Kauffman KJ, McClellan RL et al (2016) Bioinspired alkenyl amino alcohol ionizable lipid materials for highly potent in vivo mRNA delivery. Adv Mater 28:2939–2943
33. Oberli MA, Reichmuth AM, Dorkin JR et al (2017) Lipid nanoparticle assisted mRNA delivery for potent cancer immunotherapy. Nano Lett 17:1326–1335
34. He XJ, Long SA, Zheng MD et al (2010) Optimization of activated carbon preparation by orthogonal experimental design for electrochemical capacitors. Sci Adv Mater 2:545–551
35. Zhan JY, Zheng KY, Zhu KY et al (2011) Chemical and biological assessment of Angelicae Sinensis Radix after processing with wine: an orthogonal array design to reveal the optimized conditions. J Agric Food Chem 59:6091–6098
36. Meng HN, Zhang ZZ, Zhao FX et al (2013) Orthogonal optimization design for preparation of Fe_3O_4 nanoparticles via chemical coprecipitation. Appl Surf Sci 280:679–685
37. Zhao X, Wu X, Zhou H et al (2014) Optimization study of the preparation factors for argan oil microcapsule based on hybrid-level orthogonal array design via SPSS modeling. Pak J Pharm Sci 27:2129–2132

Chapter 14

Chitosan Nanoparticles for miRNA Delivery

Merve Denizli, Burcu Aslan, Lingegowda S. Mangala, Dahai Jiang, Cristian Rodriguez-Aguayo, Gabriel Lopez-Berestein, and Anil K. Sood

Abstract

RNA interference techniques represent a promising strategy for therapeutic applications. In addition to small interfering RNA-based approaches, which have been widely studied and translated into clinical investigations, microRNA-based approaches are attractive owing to their "one hit, multiple targets" concept. To overcome challenges with in vivo delivery of microRNAs related to stability, cellular uptake, and specific delivery, our group has developed and characterized chitosan nanoparticles for nucleotide delivery. This platform allows for robust target modulation and antitumor activity following intravenous administration.

Key words RNA interference, miR-34a, microRNA delivery, Chitosan nanoparticles, Prostate cancer

1 Introduction

The discovery of RNA interference (RNAi) has rapidly led to unique opportunities for the treatment of many diseases, including cancer [1, 2]. RNAi techniques represent a continuum of specific approaches for mRNA degradation, resulting in inhibition of gene expression [3]. Small interfering RNAs (siRNAs) and microRNAs (miRNAs) are small RNA molecules most frequently used for RNAi-based therapeutic applications [4].

siRNAs are exogenous double-stranded RNAs of about 20–22 nucleotides that are incorporated into RNA-induced silencing complex (RISC). SiRNA loaded into RISC is cleaved by argonaute-2 protein found in RISC, leading to the degradation of the "passenger strand" and activation of RISC with the "guide strand." Eventually, this activated RISC with single-stranded guide RNA provides recognition of target mRNA through intermolecular base pairing followed by degradation of the guide RNA. RISC can be recycled and implement numerous cleavage events [5]. The knockdown of a specific target mRNA is achieved through full complementation of siRNA to its target mRNA.

Eckart Bindewald and Bruce A. Shapiro (eds.), *RNA Nanostructures: Methods and Protocols*, Methods in Molecular Biology, vol. 1632, DOI 10.1007/978-1-4939-7138-1_14,

miRNAs are endogenous RNA molecules of about 22 nucleotides that are formed by cleavage of long primary precursors (pre-miRNAs) by Dicer. miRNAs are incorporated into RISC and bind to the 3′ untranslated regions of target mRNAs. This binding leads to either the destruction of target mRNA or translational repression [6, 7]. Unlike siRNA, miRNA regulates the expression of multiple targets owing to its partial complementation to the untranslated region in target mRNA [8]. Given the fact that the majority of oncogenes are undruggable with present strategies such as small-molecule inhibitors and monoclonal antibodies [9–11], new therapeutic approaches are needed for clinical translation. Therefore, both siRNA and miRNA have tremendous potential as therapeutic agents for multiple diseases through downregulation of the associated genes and their mRNA transcripts.

However, several challenges have to be overcome for successful clinical translation of these therapeutic agents. Susceptibility of RNAs to degradation by nucleases can hinder the RNAs from reaching the cytoplasm of target cells. Additional challenges such as stability, poor cellular uptake, or nonspecific delivery of RNAi molecules also must be addressed. The initial criteria in developing an ideal delivery system is the selection of material that is biocompatible, biodegradable, nonimmunogenic, and nontoxic. Several viral or other transfection vectors and carriers such as liposomes, polymeric nanoparticles, iron oxide nanoparticles, carbon nanotubes, and dendrimers have been developed to overcome these issues [12, 13].

Among the various options, chitosan (CH) is a particularly attractive option owing to its low immunogenicity, low toxicity, and biocompatibility [14]. CH is a natural polysaccharide that consists of repeated glucosamine and *N*-acetyl-glucosamine residues. It is derived from partial deacetylation of chitin, which is abundant in crustacean shells [15]. The cationic nature of CH enables easy and fast complex formation with negatively charged siRNA or miRNA [16], which increases the incorporation efficiency of the payload. In addition, the hydroxyl and amino groups of CH allow for chemical conjugation of specific ligands tailored for targeted therapy [17]. For instance, our group modified CH particles with an Arg-Gly-Asp (RGD) peptide via a thiolation reaction and showed that these RGD-CH-siRNA nanoparticles significantly enhanced selective intratumoral delivery in orthotopic animal models of ovarian cancer [18].

We have demonstrated several applications of siRNA/miRNA delivery using CH nanoparticles [19–26]. To assess in vivo miRNA delivery of CH nanoparticles into endothelial cells, we incorporated Cy3-labeled control miRNA into CH nanoparticles and delivered them intravenously into nude tumor-bearing mice. Forty-eight hours after delivery, mice were sacrificed and endothelial cells were assessed for colocalization of Cy3 signal. The results demonstrated endothelial and periendothelial miRNA localization in HeyA8 tumors [27].

In a recent study, we showed that CH-mediated delivery of miR-34a, a tumor-suppressive miRNA that downregulates multiple gene products involved in prostate cancer progression and metastasis, inhibited tumor growth and preserved bone integrity in a xenograft model representative of established prostate cancer bone metastasis [28]. To determine whether CH could deliver siRNAs to the tumor tissue preferentially, we incorporated Cy5.5-labeled siRNA into CH nanoparticles and administered the carriers intravenously to the mice that had been injected with PC3MM2-LG cells in the femur. An increase in Cy5.5 siRNA signal intensity was observed in the femur with the tumor compared with the femur without the tumor 3 days after siRNA delivery, suggesting that CH-incorporated siRNA was specifically delivered to the tumor tissue. Taken together, these experiments demonstrated that CH nanoparticles can accumulate in tumor tissues after intravenous administration and that miR-34a-CH nanoparticles significantly decrease prostate tumor growth in the bone.

The present chapter clarifies the experimental approaches and methods required to better understand the studies mentioned above. We detail the preparation of CH nanoparticles that were used for in vivo delivery of non-coding RNA. Cell preparation for in vivo injection, the administration, and the in vivo imaging techniques of nanocarriers are delineated. Next, we describe how the in vivo delivery is validated by in situ hybridization. Finally, immunoblotting and immunofluorescence, the procedures required to confirm the miR-34a functionality, are provided.

2 Materials

2.1 Cell Culture

1. Dulbecco's Modified of Eagle's Medium (DMEM)/Ham's F12 50/50 Mix (Invitrogen or Sigma-Aldrich).
2. Fetal bovine serum (FBS; Invitrogen).
3. Penicillin–streptomycin solution, 100× (Corning).
4. Trypsin–EDTA solution, 0.25%, sterile, filtered, suitable for cell culture, 2.5 g porcine trypsin, and 0.2 g EDTA. (Sigma-Aldrich).
5. Hank's Balanced Salt Solution (HBSS; Thermo Scientific).

2.2 Preparation of CH-miRNA or CH-siRNA Nanoparticles

1. Chitosan, low molecular weight, deacetylated chitin, poly (D-glucosamine) (Sigma-Aldrich).
2. Glacial acetic acid (Sigma-Aldrich).
3. Sodium tripolyphosphate (TPP; Sigma-Aldrich).
4. Molecular biology grade water (Lonza).

2.3 Immunoblotting

1. RIPA Lysis and Extraction Buffer (Thermo Scientific).
2. Protease inhibitor cocktail (Sigma-Aldrich).
3. Phosphatase inhibitor cocktail (Sigma-Aldrich).
4. Bicinchoninic acid assay kit (BCA; Thermo Scientific).
5. Laemmli buffer: 1.2 g SDS, 6 mg bromophenol blue, 4.7 mL glycerol, 1.2 mL Tris 0.5 M, and 2.1 mL distilled water; after everything is dissolved, add 0.93 g DTT, then aliquot and keep frozen at −20 °C.
6. Running buffer: for 10× stock, dissolve 30 g Tris-base and 144 g glycine in 1 L distilled water; for use, dilute to 1× by adding 100 mL stock and 10 mL 10% SDS to 890 mL distilled water for use.
7. Ready-to-use TGX gels.
8. Transfer buffer: for 10× stock, dissolve 30 g Tris-base and 144 g glycine in 1 L distilled water; for use, dilute to 1× by adding 100 mL stock and add 200 mL methanol to 700 mL distilled water.
9. Polyvinylidene difluoride (PVDF) membrane (Thermo Scientific).
10. Nonfat dry milk powder (Sigma-Aldrich).
11. Tris-buffered saline (TBS): for 10× stock, 31.5 g Tris–HCl and 80 g NaCl in 1 L distilled water.
12. TBS-T: add 100 mL 10× TBS and 2 mL Tween 20 (polyoxyethylene-20-sorbitan monolaurate; Thermo Scientific) to 898 mL distilled water.
13. Primary antibodies (c-Met and c-Myc from Santa Cruz Biotechnology; Axl from Cell Signaling, and GAPDH from Millipore).
14. Secondary antibodies (horseradish peroxidase-conjugated secondary goat anti-mouse or goat anti-rabbit antibody (Bio-Rad).
15. HyGLO™ Quick Spray Chemiluminescent HRP Antibody Detection Reagent (Denville Scientific).
16. Western Blot stripping buffer (Thermo Scientific).

2.4 In Situ Hybridization

1. Xylene.
2. Ethanol.
3. Hematoxylin (Harris).
4. Eosin (Harris).
5. Proteinase K (Sigma).

6. Predesigned digoxigenin-labeled miRCURY locked nucleic acid probe (Exiqon).
7. Small nuclear RNA U6 (Exiqon).
8. Polyclonal anti-digoxigenin antibody (Roche).
9. Alkaline phosphatase-conjugated secondary antibody (Roche).
10. 4-nitro-blue tetrazolium (Roche).
11. 5-bromo-4-chloro-3′-indolylphosphate (Roche).
12. Microscope imaging software for quantitative analysis of histological staining and fluorescence (Nikon NIS Elements).

2.5 Immunofluorescence (Promega DeadEnd TUNEL System)

1. Phosphate-buffered saline (PBS), 1× without calcium and magnesium (Corning).
2. Xylene.
3. Ethanol.
4. 1× antigen retrieval buffer (Dako).
5. Pap pen.
6. 3% hydrogen peroxide in methanol (Thermo Scientific).
7. Fish gelatin 4% in PBS (Sigma-Aldrich).
8. Primary antibody rat anti-mouse CD31 (BD Bioscience).
9. Secondary antibody Alexa 594 goat anti-rat (Jackson ImmunoResearch).
10. Paraformaldehyde 4% in PBS (1:4 dilution from 16% stock).
11. Triton X-100 0.2% in PBS.
12. Equilibration buffer (in Promega DeadEnd TUNEL kit).
13. Nucleotide mix (in Promega DeadEnd TUNEL kit).
14. Terminal Deoxynucleotidyl Transferase (TdT) enzyme (in Promega DeadEnd TUNEL kit).
15. 2X saline sodium citrate buffer (SSC, diluted from 20X concentrate in Promega DeadEnd TUNEL kit).
16. Hoechst 33342 mounting media to stain nuclei (Thermo Scientific).

3 Methods

3.1 Gelation of CH with Sodium Tripolyphosphate and RNA

1. Dissolve CH in 0.25% acetic acid to make the final concentration of 2 mg/mL.
2. Mix TPP (0.25% w/v) and RNA (1 μg/μL), then add this mixture to CH solution with continuous stirring to spontaneously generate the nanoparticles.
3. Incubate the mixture at 4 °C for 40 min.

4. Centrifuge the mixture at 20,817 × *g* for 50 min at 4 °C to obtain the RNA/CH particles.
5. Discard the supernatant.
6. Wash the pellet three times with distilled water to remove unbound chemicals or miRNA (*see* **Note 1**).
7. After three washes, dissolve the pellet in distilled water (5 μg/ 100 μL) and store at 4 °C.

3.2 Cell Culture

Maintain prostate tumor PC3MM2 cells in 50% Dulbecco modified Eagle medium/50% Ham F12 supplemented with 10% FBS and 1% penicillin–streptomycin at 37 °C in 5% CO_2.

3.3 Preparation of Cells for In Vivo Injections

1. Trypsinize the cells when they are 60–80% confluent.
2. Centrifuge the trypsinized cells at 290 × *g* for 5 min at 4 °C (*see* **Note 2**).
3. Wash the cells three times with PBS.
4. Count the cells with cell counter.
5. Take one million cells to a new tube and centrifuge at 290 × *g*.
6. Discard the supernatant and resuspend the cells in HBSS for each animal.
7. Inject one million PC3MM2 cells (100 μL) either subcutaneously for the subcutaneous model or into the femur of nude mice for the metastatic model.

3.4 CH-Non-Coding RNA Treatment In Vivo

3.4.1 CH-miRNA Treatment

1. One week after cell injection, divide the mice into control group and miR-34a treatment group. Administer control miRNA-encapsulated CH nanoparticles (control group; 5 μg/100 μL nanoparticles) or miR-34a-encapsulated CH nanoparticles (treatment group; 5 μg/100 μL nanoparticles) via tail vein injection every 3 days for both subcutaneous and metastatic models.
2. After 2 weeks, sacrifice the animals and harvest the tumors.
3. Fix the tumor either in formalin for paraffin embedding for in situ hybridization or snap-freeze in liquid nitrogen for Western blot analysis.

3.4.2 Delivery of CH-siRNA Nanoparticles into Bone

1. Ten days after cell injection, administer control siRNA-CH or Cy5.5-labeled siRNA-CH nanoparticles (5 μg/100 μL nanoparticles) via tail vein injection.
2. Three days after nanoparticle injection, sacrifice the mice.
3. Harvest the femurs.
4. Image the femurs using IVIS 200 (*see* Fig. 1).

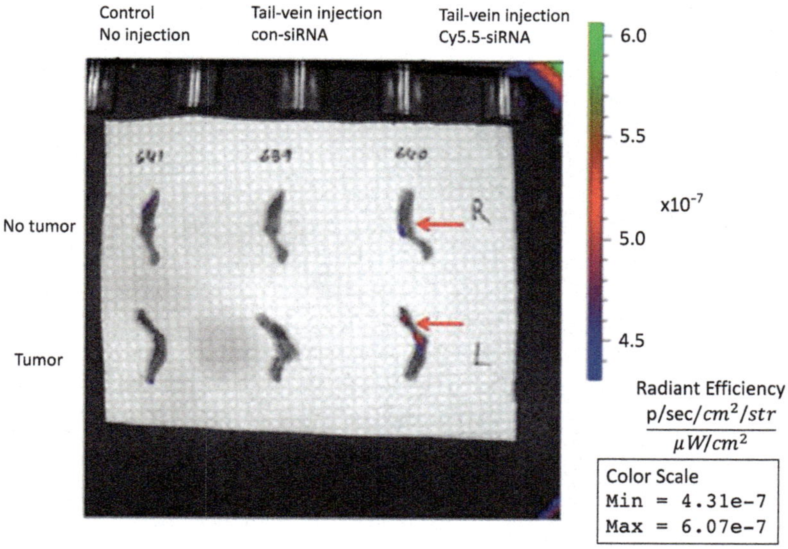

Fig. 1 Chitosan nanoparticles deliver Cy5.5-labeled siRNA to the tumor tissue preferentially. Control or Cy5.5-labeled siRNA was administered to the mice with or without a tumor in the femur and Cy5.5 signal was measured by using IVIS 200. An increase in Cy5.5 signal was observed in the femur in the tumor group compared with the nontumor group (reproduced from ref. 28 with permission from Oncotarget)

3.5 Immunoblotting

1. Homogenize the snap frozen tumor tissue in RIPA Lysis and Extraction Buffer plus protease and phosphatase inhibitor cocktails (10 μL from each for 1000 μL lysis buffer) using homogenizer. Add 200 μL from the mixture onto each tumor tissue.
2. Incubate lysate on ice for 30 min, by vortexing every 5 min.
3. Centrifuge the lysate at 15,294 × *g* for 20 min at 4 °C.
4. Transfer the supernatant to a new Eppendorf tube.
5. Perform protein quantification using the BCA assay kit.
6. Adjust an equal amount of protein (15 or 30 μg) for each sample and mix with 25 μL Laemmli sample buffer.
7. Boil protein and Laemmli sample buffer mixtures at 100 °C for 5 min.
8. Load the mixtures (25 μL) onto 8% or 12% polyacrylamide gel and transfer onto a PVDF membrane.
9. Block the membrane in 5% nonfat dry milk in 1× TBS-T for 1 h at room temperature.
10. Probe the membrane with a specific primary antibody (e.g., c-Met, c-Myc, Axl, and GAPDH) at 1:1000 dilution overnight at 4 °C.

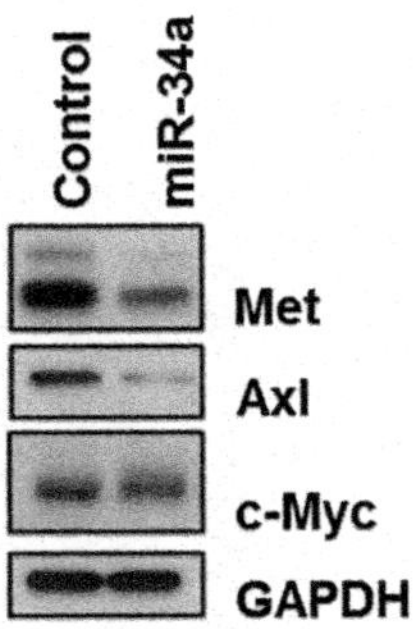

Fig. 2 Western blot results for Met, Axl, and c-Myc proteins from harvested tumor tissues treated with control miRNA or miR-34a. Treatment with miR-34a led to downregulation of Met, Axl, and c-Myc (reproduced from ref. 28 with permission from Oncotarget)

11. Remove the primary antibody and wash the membrane with TBS-T for 1 h at room temperature by changing the TBS-T every 15 min.
12. Re-probe the membrane with horseradish peroxidase-conjugated secondary goat anti-mouse or goat anti-rabbit antibody for 1 h at 1:2000 dilution at room temperature.
13. Remove the secondary antibody and wash the membrane for 1 h at room temperature, with changing the TBS-T every 15 min.
14. Remove the TBS-T.
15. Mix an equal amount (1 mL) of each ECL A and B reagent and add onto the membrane.
16. Incubate the membrane in this mixture for 1 min (*see* **Note 3**).
17. Remove the ECL reagent and blot the membrane with paper towel to remove extra ECL reagent.
18. Place the membrane into an X-ray film cassette along with a film for a few minutes.
19. Strip the membrane for 15 min using stripping buffer and wash with TBS-T for 30 min before probing with another antibody (*see* Fig. 2).

3.6 In Situ Hybridization

1. Immerse the formalin-fixed, paraffin-embedded tissue sections in xylene in a Coplin staining jar for 5 min at room temperature for deparaffinization.
2. Dehydrate the deparaffinized slides by immersing in 100% ethanol for 5 min and rehydrate in decreasing concentrations of ethanol (95%, 85%, 70%, and 50%) for 3 min each.
3. Stain one slide from each group (control and miR-34a) with hematoxylin and eosin to validate the integrity of the tissues.

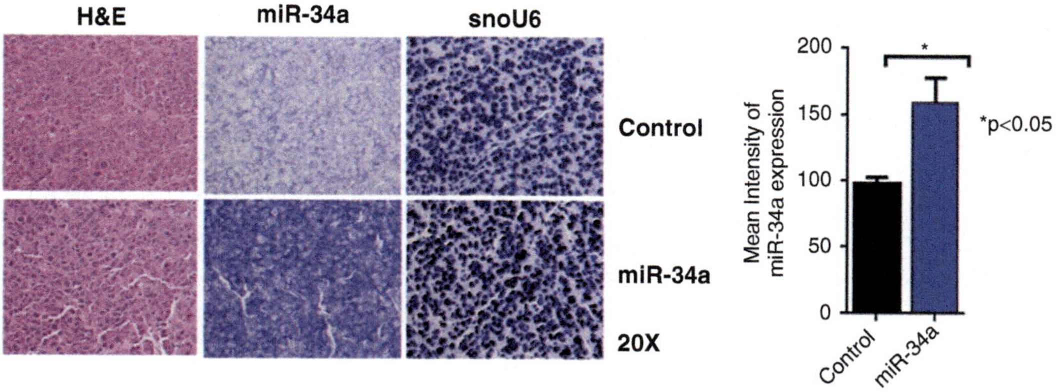

Fig. 3 In situ hybridization of miR-34a with small nuclear RNA U6 (snoU6) as endogenous control. Formalin-fixed, paraffin-embedded slides were stained with hematoxylin and eosin (H&E) or in situ hybridization was performed by using a double digoxigenin-labeled miRCURY locked nucleic acid probe designed for control miRNA or miR-34a after a digestion step with proteinase K. The mean (± standard deviation) intensities obtained from ten areas from each slide were measured using the NIS Elements software (reproduced from ref. 28 with permission from Oncotarget)

4. Digest two tissue sections from each group (control and miR-34a) with 10 μg/mL proteinase K for 10 min at room temperature and load onto the Ventana Discovery Ultra system.
5. Hybridize one of the tissue slides from each group with the double digoxigenin-labeled miRCURY locked nucleic acid probe designed for control miRNA or miR-34a for 2 h at 50 °C.
6. Use small nuclear RNA U6 as endogenous control.
7. Detect the digoxigenins with a polyclonal anti-digoxigenin antibody and alkaline phosphatase-conjugated secondary antibody using nitro-blue tetrazolium and 5-bromo-4-chloro-3′-indolyphosphate as the substrate.
8. Measure the intensities for ten areas from each slide with a microscope imaging software and compute the mean (*see* Fig. 3).

3.7 Immunofluorescence

1. Immerse paraffin-embedded sections in xylene in a Coplin staining jar for 5 min at room temperature for deparaffinization. For frozen sections, fix the sections with cold acetone for 10 min and then continue from **step** 7.
2. Dehydrate the deparaffinized slides by immersing in 100% ethanol for 5 min and rehydrate in decreasing concentrations of ethanol (95%, 85%, 70%, and 50%) for 3 min each.
3. Perform antigen retrieval with 1× Dako antigen retrieval buffer.

4. Wash with PBS three times for 3 min each.
5. Separate the slides individually, align on a slide holder placed on a humidified chamber. Draw a circle around the tissue with a Pap pen (*see* **Note 4**).
6. Block endogenous peroxidases by incubating the slides in 0.3% hydrogen peroxide in PBS for 3–5 min at room temperature.
7. Incubate the tissue sections with protein block (4% fish gelatin in PBS) for 20 min at room temperature.
8. Remove protein block and add rat anti-mouse CD31 antibody (1:500 diluted in 4% fish gelatin), incubate overnight at 4 °C.
9. Wash with PBS three times for 3 min and incubate with protein block for 5 min.
10. Remove protein blocking solution and add Alexa 594 goat anti-rat secondary antibody (Jacskon ImmunoResearch, 1:800 diluted in 4% fish gelatin), incubate 1 h at room temperature.
11. Wash with PBS three times for 3 min each.
12. Fix the tissue sections by adding 4% paraformaldehyde in PBS on the slides and incubating for 20 min at room temperature.
13. Incubate the tissues in 0.2% Triton X-100 in PBS for 15 min at room temperature.
14. Add equilibration buffer (in Promega kit) on tissue sections and incubate for 10 min at room temperature.
15. Prepare TUNEL incubation buffer by mixing 45 μL equilibration buffer, 5 μL nucleotide mix, and 1 μL TdT enzyme for each slide. Incubate the slides with the buffer for 1 h at 37 °C in the dark.
16. To terminate the reactions, wash the slides in 2× SSC for 15 min at room temperature.
17. Do the counterstaining with Hoechst mounting media and visualize under a fluorescence microscope (*see* Fig. 4).

4 Notes

1. Once CH nanoparticles are reconstituted in water, do not freeze and thaw the suspension because this process may fracture or rupture the particles. RNA/CH nanoparticles can be stored at 4 °C until they are used.
2. Use cells at 60–80% confluence for in vivo injections. Injecting overconfluent cells may lead to lower tumor take rates.
3. Rotate the membrane box by hand to make sure that the ECL mixture covers the membrane.

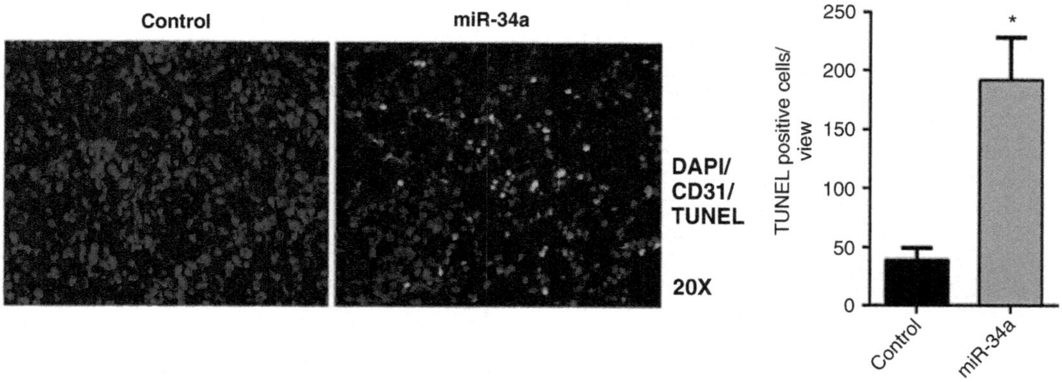

Fig. 4 Tissue sections were used for immunofluorescence reactions. Tumors treated with control miRNA and miR-34a were stained by TUNEL (*green*), CD31 (*red*), and DAPI (*blue*). Mean (± standard deviation) TUNEL-positive cells obtained from ten areas from each slide were quantified using the ImageJ software. $*p < 0.05$ (unpaired two-sided two-sample Student's *t*-test) compared with control (reproduced from ref. 28 with permission from Oncotarget)

4. Try to keep the tissues wet with PBS because letting the tissue dry during the staining procedure may cause high background.

Acknowledgments

Portions of this work were supported by the NIH (CA016672, CA109298, P50 CA083639, P50 CA098258, UH3 TR000943), the Ovarian Cancer Research Fund, Inc. (Program Project Development Grant), the Blanton-Davis Ovarian Cancer Research Program, the RGK Foundation, and the Gilder Foundation.

References

1. Bobbin ML, Rossi JJ (2016) RNA interference (RNAi)-based therapeutics: delivering on the promise? Annu Rev Pharmacol Toxicol 56:103–122
2. SY W, Lopez-Berestein G, Calin GA, Sood AK (2014) RNAi therapies: drugging the undruggable. Sci Transl Med 6 (240):240ps247–240ps247
3. Wilson RC, Doudna JA (2013) Molecular mechanisms of RNA interference. Annu Rev Biophys 42:217–239
4. Carthew RW, Sontheimer EJ (2009) Origins and mechanisms of miRNAs and siRNAs. Cell 136(4):642–655
5. Rana TM (2007) Illuminating the silence: understanding the structure and function of small RNAs. Nat Rev Mol Cell Biol 8 (1):23–36. doi:10.1038/nrm2085
6. Meister G, Tuschl T (2004) Mechanisms of gene silencing by double-stranded RNA. Nature 431(7006):343–349
7. Tijsterman M, Plasterk RH (2004) Dicers at RISC: the mechanism of RNAi. Cell 117 (1):1–3
8. Lam JK, Chow MY, Zhang Y, Leung SW (2015) siRNA versus miRNA as therapeutics for gene silencing. Mol Ther Nucleic Acids 4 (9):e252
9. Ledford H (2015) Cancer: the Ras renaissance. Nature 520(7547):278–280. doi:10.1038/520278a

10. Nero TL, Morton CJ, Holien JK, Wielens J, Parker MW (2014) Oncogenic protein interfaces: small molecules, big challenges. Nat Rev Cancer 14(4):248–262. doi:10.1038/nrc3690
11. Ramos P, Bentires-Alj M (2015) Mechanism-based cancer therapy: resistance to therapy, therapy for resistance. Oncogene 34 (28):3617–3626. doi:10.1038/onc.2014.314
12. Gharpure KM, SY W, Li C, Lopez-Berestein G, Sood AK (2015) Nanotechnology: future of oncotherapy. Clin Cancer Res 21 (14):3121–3130
13. Ozcan G, Ozpolat B, Coleman RL, Sood AK, Lopez-Berestein G (2015) Preclinical and clinical development of siRNA-based therapeutics. Adv Drug Deliv Rev 87:108–119
14. Croisier F, Jérôme C (2013) Chitosan-based biomaterials for tissue engineering. Eur Polym J 49(4):780–792
15. Shukla SK, Mishra AK, Arotiba OA, Mamba BB (2013) Chitosan-based nanomaterials: a state-of-the-art review. Int J Biol Macromol 59:46–58
16. Choi C, Nam J-P, Nah J-W (2016) Application of chitosan and chitosan derivatives as biomaterials. J Ind Eng Chem 33:1–10
17. Ragelle H, Vandermeulen G, Préat V (2013) Chitosan-based siRNA delivery systems. J Control Release 172(1):207–218
18. Han HD, Mangala LS, Lee JW, Shahzad MM, Kim HS, Shen D, Nam EJ, Mora EM, Stone RL, Lu C (2010) Targeted gene silencing using RGD-labeled chitosan nanoparticles. Clin Cancer Res 16(15):3910–3922
19. Lu C, Han HD, Mangala LS, Ali-Fehmi R, Newton CS, Ozbun L, Armaiz-Pena GN, Hu W, Stone RL, Munkarah A (2010) Regulation of tumor angiogenesis by EZH2. Cancer Cell 18(2):185–197
20. Kim H-S, Han HD, Armaiz-Pena GN, Stone RL, Nam EJ, Lee J-W, Shahzad MM, Nick AM, Lee SJ, Roh J-W (2011) Functional roles of Src and Fgr in ovarian carcinoma. Clin Cancer Res 17(7):1713–1721
21. Steg AD, Katre AA, Goodman B, Han H-D, Nick AM, Stone RL, Coleman RL, Alvarez RD, Lopez-Berestein G, Sood AK (2011) Targeting the notch ligand JAGGED1 in both tumor cells and stroma in ovarian cancer. Clin Cancer Res 17(17):5674–5685
22. Hu W, Lu C, Dong HH, Huang J, D-y S, Stone RL, Nick AM, Shahzad MM, Mora E, Jennings NB (2011) Biological roles of the delta family notch ligand Dll4 in tumor and endothelial cells in ovarian cancer. Cancer Res 71(18):6030–6039
23. Hussein YR, Sood AK, Bandyopadhyay S, Albashiti B, Semaan A, Nahleh Z, Roh J, Han HD, Lopez-Berestein G, Ali-Fehmi R (2012) Clinical and biological relevance of enhancer of zeste homolog 2 in triple-negative breast cancer. Hum Pathol 43(10):1638–1644
24. Ziebarth AJ, Nowsheen S, Steg AD, Shah MM, Katre AA, Dobbin ZC, Han H-D, Lopez-Berestein G, Sood AK, Conner M (2013) Endoglin (CD105) contributes to platinum resistance and is a target for tumor-specific therapy in epithelial ovarian cancer. Clin Cancer Res 19(1):170–182
25. Krzeszinski JY, Wei W, Huynh H, Jin Z, Wang X, Chang T-C, Xie X-J, He L, Mangala LS, Lopez-Berestein G (2014) miR-34a blocks osteoporosis and bone metastasis by inhibiting osteoclastogenesis and Tgif2. Nature 512 (7515):431–435
26. Aslan B, Monroig P, Hsu MC, Pena GA, Rodriguez-Aguayo C, Gonzalez-Villasana V, Rupaimoole R, Nagaraja AS, Mangala S, Han HD, Yuca E, SY W, Ivan C, Moss TJ, Ram PT, Wang H, Gol-Chambers A, Ozkayar O, Kanlikilicer P, Fuentes-Mattei E, Kahraman N, Pradeep S, Ozpolat B, Tucker S, Hung MC, Baggerly K, Bartholomeusz G, Calin G, Sood AK, Lopez-Berestein G (2015) The ZNF304-integrin axis protects against anoikis in cancer. Nat Commun 6:7351. doi:10.1038/ncomms8351
27. Pecot CV, Rupaimoole R, Yang D, Akbani R, Ivan C, Lu C, Wu S, Han H-D, Shah MY, Rodriguez-Aguayo C (2013) Tumour angiogenesis regulation by the miR-200 family. Nat Commun 4:2427
28. Gaur S, Wen Y, Song JH, Parikh NU, Mangala LS, Blessing AM, Ivan C, Wu SY, Varkaris A, Shi Y (2015) Chitosan nanoparticle-mediated delivery of miRNA-34a decreases prostate tumor growth in the bone and its expression induces non-canonical autophagy. Oncotarget 6(30):29161

Chapter 15

Synthesis of PLGA–Lipid Hybrid Nanoparticles for siRNA Delivery Using the Emulsion Method PLGA-PEG–Lipid Nanoparticles for siRNA Delivery

Lei Wang, Benjamin Griffel, and Xiaoyang Xu

Abstract

The effective delivery of small interfering RNA (siRNA) to tumor cells remains a challenge for applications in cancer therapy. The development of polymeric nanoparticles with high siRNA loading efficacy has shown great potential for cancer targets. Double emulsion solvent evaporation technique is a useful tool for encapsulation of hydrophilic molecules (e.g., siRNA). Here we describe a versatile platform for siRNA delivery based on PLGA-PEG–cationic lipid nanoparticles by using the double emulsion method. The resulting nanoparticles show high encapsulation efficiency for siRNA (up to 90%) and demonstrate effective downregulation of the target genes in vitro and vivo.

Key words PLGA-PEG polymer, Polymeric nanoparticles, Cationic lipid, siRNA delivery, Double emulsion solvent evaporation

1 Introduction

Gene silencing is an exciting avenue for the potential treatment of countless diseases, especially in cases where there are currently no drugs on the market. As a method for gene silencing, small interfering RNA (siRNA) is attractive because it is relatively inexpensive and it is simple to design RNA molecules specific to nearly any target gene [1–6]. However, a clinically proven siRNA-based treatment has proven to be elusive due to the inherent difficulties in delivering siRNA to their target cells without being degraded or lost [7]. siRNA are short chains of nucleic acids (compared to small molecule therapeutics) with large molecular weight and negative charge. Both the weight and charge present challenges for the delivery of siRNA to the desired targets.

In light of these issues, siRNA delivery strategies have three main goals: to protect the siRNA from the immune system and serum endonuclease degradation, to keep the siRNA in circulation

Eckart Bindewald and Bruce A. Shapiro (eds.), *RNA Nanostructures: Methods and Protocols*, Methods in Molecular Biology, vol. 1632, DOI 10.1007/978-1-4939-7138-1_15,

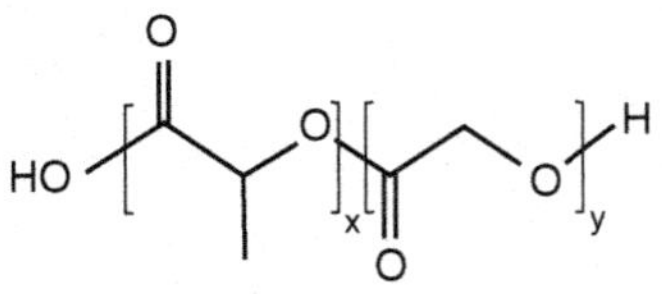

Fig. 1 Structure of PLGA. *x* refers to the number of lactide units, *y* refers to the number of glycolide units

(and avoid renal clearance), and to target the siRNA toward the cells of choice [8]. These goals are typically achieved with three strategies: (1) the modification of the naked siRNA backbone itself, (2) conjugation of the siRNA with a larger complex of molecules, or nanoparticles, and (3) encapsulation of siRNA [9]. As part of the first strategy, modification of the siRNA can provide resistance to the immune response. Further modification through conjugation of small molecules like PEG or fatty acids may also improve biocompatibility and protect against enzymatic degradation. Encapsulation of siRNA for improved biocompatibility and targeting protects the siRNA within larger biocompatible nanocomplexes. These nanocomplexes may be entirely synthetic, such as poly(lactide-co-glycolide) (PLGA) based polymers, or naturally derived.

PLGA is a commonly used biomaterial that has long been FDA approved for medical devices such as biodegradable sutures [10] (Fig. 1). Nanoparticles of PLGA-polyethylene glycol (PLGA-PEG), in a variety of formulations and variations, have been shown to be able to penetrate the blood–brain barrier and thus have been used as a means to transport drug payloads [11, 12]. Without a transport mechanism such as endocytosis, siRNAs cannot cross a cell membrane due to their negative charge. Although the cellular delivery of naked siRNA has been reported, nanoparticles can facilitate siRNA internalization into the cell via endocytosis and are highly effective vectors for the siRNA payload. Additionally, nanoparticles provide a means to protect the siRNA from nucleases that would deactivate siRNA before delivery to the target [9].

There are many methods that can be used for nanoparticle preparation using PLGA-based polymers [13]. Among them, nanoprecipitation [14], single emulsion solvent evaporation [15–17], and double emulsion solvent evaporation [17, 18] are the three most typically used. The nanoprecipitation method has the constraint that the payload should be hydrophobic. Single emulsion is another technique for nanoparticle formation that can be used when the payload is hydrophobic. In this case the payload is dissolved in an organic phase solution that is emulsified within an aqueous solution. As the solvent evaporates, nanoparticles that incorporate the payload are formed. If the payload is hydrophilic, as is the case with siRNA, then the double emulsion technique may be used. The siRNA payload is dissolved in an aqueous phase containing a surfactant while the polymer is dissolved in a solvent, forming an organic phase. The two phases are homogenized in the first emulsification step. A second emulsification is used to mix the polymer–solvent–payload suspension with another aqueous phase to generate water–oil–water nanostructures. Solvent evaporates from the water solution while stirring, which allows the nanoparticles to solidify.

Here we demonstrate that PLGA–lipid hybrid nanoparticles formed by using the double emulsion solvent evaporation technology, can be used as a versatile platform for siRNA delivery (Fig. 2). We were able to use PLGA-PEG nanoparticles incorporated with a cationic lipid G0-C14 featuring a double payload of cisplatin and siRNA targeting the REV1 and REVL genes [19]. The function of the cationic lipid is to lock negatively charged siRNA molecules inside the nanoparticles through electrostatic interactions. We found that the resulting nanoparticles are highly effective at encapsulating siRNA and drug, and demonstrated effective downregulation of target genes in vitro and vivo. The cationic lipid G0-C14 based on 0th generation poly(amidoamine) (PAMAM) dendrimer (G0) was found to give an optimal tradeoff of electrical attraction to

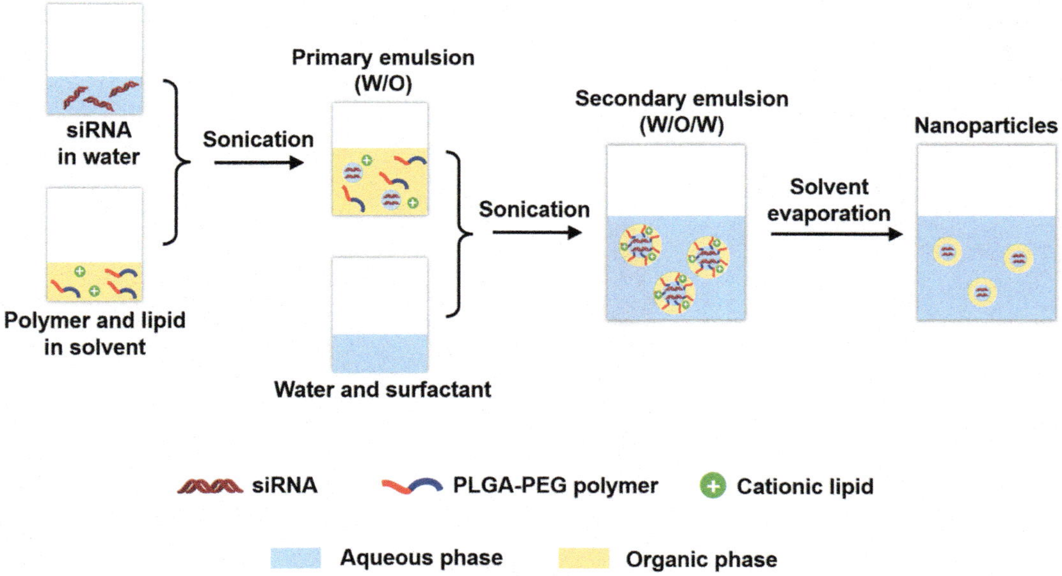

Fig. 2 Formation of PLGA-PEG–cationic lipid nanoparticles by using double emulsion solvent evaporation technique. First siRNA is dissolved in water as aqueous phase. At the same time, the polymer and lipid are dissolved in a solvent, such as DCM. This represents the organic phase. These are combined into the same vial for sonication to form the primary emulsion. The obtained emulsion is dropped into water with a surfactant, such as PVA. This creates a second two phase solution, which is then sonicated to form the secondary emulsion. The emulsion should then be transferred to water while stirring, which allows the nanoparticles to solidify and the solvent to evaporate away

siRNA molecules and biocompatibility, making them highly suitable for siRNA delivery. The synthesis of the G0 molecule is described in the protocol section of this chapter. The encapsulation efficiencies of siRNA at different weight ratios of G0-C14 to siRNA are shown in Table 1. The siRNA release from the nanoparticle can take up to 10 days in a sustained release fashion. However, the encapsulated siRNA will be released in total within a few hours without the incorporation of a cationic lipid. The sizes of nanoparticles are around 200 nm with a polydispersity index range from 0.10 to 0.23. The zeta potential increased with G0-C14–siRNA weight ratio. The nanoparticles with a fixed G0-C14–siRNA weight ratio of 20:1 showed a compact and spherical morphology with a mean diameter around 200 nm (Fig. 3).

Table 1
PLGA-PEG–G0-C14 nanoparticle characterization

G0-C14–siRNA weight ratio*	Encapsulation efficacy[a], %	Size[b], nm	Polydispersity[c]	Zeta potential[d], mV
10:1	87.0 ± 5.0	210 ± 8.5	0.23 ± 0.07	15.7 ± 4.5
20:1	95.0 ± 4.0	220 ± 12.5	0.18 ± 0.04	17.6 ± 3.1
30:1	87.0 ± 3.0	185 ± 10.0	0.10 ± 0.06	25.0 ± 2.6

a refers to siRNA encapsulation efficacy in the prepared nanoparticles; b, c, and d refer to the size, polydispersity, and zeta potential of the prepared nanoparticles, respectively. Reproduced from ref. 19 with permission from PNAS
*The initial amount of siRNA used for all weight ratios remains fixed

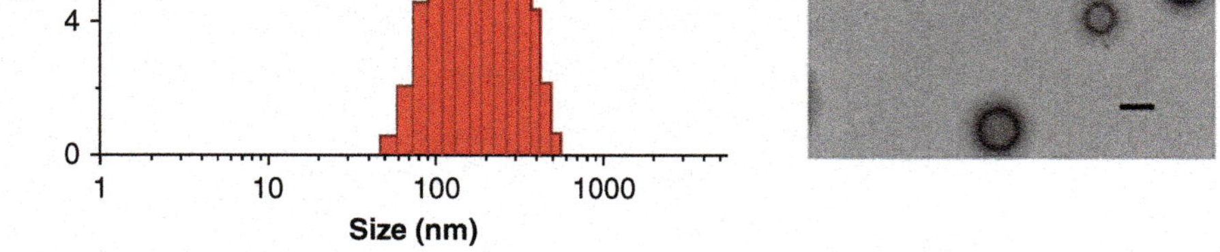

Fig. 3 Size distribution characterization of PLGA-PEG–G0-C14 nanoparticles (samples with a fixed G0-C14–siRNA weight ratio of 20:1). (**a**) Size distribution of the nanoparticles containing both cisplatin and siRNA determined by dynamic light scattering. (**b**) Representative transmission electron microscope image of the nanoparticles (Scale bar: 200 nm). Reproduced from Ref. 19 with permission from PNAS

The following procedural protocol outlines the materials and steps for synthesis of PLGA-PEG polymer through carbodiimide chemistry, synthesis of cationic lipid based on 0th generation poly (amidoamine) (PAMAM) dendrimer (G0), the synthesis of siRNA containing nanoparticles via the double emulsion method, and the steps for characterizing nanoparticles and quantifying encapsulation efficiency of nanoparticles.

2 Materials

2.1 Polymer Synthesis of PLGA-PEG

2.1.1 Materials

1. 50:50 poly(lactide-co-glycolide) with carboxylic acid groups (PLGA-COOH) (molecular weight 43,000).
2. PEG with a terminal amine and carboxylic acid group (NH_2-PEG-COOH) (molecular weight 3400).
3. N-hydroxysuccinimide (NHS).
4. 1-ethyl-3-(3-dimethylaminopropyl)-carbodiimide hydrochloride (EDC).
5. N,N-diisopropylethylamine (DIEA).
6. Anhydrous dichloromethane (DCM).
7. Cold diethyl ether–methanol washing solvent (50:50 v/v) (*see* **Note 1**).

2.1.2 Equipment

1. Magnetic stirrer.
2. Centrifuge.

2.2 Synthesis of G0-C14

2.2.1 Materials

1. Poly(amidoamine) (PAMAM) dendrimer, ethylenediamine core, generation 0.0 solution (20 wt.% in methanol).
2. 1,2-epoxytetradecane (technical grade, 85%).
3. Silica gel (technical grade, pore size 60 Å, 230–400 mesh particle size).
4. Elution solvent: DCM–methanol–NH_4OH (75:22:3 v/v)

2.2.2 Equipment

1. Rotary evaporator.

2.3 PLGA-PEG–G0-C14 Nanoparticle Preparation

2.3.1 Materials

1. Nuclease-free water.
2. 5% (w/v) polyvinyl alcohol (PVA) solution: Dissolve 25 g of PVA (molecular weight 31,000–50,000, 98–99% hydrolyzed) in 500 mL of water. Filter through a 0.45 μm Corning filter. Store at 4 °C for use.

2.3.2 Equipment

1. Magnetic stirrer.
2. Probe sonicator (QSONICA, Sonicator Q700).

2.4 Nanoparticle Characterization and Quantification of siRNA Encapsulation

2.4.1 Materials

1. Nuclease-free water.
2. Amicon Ultra-4 centrifugal filter units (MWCO 100 kDa).
3. 1% (w/v) uranyl acetate solution.
4. Quant-iT RiboGreen RNA Assay Kit.

2.4.2 Equipment

1. Zetasizer Nano-ZS (Malvern Instrument Ltd) for dynamic light scattering (DLS) measurement of nanoparticle size and zeta potential.
2. Transmission electron microscope (TEM).

3 Methods

All procedures are carried out at room temperature unless otherwise specified.

3.1 Synthesis of PLGA-PEG Polymer

1. Dissolve 250 mg of PLGA-COOH (0.005 mmol) in 2.5 mL of anhydrous DCM (*see* **Notes 2** and **3**).
2. Weigh 3.0 mg of NHS (0.025 mmol) and rinse with 0.5 mL of anhydrous DCM in a separate vial. Dissolve 4.8 mg of EDC (0.025 mmol) in 0.5 mL of anhydrous DCM in another separate vial (*see* **Note 4**).
3. PLGA-COOH is activated into PLGA-NHS by adding NHS and EDC solution to PLGA-COOH with gentle stirring for 1.5 h at room temperature (*see* **Note 5**).
4. Precipitate the PLGA-NHS with 80 mL of cold ethyl ether–-methanol washing solvent (50:50 v/v) (*see* **Note 6**).
5. Centrifuge the solution at 3000 × *g* for 10 min (4 °C) and discard the supernatant to remove residual NHS/EDC (*see* **Note** 7).
6. Repeat precipitation and centrifugation twice (*see* **Note 8**).
7. Dry the PLGA-NHS pellet under vacuum until residual ethyl ether and methanol are removed (*see* **Note 9**).
8. After drying under vacuum, dissolve 200 mg of PLGA-NHS (0.004 mmol) in 2.5 mL of anhydrous DCM (*see* **Note 3**).
9. Dissolve 20.4 mg of NH_2-PEG-COOH (0.006 mmol) in 0.5 mL of anhydrous DCM in a separate vial.
10. Add NH_2-PEG-COOH and 7.5 mg of DIEA (10 μL, 0.06 mmol) to the PLGA-NHS solution.
11. React overnight at room temperature with gentle stirring (*see* **Note 10**).
12. Precipitate the resulting PLGA-PEG polymer with 40 mL of cold ethyl ether–methanol washing solvent (50:50 v/v) (*see* **Notes 6** and **11**).

13. Centrifuge the solution at 3000 × *g* for 10 min (4 °C) and discard the supernatant to remove unreacted NH_2-PEG-COOH (*see* **Note 7**).
14. Repeat precipitation and centrifugation twice (*see* **Note 8**).
15. Dry the purified PLGA-PEG polymer under vacuum until residual solvents are removed (*see* **Note 12**).
16. Store the PLGA-PEG polymer at −20 °C.

3.2 Synthesis of Cationic Lipid G0-C14

1. Weigh 4.13 g of PAMAM (1.6 mmol) in a 7 mL vial. Evaporate methanol using a rotary evaporator for 2 h at room temperature.
2. Add 2.80 g of 1,2-epoxytetradecane (11.2 mmol) to the PAMAM. Then seal the vial.
3. Warm the vial to 90 °C in an oil bath.
4. The reaction mixture is stirred at 90 °C for 2 days.
5. The crude product is separated by column chromatography on a silica gel using a gradient elution from DCM to DCM–methanol–NH_4OH (75:22:3 v/v) (*see* **Note 13**).

3.3 Preparation of Nanoparticle Containing siRNAs via the Double Emulsion Solvent Evaporation Technique

1. Dissolve 1 mg of siRNA in 4 mL of nuclease-free water to prepare a siRNA solution (0.25 mg/mL).
2. Dissolve 10 mg of PLGA-PEG polymer and 1 or 2 or 3 mg of G0-C14 (according to a G0-C14–siRNA weight ratio of 10:1 or 20:1 or 30:1 with a fixed initial amount of siRNA) in 1 mL of DCM.
3. Prepare a mixture of 2 mL of nuclease-free water and 1 mL 5% PVA aqueous solution in a 7 mL vial (*see* **Note 14**).
4. Add 0.4 mL of siRNA solution (0.1 mg) dropwise into the polymer and lipid solution.
5. Emulsify the mixture by probe sonication at 30% power and 20% duty cycle for 15 s to form the primary emulsion (*see*-**Note 15**).
6. Add the emulsion into 3 mL of the cooling PVA aqueous solution (*see* **Note 16**).
7. Emulsify the mixture by probe sonication at 30% power and 20% duty cycle for 15 s to form the secondary emulsion.
8. Pour the final emulsion into 15 mL of nuclease-free water stirring at 1000 rpm (*see* **Note 17**).
9. Keep the emulsion stirring at room temperature for 3 h to allow the evaporation of DCM (*see* **Note 18**).
10. Wash the nanoparticle suspension three times using an Amicon Ultra-4 centrifugal filter (MWCO 100 kDa) to remove the remaining organic solvent and free molecules (*see* **Note 19**).

3.4 Nanoparticle Characterization and Quantification of siRNA

1. Mix 20 μL of the prepared nanoparticle suspension with 1 mL of water in a DLS cuvette to measure the size and zeta potential of the nanoparticles using a DLS detector (Fig. 3a and Table 1).
2. Characterize particle size and polydispersity using a TEM.
3. Nanoparticle samples are negatively stained with 1% uranyl acetate for TEM (Fig. 3b) (*see* **Note 20**).
4. Analyze siRNA nanoparticle encapsulation efficiency by using Quant-iT RiboGreen RNA Assay Kit according to the manufacturer's protocol. siRNA encapsulation efficiency is defined as the mass fraction of initial siRNA that is encapsulated by the nanoparticles (Table 1).

4 Notes

1. Prepare and store at −20 °C in advance for use.
2. DCM has a low viscosity and vapor pressure and therefore tends to drip out of pipettes. A way to help prevent this is to pre wet the inside of the pipette by pipetting some DCM and then ejecting it from your pipette before taking in the amount needed.
3. You may need to aid polymer dissolution through stirring for about 30 min.
4. Make sure to prepare the NHS and EDC solution freshly before use.
5. For maximum conjugation efficiency, dissolve NHS and EDC in anhydrous DCM immediately before adding to PLGA-COOH. The reaction time should not exceed 1.5 h for a high PLGA-NHS activity with the amine group of PEG.
6. Add the reaction solution dropwise into two of 50 mL centrifuge tubes with 40 mL of cold ethyl ether–methanol washing solvent (50:50 v/v) while gently shaking.
7. For the maximum polymer recovery, present the centrifuge at 4 °C and centrifuge immediately after polymer precipitation.
8. For the second and third precipitation and centrifugation, add 2 mL of DCM to dissolve the precipitate firstly and then add the solution dropwise into 40 mL of cold ethyl ether–methanol washing solvent (50:50 v/v) for centrifugation. Supernatant from the washing steps should be clear, especially after the second wash.
9. Normally, dry the pellet under vacuum for 2–4 h.
10. The reaction time can be shortened but a minimum of at least 12 h is needed for the best results.

11. Let DCM evaporate in a fume hood for 60–80 min before washing.
12. To achieve a more efficient polymer conjugation, use a vacuum pump at least 40–50 Torr to rapidly evaporate residual solvents. The reaction yield should be between 90% and 95%.
13. The column chromatography of the crude product is carried out on a silica gel (230–400 mesh particle sizes) at room temperature and monitored by thin-layer chromatography (TLC). Firstly, elute with DCM until residual 1,2-epoxytetradecane totally disappears (normally 2–4 column volume of eluent is needed). Then change the eluent to DCM–methanol–NH_4OH (75:22:3 v/v) until all the G0-C14 come out, which will take 1–2 h normally. The purified G0-C14 is afforded as a pale yellow viscous oil with a yield of 40–50%.
14. Mix it well and put into 4 °C at least 10 min for cooling.
15. Set the probe sonicator at 30% power (the power output recorded 12–15 W when a 7 mL vial utilized), 5 s pulse-on time and 1 s pulse-off time (That refers to 20% duty cycle), and 15 s process time. Basically, 12–15 s is workable. If the process time is longer than 15 s, the sonicated liquid will spray out because of the high energy generated during sonication. Immerse the probe into 1/2 depth of the height of the liquid in the vial. Take care to follow the next step quickly after the sonication is finished so that the particles do not coalesce.
16. The emulsion drops will go to the bottom of the vial. Do not shake or mix it with the aqueous solution.
17. Add the emulsion to the stirring water as quickly as possible. The stirring speed should not be too low or too high. A speed of approximately 1000 rpm is best.
18. During this period, the solution should turn into a semiopaque solution from a white nontransparent suspension.
19. The nanoparticles cannot be centrifuged at a speed higher than 3000 × *g*.
20. Uranyl acetate is radioactive and highly toxic if ingested. Please consult safety data sheets and local safety regulations concerning this material.

References

1. Kumar P, Wu H, McBride JL et al (2007) Transvascular delivery of small interfering RNA to the central nervous system. Nature 448:39–43
2. Díez S, Miguéliz I, Tros de Ilarduya C (2009) Targeted cationic poly (D,L-lactic-co-glycolic acid) nanoparticles for gene delivery to cultured cells. Cell Mol Biol Lett 14: 347–362
3. Wilson DS, Dalmasso G, Wang L et al (2010) Orally delivered thioketal nanoparticles loaded with TNF-α–siRNA target inflammation and

inhibit gene expression in the intestines. Nat Mater 9:923–928

4. Shi J, Xiao Z, Votruba AR et al (2011) Differentially charged hollow core/shell lipid–polymer–lipid hybrid nanoparticles for small interfering RNA delivery. Angew Chem Int Ed 50:7027–7031
5. Zhu X, Xu Y, Solis LM et al (2015) Long-circulating siRNA nanoparticles for validating Prohibitin1-targeted non-small cell lung cancer treatment. Proc Natl Acad Sci U S A 112:7779–7784
6. Shi J, Xu Y, Xu X et al (2014) Hybrid lipid-polymer nanoparticles for sustained siRNA delivery and gene silencing. Nanomedicine 10:897–900
7. Lee SJ, Kim MJ, Kwon IC et al (2016) Delivery strategies and potential targets for siRNA in major cancer types. Adv Drug Deliv Rev 104:2–15
8. Kim HJ, Kim A, Miyata K et al (2016) Recent progress in development of siRNA delivery vehicles for cancer therapy. Adv Drug Deliv Rev 104:61–77
9. Parlea L, Puri A, Kasprzak W et al (2016) Cellular delivery of RNA nanoparticles. ACS Comb Sci 18:527–547
10. Jain RA (2000) The manufacturing techniques of various drug loaded biodegradable poly(lactide-co-glycolide) (PLGA) devices. Biomaterials 21:2475–2490
11. Cheng J, Teply BA, Sherifi I et al (2007) Formulation of functionalized PLGA–PEG nanoparticles for in vivo targeted drug delivery. Biomaterials 28:869–876
12. Liu Y, Gunda V, Zhu X et al (2016) Theranostic near-infrared fluorescent nanoplatform for imaging and systemic siRNA delivery to metastatic anaplastic thyroid cancer. Proc Natl Acad Sci U S A 113:7750–7755
13. Iqbal M, Zafar N, Fessi H et al (2015) Double emulsion solvent evaporation techniques used for drug encapsulation. Int J Pharm 496:173–190
14. Farokhzad OC, Cheng J, Teply BA et al (2006) Targeted nanoparticle-aptamer bioconjugates for cancer chemotherapy in vivo. Proc Natl Acad Sci U S A 103:6315–6320
15. Lemoine D, Préat V (1998) Polymeric nanoparticles as delivery system for influenza virus glycoproteins. J Control Release 54:15–27
16. Mu L, Feng SS (2003) A novel controlled release formulation for the anticancer drug paclitaxel (Taxol®): PLGA nanoparticles containing vitamin E TPGS. J Control Release 86:33–48
17. McCall RL, Sirianni RW (2013) PLGA nanoparticles formed by single- or double-emulsion with vitamin E-TPGS. J Vis Exp e51015
18. Gref R, Minamitake Y, Peracchia M et al (1994) Biodegradable long-circulating polymeric nanospheres. Science 263:1600–1603
19. Xu X, Xie K, Zhang X-Q et al (2013) Enhancing tumor cell response to chemotherapy through nanoparticle-mediated codelivery of siRNA and cisplatin prodrug. Proc Natl Acad Sci U S A 110:18638–18643

Chapter 16

Oxime Ether Lipids as Transfection Agents: Assembly and Complexation with siRNA

Anu Puri, Serena Zampino, Mathias Viard, and Bruce A. Shapiro

Abstract

RNAi-based therapeutic approaches to combat cancer and other diseases are currently an area of great interest. However, practical applications of this approach rely on optimal tools to carry and deliver siRNA to the desired site. Oxime ether lipids (OELs) are a class of molecules among other various carriers being examined for siRNA delivery. OELs, relatively new candidates, belong to a class of non-glycerol based lipids and have begun to claim their place as an siRNA delivery carrier in the field of RNAi therapy. Chemical synthesis steps of OELs are considered relatively simple with the ability to modify the functionalities as desired. OEL–siRNA complexes can be assembled in the presence of serum-containing buffers (or cell culture media) and recent data from our and other groups have demonstrated that OELs are viable carriers for siRNA delivery in the cell culture systems. In this chapter, we provide the details of experimental protocols routinely used in our laboratory to examine OEL–siRNA complexes including their assembly, stability, and transfection efficiencies.

Key words Oxime ether lipids, RNAi therapy, Nucleic acid delivery oxime ether lipids, Structure--activity relationship, Nonsymmetric hydrophobic domain, Lipoplexes, RNA interference, Breast cancer cells

1 Introduction

Manipulations of disease-specific genes by homing small interfering RNA (siRNA) to targeted sites in vitro and in vivo are currently under intense studies. The success of RNAi therapy approaches are, however, dependent on important considerations such as the availability of siRNA molecules in their functional form in the biological system. The site of action of the RNAi machinery is intracellular and involves processes beginning from the cleavage of the long Dicer substrate RNAs (DS RNAs) by the enzyme Dicer. This reaction results in the formation of 19–21 nt siRNA duplexes. Complexation of an siRNA duplex with the Argonaute 2 protein results in the RNA induced silencing complex (RISC). Degradation of the sense strands of the double stranded RNA is catalyzed by

Eckart Bindewald and Bruce A. Shapiro (eds.), *RNA Nanostructures: Methods and Protocols*, Methods in Molecular Biology, vol. 1632, DOI 10.1007/978-1-4939-7138-1_16, © Springer Science+Business Media LLC 2017

Argonaute, whereas the antisense strand–RISC complexes bind to the target mRNAs and obliterate gene function [1]. Two main barriers that compromise the utility of siRNA are its degradation by nucleases by plasma components and its inability to cross cellular membranes due to its negative charge. Therefore, various structurally defined carriers have been developed to facilitate siRNA delivery to its site of action; cationic lipids are one of the most commonly studied nonviral agents examined to date.

A suitable lipid carrier design for siRNA delivery entails inclusion of one or more positive charges as an essential feature while maintaining its low nonspecific cytotoxicity. Other important features include its nondegradability by blood plasma components and tunable properties to ensure endosomal escape. The most popular cationic lipids, DOTAP and DOTMA contain carbonyl and ether linkages respectively. As expected, these lipids associate with negatively charged siRNA via electrostatic interactions resulting in the formation of lipoplexes (lipid–nucleic acid nanoparticles, (LNPs)). Other classes of lipids studied thus far contain various modifications in the cationic head group region or the linker domain or the hydrophobic tail region (including chain length and degree of unsaturation). Lipid hydrocarbon chains lengths are also likely to contribute to the cytotoxicity of cationic lipids [2].

It has been observed that the linking functionality of a lipid molecule contributes significantly towards the stability and observed nonspecific toxic effects [3]. To address these limitations, various chemical linkages such as orthoester [4], carbamate [5], amide [6, 7], and phosphoramidate [8] moieties have been investigated (Fig. 1). Recently, OELs that contain oxime ether bonds have been developed as nucleic acid transfection agents. The assembly of oxime linkages can be achieved by using click chemistry [9] without increasing the in vivo cytotoxicity [10–12] (Table 1). Another advantage of the molecules containing oxime ether bonds is their robustness at pH 7 and susceptibility at low pH conditions. Oxime ether lipids (containing quaternary ammonium head groups) were first reported by Biswas et al. and were tested to deliver nucleic acids (pDNA and siRNA) in cell culture assays [13]. Our recent studies have further evaluated the structure–function activity relationship of OELS to their transfection efficiencies [14]. As can be seen in Fig. 2, the examined lipids provide a versatile range of modifications in the carbon chain length from C_{12} to C_{18}, the degree of unsaturation, and modifications in the polar head groups by introducing hydroxyl groups.

The general methods and protocols employed to test OELS as siRNA delivery agents are similar to that used for other lipids. OEL liposomes are prepared by a probe sonication method. Inclusion of a destabilizing lipid such as DOPE in OEL liposomes is also critical as it facilitates endosomal escape. The formation of OEL–siRNA complexes depends on the ratios of lipids to the siRNA which is

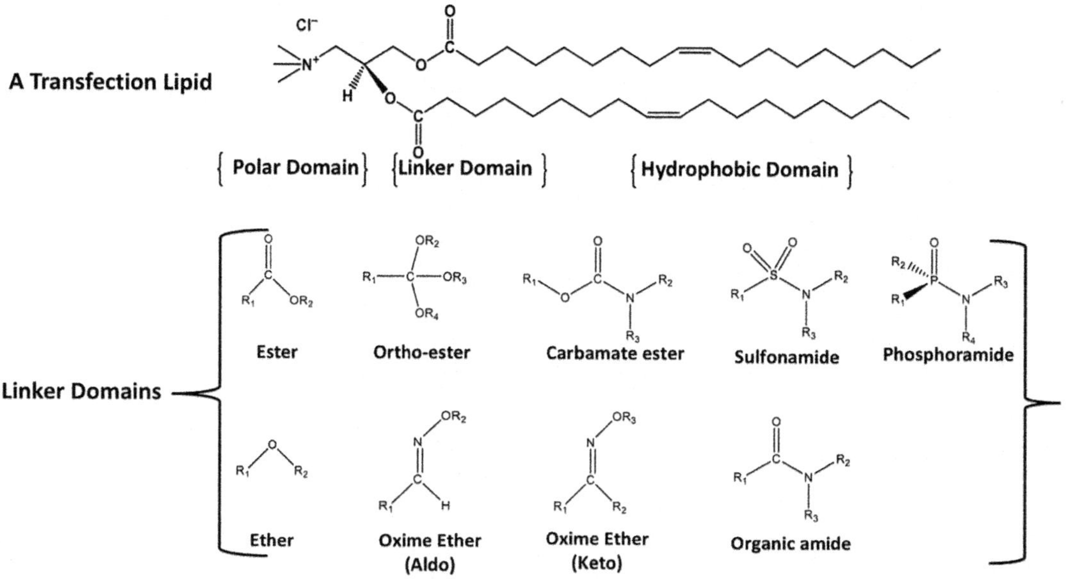

Fig. 1 General features of a transfection lipid. *Top Panel*, essential domains in a lipid molecule comprise a polar domain, linker domain and a hydrophobic domain as indicated. *Bottom Panel*, various linker domains examined to date in the context of transfection of lipids as indicated

Table 1
General properties of oxime ether lipids (OELs)

1	Robust chemical synthesis methodology
2	Ease of manipulation of structural motifs
3	Degradation of oxime ether linkages (Present in FDA-approved pharmaceuticals (Fluvoxamine))
4	Oxime ethers can serve as linking domains
5	Can transfect in the presence of serum containing media
6	Biological compatibility—well tolerated in cell culture experiments

based on electrostatic interactions. The association with siRNA results in modulation of overall charge and is determined by measurement of zeta potential. Typically, DS RNA duplexes are used for the formation of lipoplexes. However, studies that do not involve siRNA-mediated gene silencing, fluorescently labeled DNA–RNA hybrids can be used as an alternate choice for cost efficiency reasons. Typical experiments for determining the functionality of OEL–siRNA lipoplexes are monitored by two independent cell biology protocols, cellular uptake and gene silencing activity. This protocol can be used to examine the essential features of OELs including charge, hydrophobicity, and polarity of the lipid

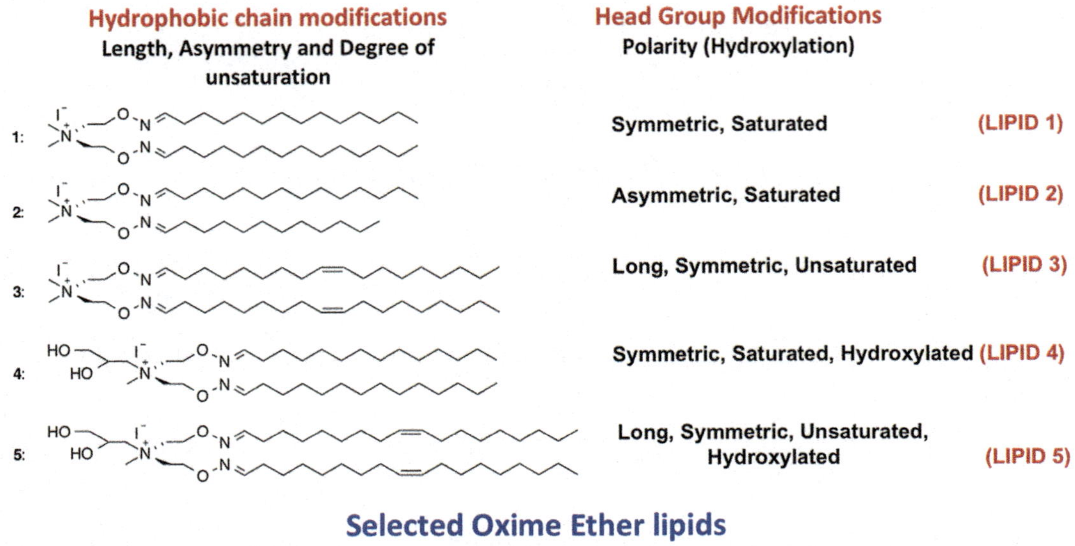

Fig. 2 Chemical structures of oxime ether lipids (OEL) examined for siRNA Delivery. The chemical modifications for each molecule shown in the figure are described on the *right*. The lipids shown here are also denominated as LIPID 1, 2, 3, 4, and 5 as indicated. (The figure is a modified version from Gupta et al. [14])

head group for efficient delivery of siRNA (*see* Figs. 1 and 2). The detailed protocols for studies of OEL–siRNA complexes followed in our laboratory are described below.

2 Materials and Equipment

2.1 Oxime Ether Lipids (OEL) Liposomes

2.1.1 Materials

1. Lipids: 1,2-dioleoyl-sn-glycero-3-phosphoethanolamine 18:1 (Δ9-Cis) PE (DOPE) (M.W. 744) from Avanti Polar Lipids, supplied as 10 mg/ml in chloroform.
2. OELs synthesized by Dr. Michael Nantz's laboratory. The chemical synthesis of OELs has been described previously by Nantz lab [13]. Stock solution prepared in chloroform at 10 mg/ml, and stored at −20 °C in crimped glass vials (1 ml capacity).
3. Water: ultrapure molecular biology grade.
4. DPBS without calcium or magnesium, sterile filtered, Cat # 17-512F (Lonza).

2.1.2 Equipment

1. Probe Sonicator. (We use Branson SLPE and SLPT Digital Sonic Dismembrator (SLPE Energy Mode 117 V, 60 Hz) Probe sonicator and Branson Ultrasonics Soundproof Enclosure for sonication of our samples. But any other brand with similar specifications can be used).
2. Molecular size and zeta potential analyzer: Zetasizer Nano ZS equipped with a He-Ne ion laser of 633 nm (Malvern Instruments) or equivalent.

2.2 Assembly of Nucleic Acid–OEL Complexes

2.2.1 Materials

1. *RNA sequences for DS RNA targeting eGFP:*

   Sense: 5′-pACCCUGAAGUUCAUCUGCACCACCG;

   Antisense: 5′-CGGUGGUGCAGAUGAACUUCAGGGUCA. The 5′ side of the sense strand is phosphorylated. Please note that the presented protocol is general and other siRNAs can be chosen instead.

   Fluorescently labeled DNA: DNA antisense labeled with Alexa 488: 5′GGAGACCGTGACCGGTGGTGCAGATGAACTTC AGGGTCAtt/3AlexF488N

2. Assembly buffer: 50 mM NaCl, 2 mM $MgCl_2$. Store at room temperature (*see* **Note 1**).

2.2.2 Equipment

1. Molecular size and zeta potential analyzer as mention under Subheading 2.1.2.
2. Dry block heater, Multiblock. Heater set to a temperature of 90 °C.

2.3 siRNA Uptake Studies

2.3.1 Materials

1. Human breast cancer cells MDA-MB-231 (ATCC).
2. Dulbecco's modified Eagle's medium (DMEM). Store at 4 °C.
3. Fetal bovine serum. Store at −20 °C.
4. Penicillin–streptomycin: 10,000 units/ml of penicillin and 10,000 μg/ml of streptomycin (GIBCO). Store at −20 °C.
5. Opti-MEM (GIBCO) transfection medium. Store at 4 °C.
6. Cell dissociation buffer (CDB) (GIBCO). Store at room temperature (*see* **Note 1**).
7. Incubator under humidification and 5% CO_2 at 37 °C.

2.3.2 Equipment

1. Cell culture cabinet.
2. CO_2 incubator.
3. Centrifuge (Sorvall Legend RT).
4. Flow cytometer (BD Biosciences).
5. Flow cytometry data analysis software such as CellQuest or equivalent.

2.4 Gene Silencing Studies

2.4.1 Materials

1. Sonicated OELs.
2. Assembled DS RNAs targeting GFP.
3. MDA-MB 231-GFP cells (MDA-MB-231/GFP) from Cell Biolabs Inc.

2.4.2 Equipment

1. Cell culture cabinet.
2. CO_2 incubator.
3. Centrifuge.

4. Flow cytometer (BD Biosciences) (*see* **Note 2**).
5. Fluorescence Microscope (Olympus IX70 model) (*see* **Note 2**).

3 Methods

3.1 OEL Liposomes

3.1.1 Preparation of Lipid Films

1. Mix lipids in a 5 or 7 ml glass tube as per requirement (*see* Table 2).
2. Remove chloroform under a stream of nitrogen or argon.
3. Resuspend lipids in 50–100 μl of chloroform–methanol (2:1 v/v) by vortex mixing.
4. Prepare films by gentle constant rotation of the tube under a stream of N_2 gas.
5. Protect the films from light (wrap in aluminum foil) and place in a desiccator attached to house vacuum for 12–24 h at room temperature (*see* **Note 1**).
6. For longer storage transfer the films to a −20 °C freezer.

3.1.2 Preparation of Liposomes (MLVs)

1. Add enzyme-free water to the lipid films (volumes per Table 1) (Fig. 3) (*see* **Note 3**).
2. Vortex the samples using a vortex mixture. Heat the samples to 50–55 °C to ensure complete suspension of the lipids and vortex again.
3. The lipid suspension at this point is called multilamellar liposomes or multilamellar vesicles (MLVs).
4. Transfer MLVs to either 5 or 14 ml capacity polypropylene round bottom tube depending on the volume of MLVs. (we

Table 2
Routinely used concentrations of oxime ether lipids for liposome formation

Lipid 4[a,b]	DOPE[a]	Total lipid (mg)	Water (ml)	Final lipid/ml (mg)
0.34 μmol (25 μl)	0.34 μmol (25 μl)	0.5	1.5	0.3
0.68 μmol (50 μl)	0.68 μmol (50 μl)	1	3	0.3
3.40 μmol (250 μl)	3.40 μmol (250 μl)	5	0.5	10.0

[a]Lipid stocks at 10 mg/ml in chloroform
[b]The values are provided for LIPID 4 as an example and are applicable to other OELs such as the lipids shown in Fig. 2

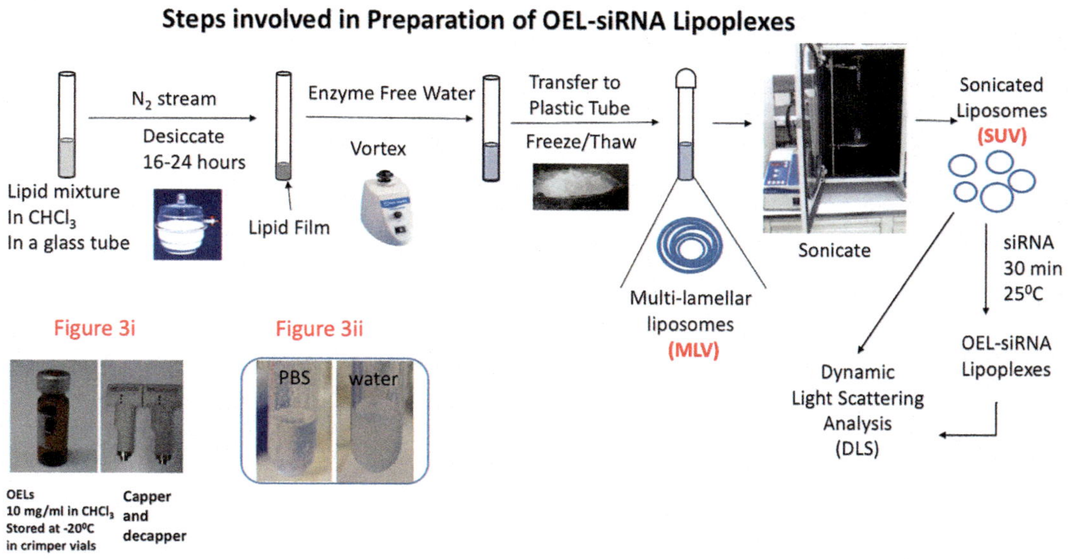

Fig. 3 Protocol to prepare oxime ether lipid–siRNA lipoplexes. Various steps and equipments involved in the preparation of OEL–siRNA lipoplexes are shown in the Figure as indicated. *Bottom panel*, *i* shows the vials and capper–decapper units we routinely use for storing the lipids. *ii* shows that reconstitution of OELS in PBS results in aggregation of the samples

have routinely used Falcon brand tubes but any brand is suitable).

5. Subject the samples to freeze–thaw cycles (5–10 times) to ensure homogeneity of the MLVs (This step is achieved by placing the tube in an ice bucket containing dry ice–ethanol and heating the frozen samples in a water bath at 45–50 °C).

3.1.3 Preparation of Sonicated Liposomes (SUVs)

1. Place MLVs (from Subheading 2.2) in a sonicator attached to a microtip probe with approx. ¼ cm of the probe inserted in the MLV suspension equipped with a beaker containing ice water.
2. Sonicate samples using the following settings with desired number of cycles: 1 min sonication with 1 min rest (time) at a pulse of 30%.
3. Typically, the MLVs ranging from 0.3 to 2 mg/ml lipid are sonicated for 7 cycles. MLVs dispersed at higher lipid concentrations (5–10 mg/ml) are sonicated using 14 cycles. The number of cycles for a given concentration of lipid in the samples is determined by examining the resulting hydrodynamic size (*see* Subheading 3.1.4 below).
4. Store sonicated liposomes at 4 °C and use for further analysis and experiments. Liposomes remain stable at least up to 3 weeks.

3.1.4 Hydrodynamic Size and Zeta Potential (Charge) Analysis of Sonicated liposomes Using Dynamic Light Scattering (DLS)

1. Dilute liposomes (from Subheading 3.1.3) in the range of 1:20 to 1:100 in water in a disposable cuvette compatible with the molecular sizer instrument in a total volume of 0.4 ml. The final concentration of liposomal lipids ranges from 3 to 15 μg/ml.
2. Perform DLS Measurements according to the instructions of the manufacturer of the utilized molecular sizer. Analyze the data using the provided software. Sonicated OEL liposomes typically exhibit a diameter ranging from 70 to 100 nm with a polydispersity index of <0.2 and a zeta potential of about 60 mV. A sample data set is shown in Fig. 4a for before and after sonication of the liposomes.

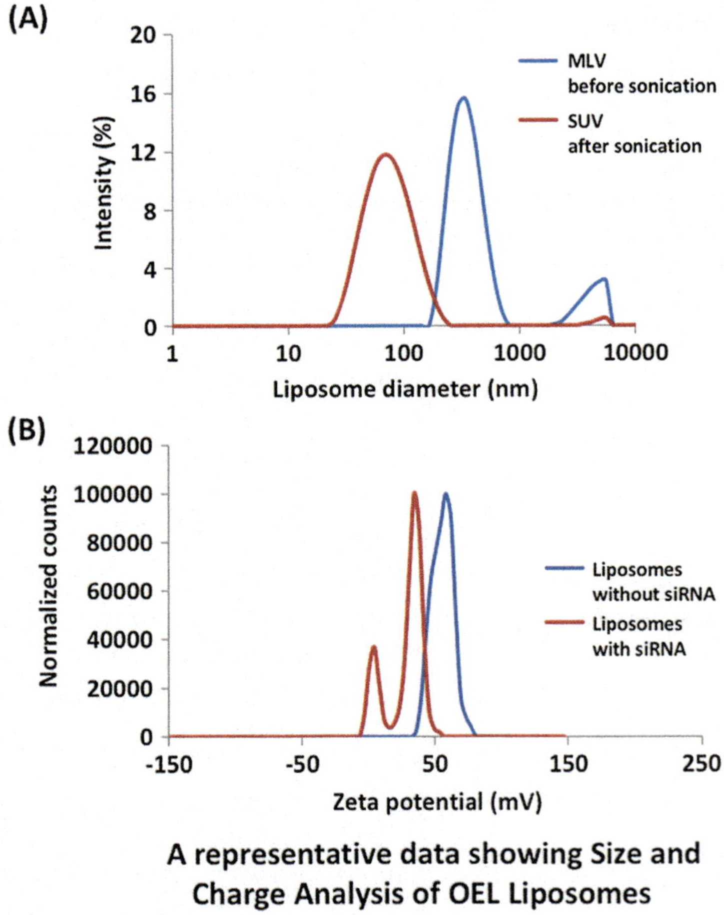

Fig. 4 Hydrodynamic size and zeta potential analysis of oxime ether lipid (OEL)–siRNA lipoplexes. (**a**) A representative data set on the average diameter of MLVs and sonicated OEL liposomes (SUVs) containing LIPID 4 is shown as indicated. The size (*X*-axis) are plotted vs. intensity in a population mode. (**b**) Zeta potential of LIPID 4 (10 μM) alone and in complexation with DS RNA (50 nM) are shown as indicated

3.2 Assembly of Nucleic Acid–OEL Complexes

Sense and antisense strands of known concentrations of nucleic acids are used for the preparation for RNA–DNA hybrids and DS RNA duplexes.

3.2.1 Preparation of siRNA Duplexes

1. Set the multiblock heater to a temperature of 90 °C.
2. Prepare appropriate concentrations of individual sense and antisense strands. Mix duplexes of either RNA–DNA hybrids or DS RNAs in assembly buffer to get 10 μM stock solution.
3. Heat duplexes prepared in the assembly buffer to 90 °C for 1 min using the preheated multiblock heater.
4. Remove the samples and allowed to anneal at room temperature (*see* **Note 1**).

3.2.2 Association of siRNA Duplexes with Sonicated OELs

1. According to the requirement of the cell culture experiments (*see* the following Subheadings 3.3 and 3.4 below) gently mix appropriate volumes of nucleic acids to the OEL liposomes to achieve the desired final concentration for the experiments (*see* **Note 4**).
2. Incubate the mixture of liposomes and nucleic acids for 30 min at room temperature (*see* **Note 1**).
3. Measure size and zeta potential of siRNA–OEL lipoplexes using the molecular sizer as described in the preceding section. An example of an analysis of the lipoplexes is shown in Fig. 4b.
4. Add appropriate amounts of nuclease/protease-free water, Opti-MEM or DMEM plus 10% FBS and 1% penicillin–streptomycin (complete media) to achieve a final concentration of lipoplexes as desired for cell biology assays (*see* Subheadings 3.3 and 3.4).

3.3 Uptake Studies (See Notes 5–7)

1. Prepare 0.3–10 mg/ml sonicated OELs (from Subheading 3.1).
2. Prepare assembled fluorescent siRNA–OEL lipoplexes hybrids where the DNA antisense strand is labeled with Alexa 488 (from Subheading 3.2.2).
3. One day before the transfection, seed MDA-MB-231 cells at a density of 100,000 cells/well using 12-well plates in 1 ml of complete medium in the CO_2 incubator.
4. Prepare siRNA–OEL lipoplexes just before starting the transfection experiment. Dilute the lipoplexes (Subheading 3.2.2) using the appropriate volume of Opti-MEM to achieve a final volume of 1 ml, containing 50 nM siRNA and various concentrations of OELs.
5. Wash each well containing the cells with 1 ml of prewarmed PBS. Remove PBS and add 1 ml DS RNA–OEL mixture (from the previous step) and incubate the cells at 37 °C for 4 h.
6. At the end of incubations, remove media and wash the cells with PBS (1 ml each, two times per well). Add 1 ml complete media per well.

7. Continue incubations for further 20–24 h at 37 °C in the CO_2 incubator.
8. At the end of incubations, dissociate the cells with CDB (0.5 ml per well) by incubating for 10–15 min at 37 °C.
9. At the end of incubations, transfer the cells to 5 ml round bottom tubes (compatible with flow cytometer use).
10. Centrifuge the cells at 1500 rpm. Discard the supernatant and resuspend the cell pellets in 500 μl of PBS and keep the samples in an ice bucket until further use.
11. Read the Alexa 488 fluorescence of the hybrid duplexes associated with cells with a flow cytometer using a 488 nm laser. Analyze the data using a flow cytometry data analysis software. A reasonable uptake would result in a major shift in the peak to high values of fluorescent cells as compared to negative controls. *See* Fig. 5a for example result.

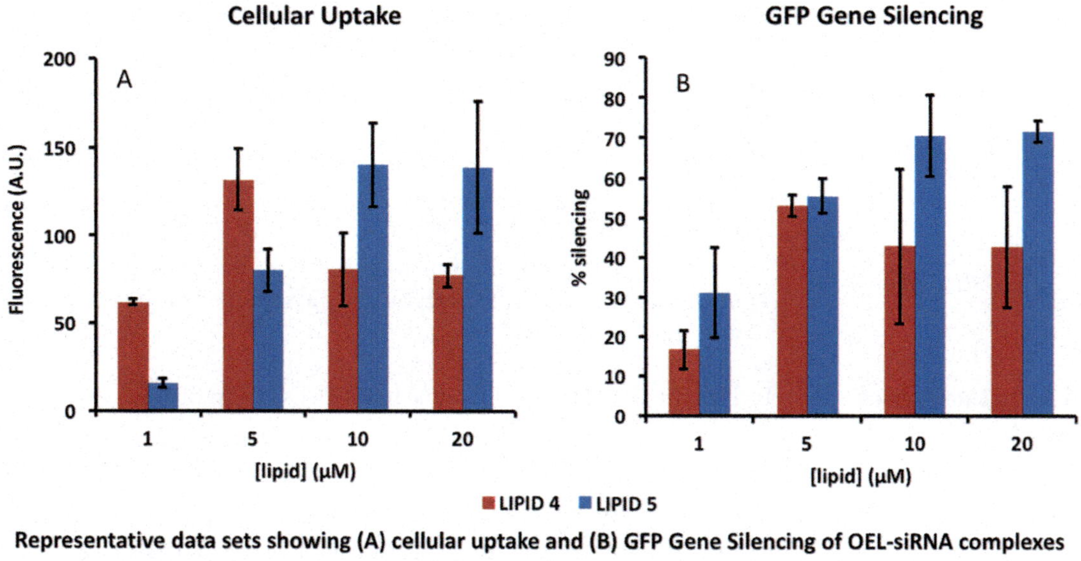

Fig. 5 Transfection studies of oxime ether lipid (OEL) liposomes–DS RNA/DNA hybrid complexes with breast cancer cells. (**a**) Cellular uptake: various concentrations of either LIPID 4 or 5 (1–20 μM) complexed with 50 nM Alexa 488 RNA–DNA hybrids were added to MDA-MB-231 cells plated on 12-well clusters and incubated for 20–24 h at 37 °C. Cellular uptake was monitored by monitoring Alexa 488 fluorescence using the flow cytometer. The values are expressed as relative fluorescence arbitrary units (A.U.) as a function of increasing concentration of the lipoplexes. The error bars represent standard error of the mean (±SEM) calculated from the values of two to three independent experiments. These data are adapted from Gupta et al. [14]. (**b**) Green fluorescent protein (GFP) gene silencing: GFP silencing efficiency of LIPID 4 or LIPID 5–DS RNA complexes was determined by adding various concentrations of the complexes containing 50 nM DS RNA to MDA-MB231-GFP cells plated on 12-well plates. The decrease in GFP fluorescence was measured with a flow cytometer. The values are expressed as percent of cells with GFP silencing. The error bars represent standard error of the mean (±SEM) calculated from the values of two to three independent experiments. These data are adapted from Gupta et al. [14]

3.4 GFP Gene Silencing Studies

1. One day before the experiment, seed MDA-MB-231/GFP at a density of 30,000 cells/well in 12-well plates in 1 ml of complete media in the CO_2 incubator.
2. To prepare OEL–siRNA lipoplexes, mix gently 5 μl of DS RNAs (10 μM) to various OEL concentrations. Incubate this mixture for 30 min at room temperature (*see* **Note 1**) prior to transfection experiments.
3. Dilute OEL–siRNA lipoplexes (from the previous step) further to a final volume of 1 ml using Opti-MEM to achieve a final concentration of DS RNA duplexes at 50 nM and OELs at 0–20 μM.
4. Wash each well containing the cells with 1 ml of prewarmed PBS. Remove PBS and add 1 ml DS RNA–OEL mixture (from **step 3** above) and incubate the cells at 37 °C for 4 h.
5. At the end of incubations, remove media and wash the cells with PBS (1 ml × 2 per well). Add 1 ml complete media per well.
6. Continue incubations for further 50–68 h at 37 °C in the CO_2 incubator.
7. At the end of incubations, dissociate the cells with CDB (0.5 ml per well) by incubating for 10–15 min at 37 °C.
8. At the end of incubations, transfer the cells to 5 ml round bottom tubes (compatible with the used flow cytometer).
9. Centrifuge the cells at 400 × *g*. Discard the supernatant and resuspend the cell pellets in 500 μl of PBS and keep the samples in an ice bucket until further use.
10. Analyze the cells for GFP fluorescence as a measure of silencing using flow cytometer in channel 1 (488 nm laser). Analyze the results using the flow cytometry data analysis software. A reasonable gene silencing would result in an increase in the population of cells associated with lower fluorescence values as compared to negative controls. *See* Fig. 5b as a sample result showing silencing of GFP by DS RNA–OEL lipoplexes.

4 Notes

1. The values of room temperature are typically considered at 25 °C. However, in the context of this chapter, a temperature range between 18 and 26 °C is appropriate.
2. We have used BD Biosciences flow cytometer and Olympus microscope, but any equipment with similar features can be used.

3. Attempts to prepare liposomes in PBS result in large aggregates, and therefore, water is used to prepare as well as store the samples. Dilutions in PBS are only done prior to a desired experiment (*see* Fig. 3).
4. This step is done just prior to the cell biology experiments.
5. The protocol provided here can also be applied to other cell lines. However, initial toxicity studies of the OELs for a given cell line is recommended. The cytotoxicity of OELs is initially tested using the standard cell viability assays. In general, OELs were found to be nontoxic up to 50 μM when examined using MDA-MB 231 cells.
6. Any other brand of tissue culture plates with similar dimensions can be used as well.
7. The scale-up and scale-down of this protocol can be done by using 6-well or 24-well clusters. The number of cells initially plated should be the same per cm^2 area of the wells).

Acknowledgments

This research has been funded in whole or in part with Federal funds from the Frederick National Laboratory for Cancer Research, National Institutes of Health, under contract HHSN261200800001E. The content of this publication does not necessarily reflect the views or policies of the Department of Health and Human Services, nor does the mention of trade names, commercial products, or organizations imply endorsement by the US Government. This work was supported by the Intramural Research Program of the National Institutes of Health, Center for Cancer Research, the National Cancer Institute. We thank Dr. Kshitij Gupta (Indian Institute of Science, Bangalore) and Dr. Michael Nantz (the University of Louisville, Kentucky) for their critical role in the exploration of OELS for siRNA delivery.

References

1. Afonin KA et al (2015) Triggering of RNA interference with RNA-RNA, RNA-DNA, and DNA-RNA nanoparticles. ACS Nano 9 (1):251–259
2. Balasubramaniam RP et al (1996) Structural and functional analysis of cationic transfection lipids: the hydrophobic domain. Gene Ther 3 (2):163–172
3. Leventis R, Silvius JR (1990) Interactions of mammalian cells with lipid dispersions containing novel metabolizable cationic amphiphiles. Biochim Biophys Acta 1023(1):124–132
4. Chen H, Zhang H, McCallum CM, Szoka FC, Guo X (2007) Unsaturated cationic ortho esters for endosome permeation in gene delivery. J Med Chem 50(18):4269–4278
5. Liu D et al (2008) Structure-function relationship research of glycerol backbone-based cationic lipids for gene delivery. Chem Biol Drug Des 71(4):336–344
6. Byk G et al (1998) Synthesis, activity, and structure–activity relationship studies of novel cationic lipids for DNA transfer. J Med Chem 41(2):229–235

7. Savva M, Chen P, Aljaberi A, Selvi B, Spelios M (2005) In vitro lipofection with novel asymmetric series of 1,2-dialkoylamidopropane-based cytofectins containing single symmetric bis-(2-dimethylaminoethane) polar headgroups. Bioconjug Chem 16(6):1411–1422
8. Le Corre SS et al (2014) Cationic lipophosphoramidates with two different lipid chains: synthesis and evaluation as gene carriers. Org Biomol Chem 12(9):1463–1474
9. Kolb HC, Finn MG, Sharpless KB (2001) Click chemistry: diverse chemical function from a few good reactions. Angew Chem Int Ed Engl 40(11):2004–2021
10. Figgitt DP, McClellan KJ (2000) Fluvoxamine. An updated review of its use in the management of adults with anxiety disorders. Drugs 60(4):925–954
11. Zhong D, Li X, Wang A, Xu Y, Wu S (2000) Identification of the metabolites of roxithromycin in humans. Drug Metab Dispos 28 (5):552–559
12. Carmona S et al (2009) Controlling HBV replication in vivo by intravenous administration of triggered PEGylated siRNA-nanoparticles. Mol Pharm 6(3):706–717
13. Biswas S et al (2011) Hydrophobic oxime ethers: a versatile class of pDNA and siRNA transfection lipids. ChemMedChem 6 (11):2063–2069
14. Gupta K et al (2015) Oxime ether lipids containing hydroxylated head groups are more superior siRNA delivery agents than their non-hydroxylated counterparts. Nanomedicine (Lond) 10(18):2805–2818

Chapter 17

Polycationic Probe-Guided Nanopore Single-Molecule Counter for Selective miRNA Detection

Kai Tian, Ruicheng Shi, Amy Gu, Michael Pennella, and Li-Qun Gu

Abstract

MicroRNAs (miRNAs) are a class of noncoding RNAs that are being explored as a new type of disease biomarkers. The nanopore single-molecule sensor offers a potential noninvasive tool to detect miRNAs for diagnostics and prognosis applications. However, one of the challenges that limits its clinical applications is the presence of a large variety of nontarget nucleic acids in the biofluid extracts. Upon interacting with the nanopore, nontarget nucleic acids produce "contaminative" nanopore signals that interfere with target miRNA discrimination, thus severely lowering the accuracy in target miRNA detection. We have reported a novel method that utilizes a designed polycationic peptide–PNA probe to specifically guide the target miRNA migration toward the nanopore, whereas any nontarget nucleic acids without the probe bound is rejected by the nanopore. Consequently, nontarget species are driven away from the nanopore and only the target miRNA can be detected at low concentration. This method is also able to discriminate miRNAs with single-nucleotide difference by using PNA to capture miRNA. Considering the significance and impact of this substantial advance for the future miRNA detection in biofluid samples, we prepared this detailed protocol, by which the readers can view the experimental procedure, data analysis, and resulting explanation.

Key words Nanopore, miRNA (miRNA), Single molecule, Biosensor, Cancer, HIV-1 TAT peptide, PNA, Probe, Diagnostics, Nucleic acids

1 Introduction

The nanopore offers a single-molecule platform for developing the next generation of biosensors and exploring various life science problems [1–14]. The nanopore technology can be utilized for rapid and low-cost gene sequencing, because individual base identity can be read out through change of the electric current during gene fragment translocation in the nanopore [15–23]. The sensitivity of the nanopore has also been applied to analyze epigenetic biomarkers, such as DNA methylation [24] and gene damage [25]. Our recent effort has been to develop the nanopore sensor for the electric detection of microRNAs (miRNAs) [26, 27]. The motivation is that miRNAs are a class of noncoding RNAs playing

Eckart Bindewald and Bruce A. Shapiro (eds.), *RNA Nanostructures: Methods and Protocols*, Methods in Molecular Biology, vol. 1632, DOI 10.1007/978-1-4939-7138-1_17, © Springer Science+Business Media LLC 2017

important roles in gene expression modulation [28–31]. Numerous research results have found that miRNAs are potential cancer biomarkers [32–40]. Aberrant expression of specific miRNAs in cells and biofluids is associated with cancer development. Our nanopore-based single-molecule counting approach for circulating miRNA detection would offer a potential noninvasive tool for cancer diagnostics.

However, current nanopore sensors have limitations with respect to detecting miRNAs directly from patient samples. One of the challenges is "contaminative signals" generated by nontarget nucleic acids in the samples. The RNA extraction from biofluids not only contains the target miRNAs, but also numerous nontarget nucleic acids species, such as miRNAs, mRNAs, and tRNAs. Due to their negative charge, all these nucleic acids species can be driven to interact with the nanopore and change its conductance under the voltage. In addition to linear-form single-stranded and hybridized double-stranded nucleic acids, RNAs can also fold into tertiary structures that may clog the nanopore. Overall, the signals produced by nontarget nucleic acids can severely interfere with the accuracy of target miRNA detection. Accurate detection of miRNA in complex biological samples demands a method that enables the nanopore to only selectively capture the target miRNA, but reject any nontarget nucleic acids from entering the nanopore.

We have recently invented a polycationic probe guiding approach that can completely overcome this challenge [41]. Briefly we designed a polycationic probe as a cargo to specifically guide the migration of the target miRNA toward an engineered nanopore, whereas any nontarget species without probe binding migrates away from the nanopore driven by the voltage. Consequently, we can only observe signature signals of the target miRNA binding with the probe, but not any signals of nontarget nucleic acids. The probe comprises a short polycationic peptide linked to a peptide nucleic acid (PNA). The PNA is used to specifically capture target miRNA, while the polycationic peptide is a leader that guides the miRNA to be captured by the nanopore. Considering the broad application value and the important nanobiophysics mechanism of this method, we prepared this short protocol to introduce the detailed experimental procedure, including nanopore formation, nanopore current recording, signal analysis, and miRNA quantification, with a discussion on accuracy, sensitivity, specificity, versatility, multiple detection, and broad impact of the approach.

2 Materials

2.1 Reagents and Consumables

1. Pretreatment solution: 1:10 hexadecane–pentane. Mix 250 μL hexadecane with 2.5 mL pentane.

2. Lipid solution: 10 mg/mL. Dissolve 25 mg 1,2-diphytanoyl-sn-glycero-phosphocholine (Avanti Polar Lipids) in 2.5 mL pentane.
3. Nucleic acids stock solution: 1 mM. Dissolve synthetic miRNAs including miRNA *let 7a*, *7b*, or *7c* in RNAase-free water to 1 mM respectively.
4. Peptide–PNA probes corresponding to the miRNAs of interest (P_{7b} etc.) (*see* **Note 1**).
5. Recording solution: 0.5, 1, or 3 M KCl, buffered with 10 mM Tris and titrated with 1M HCl to pH 7.2.
6. α-hemolysin protein: 10–100 ng per 50 μL. Dissolve the K131D protein with ddH_2O.
7. Restriction enzymes: *EcoN* I or *Hind* III.
8. Expand Long Template PCR Synthesis Kit (Roche Diagnostic GmbH).
9. Coupled in vitro Transcription and Translation kit (IVTT, Promega).
10. XL10-gold super-competent *E. coli* cells.
11. Rabbit blood cell membrane (rabbit blood from Pel-Freez Biologicals).
12. Micron centrifugal filters that enable the concentration of biomolecules with a with a membrane nominal molecular weight limit (NMWL) of 10 kDa.

2.2 Setup and Instrument

1. Preparation of the chamber.
   (a) Compartments. Fabricate two Teflon compartments. Each compartment has a volume of 0.2–2 mL depending on specific design.
   (b) Partition. Prepare a piece of 25-μm thick Teflon film (Goodfellow) as partition and puncture a 100–150 μm wide tiny aperture in the center of the film. The lipid bilayer membrane will span over the aperture.
   (c) Chamber assembling. The chamber is assembled by tightly clicking the partition with two compartment blocks as shown in Fig. 1a. Seal the partition–compartment interface with vacuum grease. The cells on both sides of the partition are named *cis* and *trans*. The chamber is placed on the chamber holder, which has been preinstalled with two stirring motors under the bottom of the two cells for stirring. The chamber and the holder are placed in a Faraday box for shielding.
2. A pair of electrodes. Fabricate the Ag–AgCl electrode by merging a 0.5- or 1-mm wide silver wire in bleach overnight. A qualified electrode should have a dark brown surface. Fill 1.5%

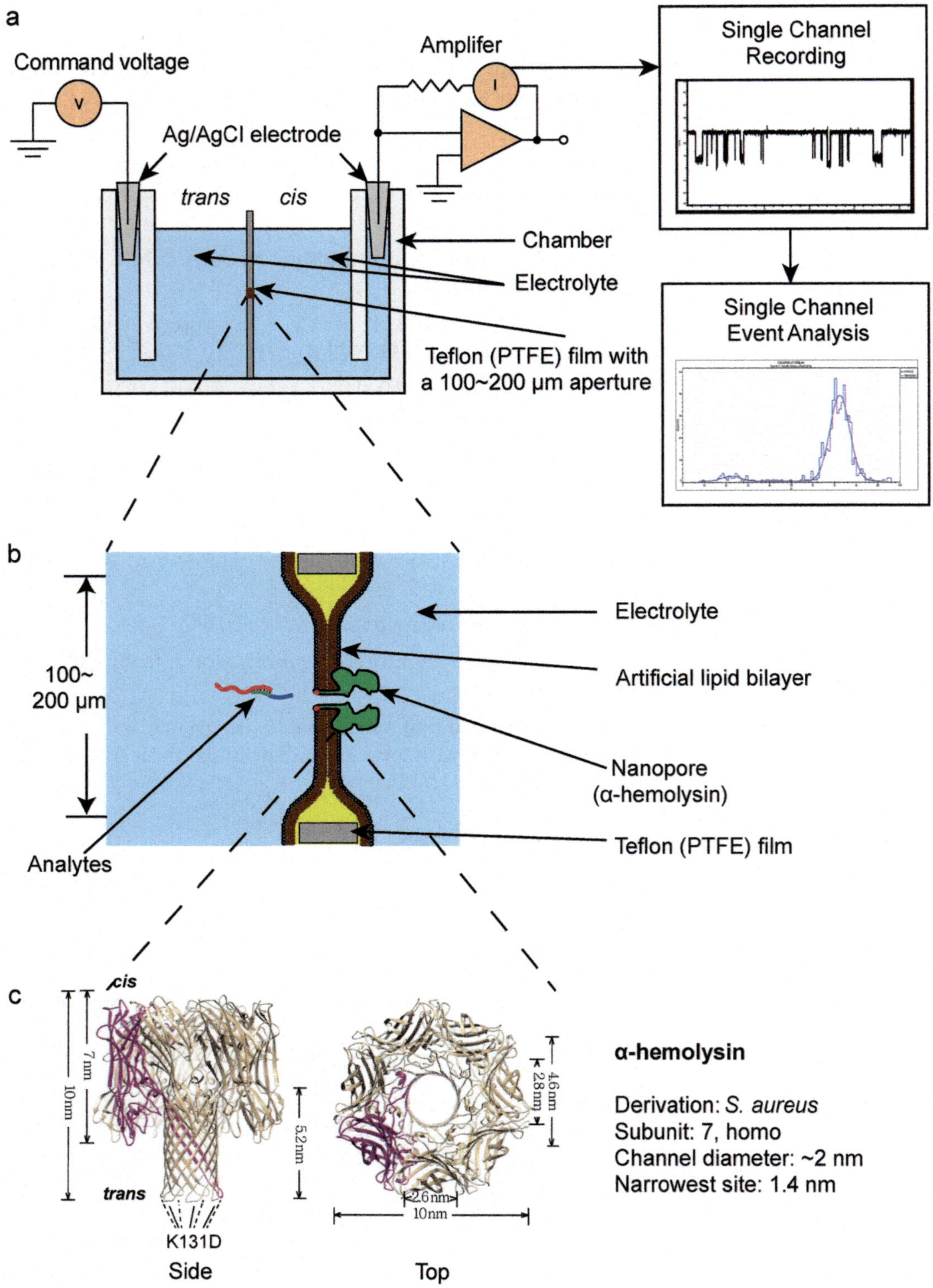

Fig. 1 Diagrams showing the lipid bilayer/protein pore experimental setup and components for electrical detection of miRNA in the engineered nanopore. (**a**) Schematic of experimental setup, including chamber, electrode, and amplifier (*left*). Nanopore current trace and analysis are also shown on the *right*. (**b**) Artificial lipid bilayer with a single nanopore embedding covers the aperture of Teflon film. (**c**) The structure of α-hemolysin nanopore and the engineered site in this method

agarose resolved in 3 M KCl in a 50 μL plastic pipette tip and fix the electrode in the gel.

3. Pico-Ampere current amplifier, used for recording ion current through the nanopore. The amplifier should be able to record pico-Ampere (ideally sub-pico-Ampere) currents.
4. Analog-to digital (AD) converter, used for acquiring the nanopore current data from the amplifier and for communicating the data to a computer. The AD converter should have a resolution of at least 16 bits and be capable of utilizing a sample rate of 250 kHz.
5. Computer that is compatible with the AD converter as well as software for data acquisition and analysis.

3 Methods

3.1 αHL Mutagenesis and Nanopore Purification

1. The K131D mutant αHL can be constructed using site-directed mutagenesis [42]. Digest the wild-type αHL plasmid by *EcoN* I or *Hind* III, respectively, into a linear template for PCR.
2. Add 20–30 ng linearized plasmid DNA and 200 pmol each of a nonmutagenic and a mutagenic primer with Expand Long Template PCR Synthesis Kit to 50 μL PCR mixture.
3. Perform PCR in the following temperature condition: 94 °C >5 min; 20 cycles of 94 °C 1 min, 50 °C 1 min, and 72 °C 3 min; 72 °C 6 min.
4. The two long PCR fragments, each from an enzyme digestion product, contain two overlapping sequences. Extract PCR fragments from the agarose gel.
5. Mix extracted PCR fragments.
6. Transform into XL10-gold super-competent *E. coli* cells, to form recombinant plasmid.
7. Extract mutant plasmid from *E. coli* cells.
8. Synthesize the K131D αHL protein using Coupled in vitro Transcription and Translation kit by following the protocol of the kit.
9. A 25 μL reaction contains 4 μL DNA template (400 ng/μL in stock solution) and complete amino acid mixture (including [^{35}S]methionine) in IVTT components, and is incubated at 37 °C for 1 h to produce protein monomers.
10. Mix the supernatant of the reaction with rabbit blood cell membrane to assemble the protein heptamer.
11. Separate the assembled heptameric protein by electrophoresis on a 7.5% SDS–polyacrylamide gel.

12. Cut out the oligomeric protein band.
13. Dissolve the gel band in 500 μL water.
14. Collect the protein using a micron centrifugal filter.
15. Divide the protein solution into ten aliquots, and store at −20 °C.

3.2 Nanopore Formation

1. Formation of lipid bilayer. Drop 2–5 μL of pretreatment solution by a 10 μL micropipette over the aperture of Teflon film, and immediately blow to dry. This aids in the formation a stable lipid bilayer.
2. Put a stirring bar into each cell of the chamber and place the chamber on the holder.
3. Connect the chamber to the amplifier through a pair of Ag–AgCl electrodes. Voltage provided by the amplifier is applied through the *trans* electrode; the *cis* cell is grounded.
4. Form the lipid bilayer according to the monolayer folding process described in ref. 43: inject 0.1–1 mL recording solution into each cell (0.2–2 mL), then drop 5 μL of lipid solution by a micropipette onto the surface of the solution.
5. Wait for 2–3 min to evaporate the pentane so as to form a molecular monolayer on the solution surface.
6. Gently inject another 0.1–1 mL recording solution from the bottom of each cell. The liquid surfaces on both sides are raised over the aperture of the partition. Immediately, the lipid monolayers on both sides automatically hybridize and form a bilayer membrane covering the aperture.
7. Use the data acquisition software to monitor capacitance and current thermal noise during membrane formation. A qualified lipid bilayer for a single channel recording should have a membrane resistance of ~100 GΩ, a capacitance of ~100–200 pF, and a current noise of 1.2–1.8 pA (the value of I_{RMS} on the amplifier panel).
8. Insertion of nanopore in the lipid bilayer. Upon bilayer formation, release 0.5–2 μL of protein solution to the *cis* solution close to the bilayer.
9. Gently stir the recording solution, while monitoring the current using data acquisition software. After a short time (tens of seconds to several minutes), the current suddenly increases sharply from zero to a specific level at a holding voltage, such as ~180 pA at +120 mV for K131D αHL, which indicates the insertion of a single nanopore in the lipid bilayer. Upon insertion, the wide entrance (nanocavity) of the pore faces the *cis* solution, while the narrow entrance (β-barrel) faces the *trans* solution (*see* the nanopore orientation in Fig. 1b).

3.3 Simultaneous Enrichment and Detection of miRNAs with a Polycationic Probe

1. Probe–miRNA hybridization. The probe P_{7b} contains a 10-base PNA sequence designed to specifically hybridize with miRNA Let-7b. The PNA was extended at the N-terminal with an 11-amino acid HIV-1 TAT polycationic peptide [44], which includes eight positively charged amino acids (six arginines and two lysines) (Fig. 2a). The P_{7b} and Let-7b mixture is added in *trans* solution (*see* **Notes 1** and **2**).
2. Observation of single probe–miRNA molecules. Apply a positive voltage, such as +180 mV, from the *trans* side to the grounded *cis* side and record the current using the data acquisition software. This voltage polarity can drive cationic molecules toward the pore while repelling anionic molecules away from the pore (*see* **Notes 3** and **4**).
3. Analysis of the signals. P_{7b} alone in *trans* solution shows a large number of Level-1 blocks (Fig. 2b). The duration of these blocks (τ_{off}) was 4.8 ms and their relative conductance I_R/I was 8.2% (I_R and I are currents of the block and the empty pore Fig. 2b). When the P_{7b} and miRNA Let-7b mixture is added in *trans* solution, a large number of distinct Level-2 blocks appear, while Level-1 blocks diminish. Compared with the Level-1 blocks, the Level-2 blocks were sixfold longer with a mean duration of 28 ± 4 ms and featured higher relative conductance at 26% (Fig. 2b). Therefore the Level-2 blocks serve as signatures for Let-7b identification. The occurring frequency of the signature signals generated by the P_{7b}·Let-7b complex (f_{sig}, the number of signature blocks per second) is linearly correlated with the Let-7b concentration, i.e., $f_{sig} \approx k_{on}\cdot[\text{Let-7b}]$, where k_{on} is the capture rate of the P_{7b}/ Let-7b complex and [Let-7b] is the target miRNA concentration. Thus, f_{sig} is used to quantify miRNA concentration. f_{sig} corresponds to measuring the intervals between adjacent signature blocks using the software. τ_{sig} is obtained by fitting the histogram of all interval values to an exponential distribution function:

$$f(t) = Ae^{-t/\tau}$$

   The inverted τ_{sig} is f_{sig}, i.e., $f_{sig}=1/\tau_{sig}$ (*see* **Note 5**).

4. Quantification of miRNAs. Repeat the P_{7b}/Let-7b experiments described above in a series of miRNA concentrations, such as 50 pM, 500 pM and 5 nM. Make a calibration curve using the set of f_{sig} at each concentration. The concentration of an unknown Let-7b sample can be determined according the calibration curve (*see* **Note 6**).

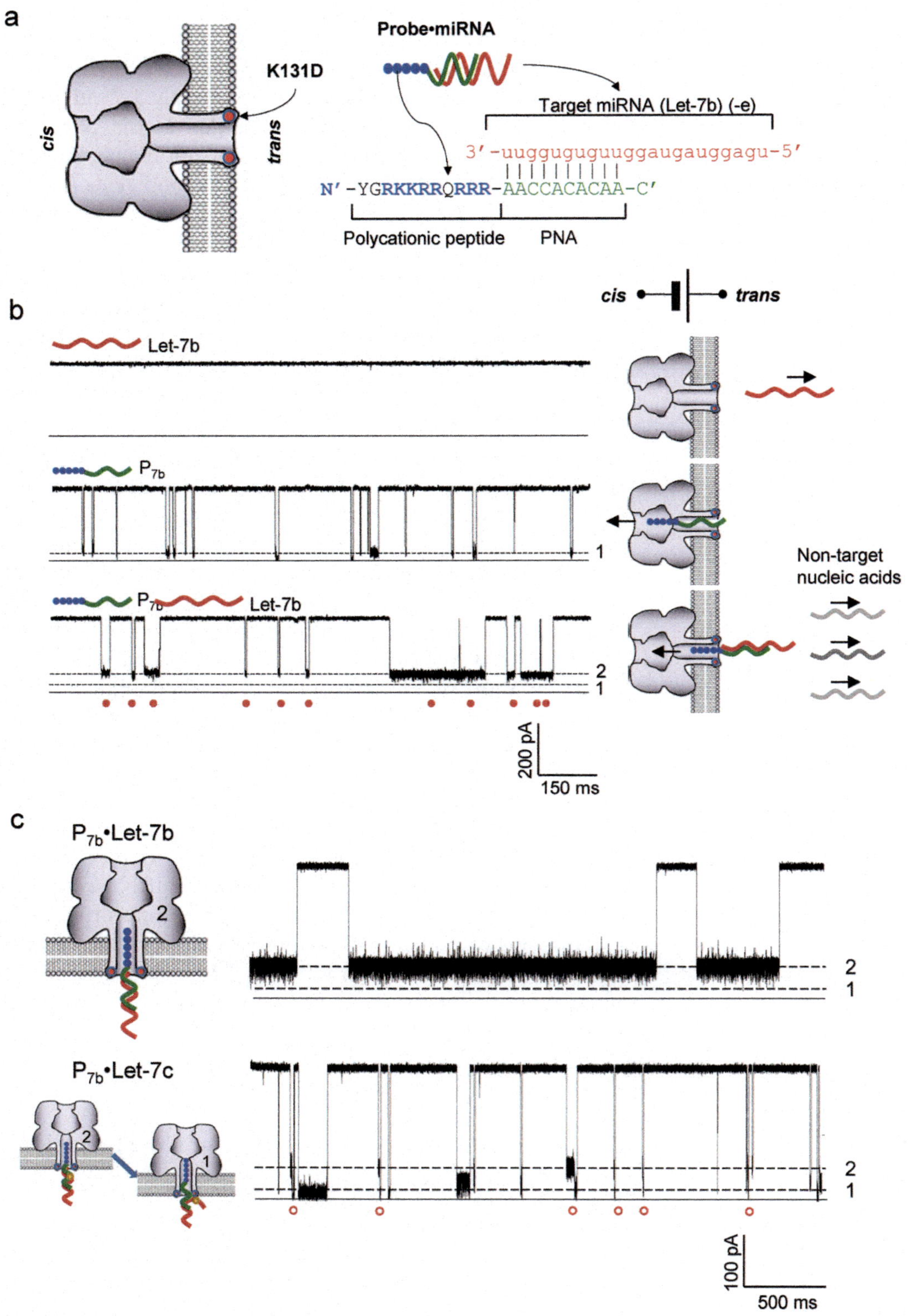

Fig. 2 Polycationic probe-guided specific detection of miRNA using the K131D protein nanopore. (**a**) Diagram showing cartoon of the *trans* opening of K131D with a group of negatively charged residues engineered around the *cis* entrance (*red dots* at the end of stem represent the negatively charged aspartic acids), and the

4 Notes

1. Polycationic probe. The polycationic probe is the key to guide the selective capture of target miRNA species. As shown in Fig. 2a, the probe is a peptide–PNA conjugate. PNA is used to capture miRNA, and the polycationic peptide is the leader that enables the nanopore to attract the probe·miRNA complex and guide its capture by the nanopore. The peptide leader is not necessary long, but need contain high density of positively charged amino acids to generate driving force near the nanopore entrance. The peptide–PNA copolymer can be synthesized (commercially available) at once without additional chemical cross-linking, as peptide and PNA have the same backbone. The sequence of the peptide leader is programmable. Its properties can be tuned by changing the peptide length and the number, position, and type of charged amino acids. The peptide can also be functionalized in any position of the sequence by chemical modification (such as cross-linking at cysteine), making it possible to generate unique signatures for multiplex detection.
2. Multiplex detections. This method has potential to be developed for multiplex detections, in which multiple target miRNAs as a biomarker panel can be simultaneously detected using one nanopore. Specific cancer diagnostics requires accurate detection of a biomarker panel that consists of multiple miRNAs. But currently one nanopore is only used to detect one miRNA. We have developed barcoded DNA probes for simultaneous detection of multiple miRNAs in one nanopore [27]. We designed a series of barcode probes to encode different targets. When the probe binds with the target, the barcode group such as polyethylene glycol attached on the probe can

Fig. 2 (continued) structures (sequences) of the polycationic probe, miRNA (Let-7b), and the probe–miRNA hybrid. The probe contains a HIV-1 TAT polycationic peptide (cationic Lys and Arg residues are marked in *blue*) and PNA sequence complementary to Let-7b sequence (marked in *green*). The sequence of miRNA Let-7b is marked in *red*; (**b**) (*Upper*) current trace of K131D pore showing that no block was observed in the presence of miRNA (Let-7b). (*Middle*) Current trace of K131D pore showing that the polycationic probe (P_{7b}) alone generated a large number of Level-1 blocks. (*Lower*) Current trace showing a large number of new Level-2 blocks (*solid red dots*) that were generated by the probe·miRNA hybrid (P_{7b}·Let-7b) in the presence of the miRNA (Let-7b) and its probe (P_{7b}). Models of suggested molecular processes are illustrated on the *right* of traces. Traces were recorded at +180 mV in 1 M KCl and 10 mM Tris (pH 7.2); (**c**) (*Upper*) Current traces and cartoon model showing characteristic blocks generated by the P_{7b}·Let-7b hybrid that is fully matched. (*Lower*) Current traces and cartoon model showing characteristic blocks generated by the P_{7b}·Let-7c hybrid that contains a single mismatched base pair (*lower*). Both blocks can be discriminated from the Level-1 block duration and the unzipping signature of the Level-2 terminal of the block (*hollow red dots*). Traces were recorded at +130 mV in 3 M/0.5 M (*cis*/*trans*) asymmetric solutions specifically modulate the pore's ion flow. The resulting signature serves as a marker for the encoded target, and counting different signatures allows simultaneous analysis of multiple targets in one pore. This approach was verified by using a panel of lung cancer-derived miRNAs as the target. Supported by these result, we can develop barcoded polycationic peptide–PNA probes to simultaneously detect multiple cancer-derived miRNAs in one nanopore. Specifically, the polycationic peptide sequence can contain a cysteine at a designated position, and various chemical groups can be attached to cysteine to generate specific blocking levels when the probe is trapped in the nanopore. Each miRNA will be driven by a barcoded carrier probe that can generate a distinct signature for target recognition. As a result, multiple miRNAs can be quantified by counting their distinct signatures in one pore.

3. Versatility. We have developed the polycationic probe guiding method to selectively detect miRNA in a complex mixture also containing nontarget species. The biophysical mechanism behind this nano-phenomenon is not the electrophoretic and the electroosmotic effect. The specific driving force comes from a short range (several nanometers) electrical field formed by a group of negatively charged residues engineered around the *trans* entrance of the K131D pore. This field can only generate an attractive driving force on the polycationic peptide of the probe, but has much less repulsive force acting on the attached miRNA. Therefore the net force on the probe·miRNA complex remains attractive. Our unpublished data indicates that this method is also applicable to long nucleic acids by using a probe that has the same polycationic peptide head as we used for miRNA detection. We expect that this biophysical mechanism can be utilized to selectively capture any gene fragment, long or short, in complex samples that contain a large number of nontarget species.

4. Significance and broad impact. (1) The single-molecule counting method described here has the potential to become a new player to meet clinical needs for the detection of miRNA and other noncoding RNAs. The system is highly sensitive, specific, label-free, without need of amplification or secondary detection, fast (<1 h), inexpensive, simple to operate, and offers a multiplex detection. It may provide higher accuracy than PCR in assaying of patient samples [26]. For example, it could potentially become a new platform to develop a confirmatory test when the lung nodule is identified by low-dose helical computed tomography (LDCT) in high risk populations for lung cancer screening. The validated nanopore technology could be applied to detect any pathogenic DNA/RNA, single nucleotide polymorphism (SNP), and epigenetics such as DNA

methylation, for early diagnosis, prediction of cancer metastasis, and monitoring of response to therapy and thus could have a broad impact. Notably, the nanopore's SNP discrimination capability can be used not only to distinguish the family members of miRNA but also to detect a point mutation of the oncogene DNA in blood (circulating nuclear acids) that holds a huge potential for the early detection and monitoring of cancer. (2) The method we described in this report represents a substantial step toward medical applications of the nanopore. Unlike traditional electrophoretic and electroosmotic driving that lack selectivity, the polycationic guiding approach we invented provides a new route to nanopore selective capture of any nucleic acids target in a mixture for noise-free detection. Therefore for the first time it becomes possible to translate the nanopore sensor into a clinically usable tool. (3) The lipid bilayer supported protein nanopore system is not stable enough for clinical samples. There are three strategies to resolve the problem. First, whole DNAs or RNAs can be extracted from samples to avoid the contact between lipid bilayer and "membrane breakers" in the samples. Second, more solid supported systems are available now in several companies, such as Oxford Nanopore Technologies. Third, hybrid nanopores, i.e. combined solid and protein nanopores, may provide a new way to improve the stability of nanopore systems in clinical use [4]. Our long-term vision is to provide a compact, durable, and reliable nanopore device for performing rapid, accurate, low-cost, and high-throughput assays in a clinical diagnostic setting. If this method can be validated, we would be able to selectively and sensitively detect any type of nucleic acids, DNAs or RNAs, long or short, in complex clinical samples. The outcome will open a new avenue with significance not only in cancer detection and other human disease molecular diagnostics, but more broadly in fields such as plant science and foodborne detection where rapid genetic detection is required.

5. Accuracy and specificity. The most unique and useful advantage of this polycationic probe guiding method is that the probe (peptide–PNA) can guide the target miRNA to migrate toward the nanopore and is detected, whereas any nontarget nucleic acid species that are not binding with the probe migrate away from the nanopore without generating any "contaminative" or interference signals. As a result of this "noise-free" detection, only target miRNA signals are specifically generated and accurately recognized. We have used this method to test the mixture of target miRNA and various nontarget RNA and DNA. It has been found that the capture rate for the target miRNA alone and that in the mixture are the same, suggesting the

accuracy of this method. The specificity of this method is reflected by the capability to discriminate single-base differences between miRNAs with similar sequence, such as the Let-7 miRNA family. For example, the probe P_{7b} has been designed to detect Let-7, and can form fully matched hybrid P_{7b}·Let-7b. When P_{7b} hybridizes with another miRNA Let-7c, which has one-base difference from Let-7b, the hybrid P_{7b}·Let-7c forms a single mismatched base pair. Recorded in 3 M/ 0.5 M *cis/trans* KCl (pH 7.2) asymmetric solution at +130 mV. The duration of Level-2 signature of fully hybridized P_{7b}·Let-7b was 2.3 s (Fig. 2c), suggesting that P_{7b}·Let-7b is so stable that it cannot be unzipped prior to release from the pore. In contrast, the P_{7b}·Let-7c hybrid containing one mismatch generates a distinct type of two-level block from Level 2 to Level 1 (Fig. 2c). The Level-2 signature was only 19 ms, about 120 times shorter than Level-2 blocks for P_{7b}·Let-7b; and the terminal Level-1 block corresponds to the unzipping of the hybrid followed by translocation of the probe through the pore, suggesting that one mismatch induced in a RNA–PNA duplex can greatly reduce its stability. Overall, different stabilities of RNA–PNA hybrids without and with mismatch can generate distinct signatures that can be utilized to discriminate single-base difference in miRNAs.

6. Sensitivity. The K131D mutant pore, which has been engineered with a ring of anionic aspartic acids at the *trans* opening, plays an important role in attracting cationic molecules. It greatly increases the target capture rate k_{on}. The P_{7b}·Let-7b complex was rarely trapped in the wild type, while its k_{on} was $80 \pm 9\ \mu M^{-1}\ s^{-1}$ in the K131D. Also, it is much higher than the k_{on} for the miRNA in complex with a DNA probe, which was reported as $1.4\ \mu M^{-1}\ s^{-1}$ [26]. Therefore the combination of the polycationic probe and the engineered nanopore enhance k_{on} over 50-fold. The frequency of P_{7b}/Let-7b signatures consistently increases with increasing Let-7b concentrations ranging from 50 pM to 5 nM at +180 mV. It can be fitted to a straight line in the log–log scale. Therefore, Let-7b can be detected as low as tens of picomolar. Note that the K131D pore itself may spontaneously generate blocks [45], but the occurrence is very low ($\sim 6 \times 10^{-3}\ s^{-1}$ above +140 mV), and its current pattern (amplitude) is different from the signature of the probe·miRNA complex. Therefore, the background noise signal from the pore itself can be excluded.

Acknowledgment

We appreciate Amy Gu and Michael Pennella who contributed to the nanopore experiments as highschool student research trainees.

This work was supported through initial NSF CAREER award 0546165 and NIH R01-GM079613, and current NIH R01-GM114204. This investigation was conducted in a facility constructed with support from the Research Facilities Improvement Program Grant C06-RR-016489-01 from the National Center for Research Resources, National Institutes of Health.

References

1. Bayley H et al (2008) Droplet interface bilayers. Mol Biosyst 4(12):1191–1208
2. Bayley H, Jayasinghe L (2004) Functional engineered channels and pores (Review). Mol Membr Biol 21(4):209–220
3. L-Q G, Shim JW (2010) Single molecule sensing by nanopores and nanopore devices. Analyst 135(3):441–451
4. Hall AR et al (2010) Hybrid pore formation by directed insertion of [alpha]-haemolysin into solid-state nanopores. Nat Nanotechnol 5 (12):874–877
5. Hornblower B et al (2007) Single-molecule analysis of DNA-protein complexes using nanopores. Nat Methods 4(4):315–317
6. Howorka S, Siwy Z (2009) Nanopore analytics: sensing of single molecules. Chem Soc Rev 38(8):2360–2384
7. Langecker M et al (2012) Synthetic lipid membrane channels formed by designed DNA nanostructures. Science 338(6109):932–936
8. Ma L, Cockroft SL (2010) Biological nanopores for single-molecule biophysics. ChemBioChem 11(1):25–34
9. Majd S et al (2010) Applications of biological pores in nanomedicine, sensing, and nanoelectronics. Curr Opin Biotechnol 21(4):439–476
10. Movileanu L (2009) Interrogating single proteins through nanopores: challenges and opportunities. Trends Biotechnol 27(6):333–341
11. Olasagasti F et al (2010) Replication of individual DNA molecules under electronic control using a protein nanopore. Nat Nanotechnol 5 (11):798–806
12. Venkatesan BM, Bashir R (2011) Nanopore sensors for nucleic acid analysis. Nat Nanotechnol 6(10):615–624
13. Wanunu M, Morrison W, Rabin Y, Grosberg AY, Meller A (2010) Electrostatic focusing of unlabelled DNA into nanoscale pores using a salt gradient. Nat Nanotechnol 5(2):160–165
14. Wendell D et al (2009) Translocation of double-stranded DNA through membrane-adapted phi29 motor protein nanopores. Nat Nanotechnol 4(11):765–772
15. Branton D et al (2008) The potential and challenges of nanopore sequencing. Nat Biotechnol 26(10):1146–1153
16. Cherf GM et al (2012) Automated forward and reverse ratcheting of DNA in a nanopore at 5-A precision. Nat Biotechnol 30(4):344–348
17. Kasianowicz JJ, Brandin E, Branton D, Deamer DW (1996) Characterization of individual polynucleotide molecules using a membrane channel. Proc Natl Acad Sci U S A 93(24):13770–13773
18. Manrao EA et al (2012) Reading DNA at single-nucleotide resolution with a mutant MspA nanopore and phi29 DNA polymerase. Nat Biotechnol 30(4):349–353
19. Ashton PM et al (2015) MinION nanopore sequencing identifies the position and structure of a bacterial antibiotic resistance island. Nat Biotechnol 33(3):296–300
20. Bolisetty MT, Rajadinakaran G, Graveley BR (2015) Determining exon connectivity in complex mRNAs by nanopore sequencing. Genome Biol 16(1):1–12
21. Laszlo AH et al (2014) Decoding long nanopore sequencing reads of natural DNA. Nat Biotechnol 32(8):829–833
22. Norris AL, Workman RE, Fan Y, Eshleman JR, Timp W (2016) Nanopore sequencing detects structural variants in cancer. Cancer Biol Ther 17(3):246–253
23. Quick J et al (2016) Real-time, portable genome sequencing for Ebola surveillance. Nature 530 (7589):228–232
24. Wallace EVB et al (2010) Identification of epigenetic DNA modifications with a protein nanopore. Chem Commun 46(43):8195–8197
25. An N, Fleming AM, White HS, Burrows CJ (2012) Crown ether–electrolyte interactions permit nanopore detection of individual DNA abasic sites in single molecules. Proc Natl Acad Sci U S A 109(29):11504–11509
26. Wang Y, Zheng D, Tan Q, Wang MX, L-Q G (2011) Nanopore-based detection of circulating microRNAs in lung cancer patients. Nat Nanotechnol 6(10):668–674
27. Zhang X, Wang Y, Fricke BL, L-Q G (2014) Programming nanopore ion flow for encoded

multiplex microRNA detection. ACS Nano 8 (4):3444–3450

28. Carthew RW, Sontheimer EJ (2009) Origins and mechanisms of miRNAs and siRNAs. Cell 136(4):642–655
29. Kim VN, Han J, Siomi MC (2009) Biogenesis of small RNAs in animals. Nat Rev Mol Cell Biol 10(2):126–139
30. Lee RC, Feinbaum RL, Ambros V (1993) The C. elegans heterochronic gene lin-4 encodes small RNAs with antisense complementarity to lin-14. Cell 75(5):843–854
31. Wightman B, Ha I, Ruvkun G (1993) Post-transcriptional regulation of the heterochronic gene lin-14 by lin-4 mediates temporal pattern formation in C. elegans. Cell 75(5):855–862
32. Boeri M et al (2011) MicroRNA signatures in tissues and plasma predict development and prognosis of computed tomography detected lung cancer. Proc Natl Acad Sci U S A 108 (9):3713–3718
33. Hu Z et al (2010) Serum microRNA signatures identified in a genome-wide serum microRNA expression profiling predict survival of non-small-cell lung cancer. J Clin Oncol 28 (10):1721–1726
34. Hunt EA, Goulding AM, Deo SK (2009) Direct detection and quantification of micro-RNAs. Anal Biochem 387(1):1–12
35. Iorio MV, Croce CM (2009) MicroRNAs in cancer: small molecules with a huge impact. J Clin Oncol 27(34):5848–5856
36. Landi MT et al (2010) MicroRNA expression differentiates histology and predicts survival of lung cancer. Clin Cancer Res 16(2):430–441
37. Mitchell PS et al (2008) Circulating microRNAs as stable blood-based markers for cancer detection. Proc Natl Acad Sci U S A 105 (30):10513–10518
38. Shen J et al (2011) Plasma microRNAs as potential biomarkers for non-small-cell lung cancer. Lab Invest 91(4):579–587
39. Sozzi G et al (2009) Plasma DNA quantification in lung cancer computed tomography screening. Am J Respir Crit Care Med 179 (1):69–74
40. Zheng D et al (2011) Plasma microRNAs as novel biomarkers for early detection of lung cancer. Int J Clin Exp Pathol 4(6):575–586
41. Tian K, He Z, Wang Y, Chen SJ, LQ G (2013) Designing a polycationic probe for simultaneous enrichment and detection of microRNAs in a nanopore. ACS Nano 7(5):3962–3969
42. Howorka S, Bayley H (1998) Improved protocol for high-throughput cysteine scanning mutagenesis. Biotechniques 25(5):764–766, 768, 770 passim
43. Montal M, Mueller P (1972) Formation of bimolecular membranes from lipid monolayers and a study of their electrical properties. Proc Natl Acad Sci U S A 69(12):3561–3566
44. Takeshima K, Chikushi A, Lee K-K, Yonehara S, Matsuzaki K (2003) Translocation of analogues of the antimicrobial peptides magainin and buforin across human cell membranes. J Biol Chem 278(2):1310–1315
45. Mohammad MM, Movileanu L (2010) Impact of distant charge reversals within a robust beta-barrel protein pore. J Phys Chem B 114 (26):8750–8759

Chapter 18

Intracellular Reassociation of RNA–DNA Hybrids that Activates RNAi in HIV-Infected Cells

Angelica N. Martins, Weina Ke, Vaishnavi Jawahar, Morriah Striplin, Caryn Striplin, Eric O. Freed, and Kirill A. Afonin

Abstract

Human immunodeficiency virus Type 1 (HIV-1) is the major cause of acquired immune deficiency syndrome (AIDS). In 2014, it was estimated that 1.2 million people died from AIDS-related illnesses. RNA interference-based therapy to block HIV replication is a field that, as of now, is without any FDA-approved drugs available for clinical use. In this chapter we describe a protocol for testing and utilizing a new approach that relies on reassociation of RNA–DNA hybrids activating RNAi and blocking HIV replication in human cells.

Key words RNA–DNA hybrids, RNA interference, HIV-1, Hybrids reassociation, Dicer substrate RNA, RNA nanotechnology

1 Introduction

RNA-based therapy is a new frontier in cancer research, creating an intriguing alternative to small molecule based treatments [1]. In a posttranscriptional, sequence-specific gene silencing process known as RNA interference [2, 3], longer double-stranded RNAs or self-folded, single-stranded RNAs are processed by an enzyme known as Dicer into shorter duplexes, called small interfering RNA (siRNA), or micro RNAs (miRNAs) [4–7]. The processed duplexes are then loaded onto an RNA-induced silencing complex (RISC) which uses, based on the thermodynamic asymmetry, one of the RNA strands as a guide to target the messenger RNA (mRNA) sequence [8, 9]. The targeted mRNA is then cleaved, thereby preventing it from undergoing translation [10], effectively curtailing the corresponding protein production. This method of cell regulation has a myriad of possible applications in the biomedical field.

RNA-based therapy provides an advantage over traditional pharmacological approaches by potentially being able to effectively

Eckart Bindewald and Bruce A. Shapiro (eds.), *RNA Nanostructures: Methods and Protocols*, Methods in Molecular Biology, vol. 1632, DOI 10.1007/978-1-4939-7138-1_18,

target any protein in the diseased cells. However, there are still obstacles in making RNA therapy a viable clinical treatment [7]. First, RNAs are quickly degraded in the bloodstream by nucleases, which provide a natural defense against foreign nucleic acids. Secondly, analyzing the pharmacodynamics of RNA-based therapies is difficult because adding functionalities, such as fluorescent probes for visualization, to the RNA strands has the potential to alter the RNA's activity. Lastly, ensuring conditional activation of the RNAi in the targeted diseased cells alone may be essential for minimizing collateral damage in healthy cells. The recently developed technique that is based on the reassociation of RNA–DNA hybrids [11] is a novel approach that may offer some solutions to the problems encountered in the field of nucleic acid-based therapeutic technologies.

The concept behind the hybrids is similar to split-protein systems [12] which involve the division of functional proteins into nonfunctional pieces that under certain conditions reassociate, restoring the original function. This permits tighter control over the functionality of the proteins, as well as an increase in molecule detectability. The RNA–DNA hybrids are used to split the RNAi triggers into two pieces, by having one strand of the Dicer substrate RNA [13] (DS RNA) in each of the two hybrids, and letting the DNA serve as the other half of each duplex. RNA–DNA hybrids cannot be processed by Dicer and as such are inactive. When cognate hybrids are present in the same cell, the complementary sequences on the hanging ends of their DNA strands, known as toeholds, recognize each other and initiate the interactions. The DNA strands of hybrids then reassociate into the longer DNA duplexes, releasing, through branch migration, the RNA fragments which assemble into the functional DS RNAs, thus activating RNAi (Fig. 1). The use of RNA toeholds was also shown to be functional [14]. It is important to mention that the use of conventional 21-mer siRNAs in place of DS RNAs for RNA/DNA hybrids does not entirely deactivate the individual hybrids and some gene silencing is still observed for the siRNA antisense-carrying hybrids [15]. Novel computational algorithms were later developed to follow, in silico, the process of hybrid formation and reassociation, thereby automating hybrid design [16].

The simultaneous release of multiple DS RNAs (up to seven at once) on reassociation of two hybrids has been previously demonstrated [15]. However, the resulting long DNA duplexes showed some immunostimulatory effects. To avoid this, while still simultaneously activating multiple DS RNAs, previously characterized RNA-, RNA/DNA-, and DNA-based nanoparticles, such as nanorings [17–19] and nanocubes [20, 21], were decorated with multiple nonfunctional RNA–DNA hybrids. During the addition of cognate hybrids, the activation of nanoparticles and triggering of specific RNAi were promoted [22, 23]. Additional functionalities

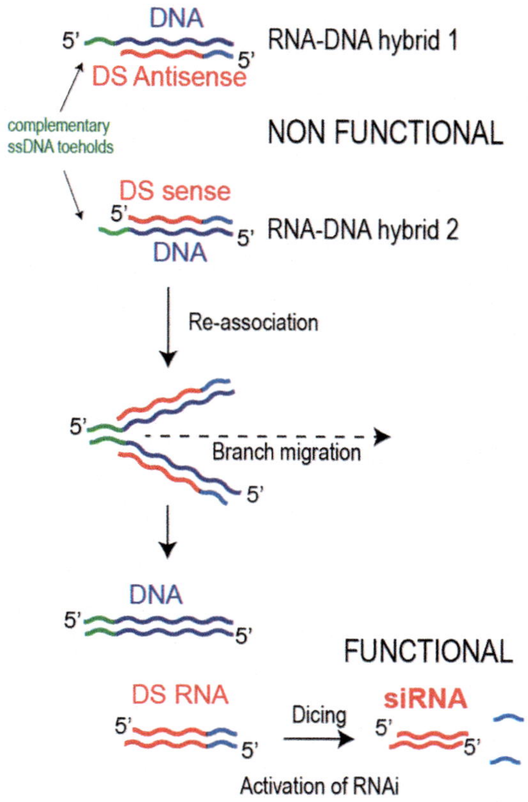

Fig. 1 Schematic illustration of the reassociation of RNA–DNA hybrids through the ssDNA toehold interactions that results in release of Dicer substrate RNAs and further RNAi activation through dicing

such as fluorescent dyes and targeting agents can be added onto both the 5′- and 3′-ends of the DNA strands to allow a more detailed analysis of the treatment effects and delivery rates without interfering with the RNA function. The RNA–DNA hybrid approach allows for greater control over RNA interference activation, while allowing the direct visualization of intracellular reassociation in real time. RNA–DNA hybrids are also more stable in human blood compared to conventional RNAs [11]. The RNA–DNA hybrid approach also allows for simultaneous activation of other split functionalities such as, for example, FRET and functional aptamers [15, 24]. Recently, the use of modified RNAs, in place of DNA strands, has been shown to offer certain advantages [25].

According to UNAIDS, it is estimated that more than 37 million people are currently living with HIV, and the number of people with HIV continues to increase. Although effective treatment against HIV is available, the virus continues to evolve and acquire drug resistance mutations, becoming resistant to available therapies. Therefore, it is important to continue research for new drugs as well as new targets. Blocking HIV replication using RNAi is a field well studied using cell culture assays. Several regions within the HIV genome have been tested in order to find potential effective targets. Berkhout's group has worked extensively in this field, and to test the hypothesis that RNA–DNA hybrids could inhibit HIV replication we selected two HIV targets, viral protease (PR) and surface envelope glycoprotein (gp120), validated by their published work [26].

2 Materials

All solutions should be prepared using double-deionized ultrapure water (18 MΩ at 25 °C) and biological grade reagents. It is recommended to have all reagents freshly made and filtered. Once prepared, all reagents and buffers can be stored at room temperature (unless indicated otherwise). All waste disposal regulations should be followed when disposing of waste materials. RNAs, DNAs, and fluorescently labeled DNAs for hybrid duplexes can be purchased from commercial vendors such as Integrated DNA Technologies, Inc. or others. In the case of the fluorescently labeled DNAs, we recommend the use of an additional linker of two nucleotides (e.g., TT) prior to the fluorescent tags. Different software tools such as, for example, IDT OligoAnalizer or Scripps Oligo Extinction Coefficient Calculator can be used to determine the extinction coefficients for each oligo.

The following equipment is required for the described experiments: vortex mixer, SpeedVac concentrator, horizontal gel electrophoresis, PCR thermocycler, water bath, ChemiDoc XS, fume hood, CO_2 incubator, biosafety cabinet type II A2, microfuge centrifuge, oven, Shaker, Personal Molecular Imager™ (PMI™) System and correspondent software and Geiger counter.

2.1 Preparation and Analysis of RNA–DNA Hybrids

1. 10× TBE: 0.89 M Tris, 0.89 M boric acid, 20 mM EDTA, pH ~8.3.
2. 10× TB: 0.89 M Tris, 0.89 M boric acid, pH ~8.3.
3. Eluting buffer: 10 mM Tris pH 7.5, 300 mM NaCl, 0.5 mM EDTA.
4. 5× assembly buffer: 5×TB (pH 8.3), 10 mM $Mg(OAc)_2$.
5. Denaturing gel polyacrylamide gel electrophoresis (PAGE): 8 M urea, 8% acrylamide with 19:1 acrylamide–bis-acrylamide ratio, 1× TBE.
6. Nondenaturing native PAGE: 8% acrylamide with 19:1 acrylamide–bis-acrylamide ratio, 1× TB, 2 mM $Mg(OAc)_2$.

2.2 Cell Culture and Transfection Reagents

1. Plasmid: Full-length HIV-1 molecular clone pNL4-3 [27] (Although pNL4-3 is not a harmful reagent, it is still a recombinant DNA that can generate full length virus, and therefore should be handled with caution according to NIH Guidelines for recombinant or synthetic nucleic acids).
2. RNAi inducers: PR RNA–DNA hybrids, Env RNA–DNA hybrids, PR DS RNA, Env DS RNA (5, 10, and 20 nM final concentrations) (*see* Subheading 3—Hybrid Design).
3. Transfection reagent: Lipofectamine 2000™ (Invitrogen).
4. Medium: Dulbecco's modified Eagle's medium (DMEM) High Glucose supplemented with 100 U/mL penicillin, 100 μg/mL streptomycin, 2 mM L-glutamine and 5% fetal bovine serum (FBS), Opti-MEM-I reduced serum, Roswell Park Memorial Institute 1640 Medium lacking the amino acids methionine (Met) and cysteine (Cys) (RPMI 1640 Cys^{-}/Met^{-}), supplemented or not with 2% (vol/vol) dialyzed (dFBS).

2.3 Metabolic Labeling

1. Protein Labeling Mix: [^{35}S]Met/Cys (a PerkinElmer mixture containing both ^{35}S-L-methionine and ^{35}S-L-cysteine).
2. Cell lysis buffer: 300 mM NaCl, 50 mM Tris–HCl (pH 7.5), 0.5% Triton X-100 (TX-100), 10 mM iodoacetamide, and two tablets of cOmplete protease inhibitor cocktail tablets per 100 mL of solution.
3. Triton wash buffer: 300 mM NaCl, 50 mM Tris–HCl (pH 7.5), 0.5% Triton X-100.
4. Protein A Agarose for Immunoprecipitation (ThermoFisher).
5. Bovine serum albumin (BSA).
6. Antibody: HIV-IgG (pool of purified serum from HIV-1-infected patient obtained from AIDS Research and Reference Reagent Program. Although the reagent is not considered to be biohazardous, it should be handled with caution).
7. SDS/DOC wash buffer: 300 mM NaCl, 50 mM Tris–HCl (pH 7.5), 0.1% SDS, 0.1% deoxycholic acid.
8. 2× sample buffer: 2× Laemmli buffer: 120 mM Tris–HCl (pH 6.8), 4% SDS, 20% glycerol, 10% β-mercaptoethanol, 0.02% bromophenol blue.

2.4 Radioimmunoprecipitation SDS-PAGE Gel

1. Acrylamide–AcrylAide lower gel: 29% acrylamide–1% AcrylAide gel reagent for SDS-PAGE, 1.5 M Tris–HCl, (pH 8.7), 0.4% SDS, 0.5 mL 2% ammonium persulfate (APS), 40 μL *N*,*N*,*N*′,*N*′-tetramethylethylenediamine (TEMED).

   Important: AcrylAide Cross-Linker can be obtained as a special order to Lonza, Cat No. 51015.

2. Stacking upper gel: 30% acrylamide–0.8% bis-acrylamide solution for SDS-PAGE, 0.5 M Tris–HCl, pH 6.8, 0.4% SDS, 0.2 mL 2% APS, 20 μL TEMED.
3. Gel bond paper.
4. Tris–Glycine–SDS running buffer: 25 mM Tris, 192 mM glycine, 1% (w/v) SDS.
5. Fixative: 40% (vol/vol) methanol, 10% (vol/vol) acetic acid in dH_2O.
6. Enhancer solution: 1 M salicylic acid with 2% glycerol (vol/vol).
7. Storage Phosphor screen for gel exposure.
8. Radiacwash decontaminant to cleanup isotope contamination.

3 Experimental Procedure

Hybrid design

General designing strategy:

1. Identify target gene and find corresponding siRNAs (21/21 nts) based on the published literature.
2. Using the sequences of known siRNAs, design asymmetric (25/27 nts) Dicer substrate RNAs (DS siRNAs) through elongation of the siRNA sense strands with corresponding sequences identified and extracted from mRNAs, as shown in Fig. 2.
3. Design DNA strands by taking the reverse complements of individual DS RNA strands.
4. Each of the DNA strands in the hybrids is decorated with ssDNA toehold. The toeholds in the hybrids are complementary to each other for hybridization. The standard toehold sequence [11] that has been used in multiple designs is as follows: 5′-GGAGACCGTGAC-hybrid (and its complement hybrid - GTCACGGTCTCC-3′), and can be routinely used for any new hybrid designs.

Hybrid design for HIV targets:

1. Identify gene locations where previous studies have shown that siRNAs targeting that sequence inhibit HIV replication.

   Important: a single nucleotide difference, upstream or downstream of the siRNA sequence can abrogate the ability to block HIV replication.
2. Use a sequence location where multiple siRNAs are tested.
3. Two viral targets are used, viral protease (PR) and surface envelope glycoprotein (gp120). More precisely, positions 2332 to 2356 for PR and 7642 to 7665, according to the virus strain

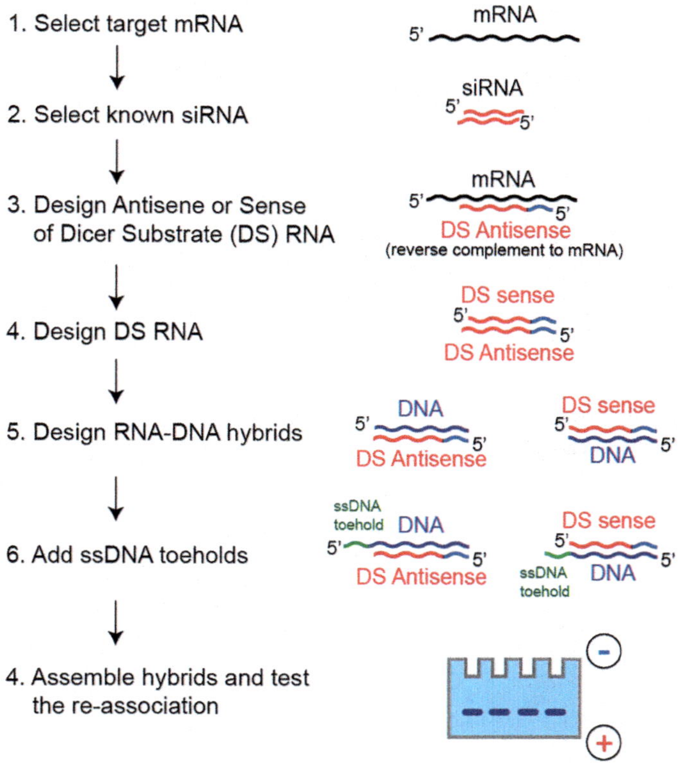

Fig. 2 Main steps required for RNA–DNA hybrids design and characterization: First target mRNA should be identified, then Dicer substrate (DS) RNA should be designed by elongating the known siRNA sequences, and the corresponding DNA strands with ssDNA toeholds should be added to form the hybrids. The resulting hybrids can be assembled and their reassociation confirmed by gel electrophoretic techniques

NL43. The sequence accession number AX032749 can be obtained from NCBI GenBank. The following DS RNAs and DNAs for RNA–DNA hybrids are designed as described above.

Hybrids designed against Protease:

DS sense

5′-pGAGCAGAUGAUACAGUAUUAGAAGA

DNA that binds DS sense

5′-TCTTCTAATACTGTATCATCTGCTCCTGTCACGGTC TCC

DS antisense

5′-UCUUCUAAUACUGUAUCAUCUGCUCCU

DNA that binds DS antisense

5′-GGAGACCGTGACAGGAGCAGATGATACAGTATTAG AAGA

Hybrids designed against Envelope:
DS sense
5′-GGACAAUUGGAGAAGUGAAUUAUAUU
DNA that binds DS sense:
5′-AATATAATTCACTTCTCCAATTGTCCGTCACGGTCT
CC
DS antisense
5′-pUAUAAUUCACUUCUCCAAUUGUCC
DNA that binds DS antisense
5′-GGAGACCGTGACGGACAATTGGAGAAGTGAATTAT
ATT

3.1 Purification of DNAs and RNAs with Urea Gels

Gel electrophoresis

1. It's recommended to purify all the purchased DNAs and RNAs prior to further experiments with RNA–DNA hybrids. To purify oligos, prepare a denaturing gel with TBE. Polymerization is initiated by mixing APS solution and TEMED.
2. Run a vertical gel with spacer thickness of 0.75 mm, TBE is used as a running buffer solution (*see* **Notes 1–4**).

   Important: Use a syringe to wash the loading wells several times with running buffer prior to loading your samples.
3. Separate the glass plates and wrap the gel in Saran Wrap.
4. Visualize the nucleotide bands on TLC plate using a UV lamp at short wavelength (254 nm).
5. Outline the nucleic acid bands and cut out the DNA and RNA bands.
6. Recover the DNA and RNA strands by incubating the gel pieces overnight at 4 °C in eluting buffer.
7. Recover as much supernatant as possible, without collecting gel, and place into new tube.
8. Add equal volumes of 100% ethanol to each tube containing eluted oligos.
9. Place oligos in a −20 °C freezer for 3 h or overnight.
10. Centrifuge for 30 min at 11,000 × *g* and 4 °C.
11. Discard supernatant.
12. Rinse with of 90% ethanol and repeat **steps 10** and **11**.
13. Discard supernatant and dry the oligos in the vacuum concentrator.
14. Suspend dry pellets in water.

    Important: Resuspended oligos should be always kept on ice or frozen. If frozen, oligos must be vortex-mixed for 5 sec prior to each use.

15. Measure the absorbance at 260 nm (A260) for each purified oligo using a UV spectrometer.
16. Based on the sequence, determine the extinction coefficients for each oligo using IDT OligoAnalyzer tools (https://www.idtdna.com/calc/analyzer).
17. Use the Beer–Lambert law to calculate the concentration of purified oligos:
    $A = a \times b \times c$, where A is the measured absorbance at 260 nm; a—extinction coefficient; b—path length; c—concentration.

3.2 Assembly of RNA–DNA Hybrids

1. Mix cognate RNA and DNA strands of designed hybrids at equimolar concentrations in water.
2. Incubate the mixture in a heat block at 95 °C for 2 min to melt all hydrogen bonds.

   Important: Do not incubate the mixture longer, as it can promote the degradation of RNAs.
3. Snap-cool the mixture by rapidly transferring it to the heating block set at 30 °C. As an alternative, snap-cooling to RT can be used.
4. Add 5× assembly buffer (to the final 1×) and incubate for 20 min.
5. Run quality control experiments using native polyacrylamide gel electrophoresis (native PAGE) [15]. For visualization, use total staining with ethidium bromide. As an alternative, assemblies containing one fluorescently labeled DNA strand (e.g., with Alexa 488) together with nonlabeled RNA can be directly visualized on the gels using ChemiDoc MP system (*see* **Notes 5–7**).

3.3 Native PAGE Experiments for RNA–DNA Assemblies and Visualization of Reassociation

1. Mix two cognate hybrids at equimolar concentrations and incubate for 30 min at 37 °C.
2. Prepare a sequencing gel for vertical electrophoresis for high-resolution native PAGE, i.e., 10% (19:1) acrylamide, 1× assembly buffer.
3. For vertical gel with the spacer thickness of 0.75 mm, load samples in individual lanes of a gel (5 μL per lane), perform electrophoresis for 2 h at 20 W at 4 °C.

   Important: Use a syringe to wash the loading wells several times with running buffer before loading.
4. For EtBr-stained gels, a ChemiDoc MP can be used for visualization.

3.4 HIV-1 Inhibition Experiments

Cotransfection of HIV and siRNAs

HeLa cells are cultured with DMEM High Glucose, and are maintained at 37 °C in 5% CO_2 incubator humid.

1. Day 1: Plate 120,000 HeLa cells per well in a 12-well plate.

2. Day 2: Cotransfect 800 μg of pNL4-3 molecular clone [27] plus 5, 10, and 20 nM (final concentrations) of RNA–DNA hybrids, or pNL43 plus DS RNAs as positive controls. For knockdown, negative control RNA–DNA hybrids and DS RNAs targeting other genes (e.g., eGFP) can be used. Virus control is measured only with pNL43 transfection and Mock only transfection reagent is used. Use 1 μL of Lipofectamine 2000 (L2K) per well. For RNA–DNA hybrids, incubate DS sense and DS anti-sense containing hybrids in separate tubes using half of the volume of L2K, and 400 μg of pNL4-3 for each tube.
   (a) Add to each tube Opti-MEM I (to dilute L2K and nucleic acids, to a maximum final volume of 50 μL).
   (b) Add RNA–DNA hybrids (containing sense or antisense) and DS RNAs. For virus control do not add RNA–DNA hybrids or DS RNAs.
   (c) Add 400 or 800 μg of pNL4-3.
   (d) Add 0.5 or 1 μL of L2K.
   (e) Incubate for 30 min.
   (f) Discard old DMEM and add 450 μL of fresh DMEM without penicillin and streptomycin to each well from plated cells.
   (g) Add dropwise the complexes of L2K/pNL4-3; L2K/pNL4-3/RNA–DNA hybrid sense+L2K/pNL4-3/RNA–DNA hybrid antisense; or L2K/pNL4-3/DS RNA.
   (h) Discard the media 5 h post transfection.
   (i) Add 1 mL of DMEM 5% FBS, 100 U/mL penicillin, 100 μg/mL streptomycin, and 2 mM L-glutamine.
   (j) Incubate for 2 days before metabolic labeling (Day 4).

Metabolic Labeling

During the metabolic labeling assay culture cells with RPMI 1640 Medium lacking the amino acids methionine (Met) and cysteine (Cys), supplemented or not with 2% (vol/vol) dFBS [28].

1. Discard the DMEM media.
2. Starve the cells with 1 mL of RPMI 1640 Cys^-/Met^- and incubate for 30 min at 37 °C.
3. Discard starvation media.
4. Add 1 mL of RPMI 1640 Cys^-/Met^- supplemented with 2% dFBS and 250 μCi[^{35}S] Met/Cys (a mixture containing both ^{35}S-L-methionine and ^{35}S-L-cysteine) (*see* **Notes 8–11**).
5. Incubate at 37 °C for 4 h.

   Important: Handling isotopes requires special training and approval. All solid and liquid waste should be discarded separately following the rules of your institution. If any questions rise, consult the radiation safety program at your Institution.

6. Dispose the supernatant if you will not measure the virus production.
7. Lyse the cells using 250 μL of Cell lysis buffer at room temperature.

Immunoprecipitation

Important: Preparation of cell lysates and all subsequent steps use screw-cap tubes with O-rings (Sarstedt) to reduce aerosol production and contamination of equipment with isotopes.

1. Preclearance of cell-associated material.
   (a) Spin the cell lysate at 11,000 × *g* for 5 min using a microfuge.
   (b) Transfer 50 μL of the cell lysate supernatant to a new tube.
   (c) Add 500 μL of Triton wash buffer.
   (d) Mix with 50 μL Protein A agarose beads/tube and 50 μL 1% BSA/tube.
   (e) Incubate in low rotation at 4 °C for 1 h.
   (f) Spin samples for 1 min at 11,000 × *g* using a microfuge.
   (g) Collect 550 μL from the supernatant, avoiding disturbing the pellet (This is the precleared lysate).
2. Couple HIV-IgG with protein A agarose.
   (a) Prepare a mixture of 50 μL protein A agarose beads + 30 μL 1% BSA + 1 μL of HIV-1 IgG in 1 mL of Phosphate-buffered saline (PBS) per sample.
   (b) Incubate in low rotation at 4 °C for 1 h.
   (c) Spin beads now coupled with HIV-IgG for 5 min at 500 × *g* using a microfuge.
   (d) Wash beads 2× with 500 μL of PBS, repeating the spinning from **step c**.

   Important: Instead, the coupling of the beads and the washing steps can be performed in a pool by mixing enough solution to all samples in 15 mL tubes (500 μL/ per sample). Then before the last spin transfer to a tube and waste the supernatant.
   (e) After the last wash aspirate all the PBS leaving the beads coupled with HIV-IgG.
   (f) Add the precleared lysate to the tube with the beads pellet.
   (g) Immunoprecipitate the samples by incubating in slow rotation at 4 °C overnight.

3. Washing the beads complexed with HIV-IgG and HIV proteins from the cell lysate (*see* **Note 12**)
   (a) Spin samples for 1 min at 14,000 RPM.
   (b) Wash beads 2× with 1 mL of Triton wash buffer room temperature (RT) (Aspirate the supernatant, and inverting the tube 2× after addition of 1 mL of Triton wash buffer, then repeat the spin for 1 min at 11,000 × *g*).
   (c) Add 1 mL of SDS/DOC wash buffer to the beads pellet.
   (d) Incubate at RT for 5 min.
   (e) Spin samples for 1 min at 11,000 × *g*.
   (f) Aspirate the SDS/DOC wash buffer.
   (g) Resuspend the beads in 100 μL 2× Sample buffer.
   (h) Vortex briefly.
   (i) Incubate in dry heating block at 99 °C for 5 min.
   (j) Spin samples for 1 min at 11,000 × *g*.
   (k) Transfer 80 μL of the samples supernatant to a new tube.

Radioimmunoprecipitation SDS-PAGE

1. Gel preparation
   (a) Acrylamide–AcrylAide lower gel.
   (b) Stacking upper gel.
   (c) Add gel bond paper to the smaller glass, with the hydrophobic side facing the glass.
   (d) Once the gel is casted and polymerized load the samples.
2. Run the gel using 1× Tris–glycine–SDS buffer.
3. Gently remove the gel bond paper with the gel attached.
4. Fix the gel: Add the gel to a container with gel fixative solution.
5. Incubate by slow shaking at room temperature for 2 h.
6. Discard the gel fixative.
7. Add Enhancer solution to the container.
8. Incubate in slow shaking at RT for 2 h.
9. Place the gel on paper towel.
10. Incubate in dry 100 °C oven until the gel dries (*see* **Note 13**).
11. Place the gel in an X-ray film cassette.
12. Add a phosphorimage screen for quantification.

    Important: Face the gel to the white side of the phosphorimage screen. The time of exposure will depend on the signal. We recommend the first exposure to be overnight, or use the Geiger counter for measuring signal intensity.
13. Read the screen in any Phosphorimager (**Fig. 3**).

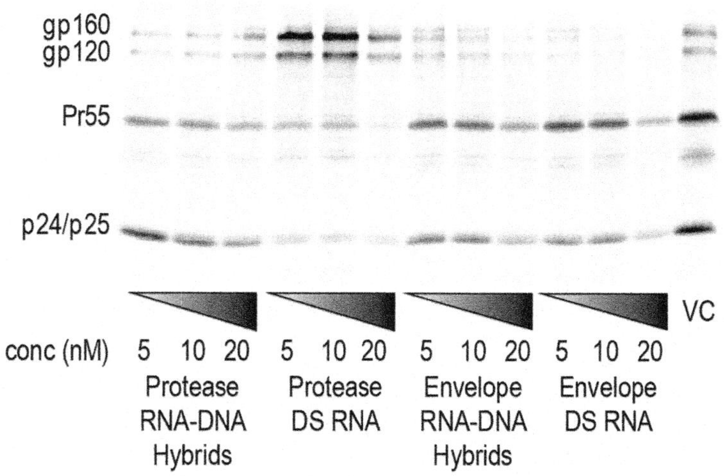

Fig. 3 Inhibition of HIV-1 expression and production by intracellular reassociation of RNA–DNA hybrids and activation of RNAi. HeLa cells were transfected with pNL4-3, without (VC virus control) and with RNA–DNA hybrids or DS RNAs. At 48 h posttransfection, HeLa cells were metabolically labeled with [^{35}S] MetCys for 4 h. Cell lysates were radioimmunoprecipitated. Positions of envelope glycoprotein precursor (gp160), and surface glycoprotein (gp120); Pr55Gag (Pr55), CA/CA-SP1 (p24/p25). (Original figure from Afonin et al., Nat Nano 2013)

Quantifying HIV proteins inside the cell

Important: Use the software from the Phosphorimager used to quantify the bands of interest. Always subtract the background value from each band. Since we target PR, and PR is located in the full length mRNA, we quantify the expression of Gag. Although the siRNA targeting Envelope binds to the Envelope mRNA, it also binds the full-length mRNA.

14. Quantification of total cell-associated Gag Pr55Gag (Pr55), matrix-capsid (p41), and capsid/capsid-SP1 (p24/p25) (Pr55 + p41 + p25 + p24).
15. Plot a column graph on Excel normalizing the relative amount of Gag Pr55 + p41 + p25 + p24 from at least three repetitions.

4 Notes

1. Avoid overloading samples to prevent less efficient separation in native PAGE gel.
2. Prior to pipetting, mix well to avoid unequal concentrations.
3. Prerun the native PAGE gel as a calibration and quality control step to remove magnesium ions.

4. After loading of the samples, run the gel at a lower voltage to allow all nucleic acid strands of all lengths to begin separation at the same point.
5. In order to obtain the best result for strand association, make sure each DNA–RNA duplex is in exactly equal concentrations.
6. Cool tubes before opening after incubating at 95 °C to avoid loss of the volatilized sample so that the concentration remains unchanged.
7. For native PAGE visualization use total staining with ethidium bromide. As an alternative, assemblies containing fluorescence labeled RNA/DNA strands together with nonlabeled RNA/DNAs can be visualized using a gel imaging system such as ChemiDoc.
8. Use the Geiger counter to monitor any contamination with ^{35}S during the work with radioisotopes.
9. Check your gloves with the Geiger counter before disposing of them.
10. Radiacwash decontaminant to cleanup isotope contamination works better than regular soap.
11. For more difficult isotope contamination use household bleach, then rinse with ethanol 70%.
12. When removing or aspirating supernatants, avoid disturbing the pellet.
13. The gel will change from clear to opaque white. Do not over-dry to avoid breaking of the gel. Use an oven connected to a vacuum line for shorter incubation time.

References

1. Bramsen JB, Kjems J (2012) Development of therapeutic-grade small interfering RNAs by chemical engineering. Front Genet 3:154
2. Fire A, Xu S, Montgomery MK et al (1998) Potent and specific genetic interference by double-stranded RNA in Caenorhabditis elegans. Nature 391(6669):806–811
3. Parrish S, Fleenor J, Xu S et al (2000) Functional anatomy of a dsRNA trigger: differential requirement for the two trigger strands in RNA interference. Mol Cell 6(5):1077–1087
4. Lund E, Dahlberg JE (2006) Substrate selectivity of exportin 5 and Dicer in the biogenesis of microRNAs. Cold Spring Harb Symp Quant Biol 71:59–66
5. Ji X (2008) The mechanism of RNase III action: how dicer dices. Curr Top Microbiol Immunol 320:99–116
6. Macrae IJ, Zhou K, Li F et al (2006) Structural basis for double-stranded RNA processing by Dicer. Science 311(5758):195–198. doi:10.1126/science.1121638
7. Dao BN, Viard M, Martins AN et al (2015) Triggering RNAi with multifunctional RNA nanoparticles and their delivery. DNA RNA Nanotechnol 1(1):27–38
8. Schwarz DS, Hutvagner G, Du T et al (2003) Asymmetry in the assembly of the RNAi enzyme complex. Cell 115(2):199–208
9. Khvorova A, Reynolds A, Jayasena SD (2003) Functional siRNAs and miRNAs exhibit strand bias. Cell 115(2):209–216
10. Pratt AJ, MacRae IJ (2009) The RNA-induced silencing complex: a versatile gene-silencing machine. J Biol Chem 284(27):17897–17901

11. Afonin KA, Viard M, Martins AN et al (2013) Activation of different split functionalities on re-association of RNA-DNA hybrids. Nat Nanotechnol 8(4):296–304
12. Shekhawat SS, Ghosh I (2011) Split-protein systems: beyond binary protein-protein interactions. Curr Opin Chem Biol 15(6):789–797
13. Rose SD, Kim DH, Amarzguioui M et al (2005) Functional polarity is introduced by Dicer processing of short substrate RNAs. Nucleic Acids Res 33(13):4140–4156
14. Afonin KA, Viard M, Tedbury P et al (2016) The use of minimal RNA toeholds to trigger the activation of multiple functionalities. Nano Lett 16(3):1746–1753
15. Afonin KA, Desai R, Viard M et al (2014) Co-transcriptional production of RNA-DNA hybrids for simultaneous release of multiple split functionalities. Nucleic Acids Res 42 (3):2085–2097
16. Afonin KA, Bindewald E, Kireeva M et al (2015) Computational and experimental studies of reassociating RNA/DNA hybrids containing split functionalities. Methods Enzymol 553:313–334
17. Afonin KA, Grabow WW, Walker FM et al (2011) Design and self-assembly of siRNA-functionalized RNA nanoparticles for use in automated nanomedicine. Nat Protoc 6 (12):2022–2034
18. Grabow WW, Zakrevsky P, Afonin KA et al (2011) Self-assembling RNA nanorings based on RNAI/II inverse kissing complexes. Nano Lett 11(2):878–887
19. Yingling YG, Shapiro BA (2007) Computational design of an RNA hexagonal nanoring and an RNA nanotube. Nano Lett 7 (8):2328–2334
20. Afonin KA, Bindewald E, Yaghoubian AJ et al (2010) In vitro assembly of cubic RNA-based scaffolds designed in silico. Nat Nanotechnol 5 (9):676–682
21. Afonin KA, Kasprzak W, Bindewald E et al (2014) Computational and experimental characterization of RNA cubic nanoscaffolds. Methods 67(2):256–265
22. Afonin KA, Viard M, Kagiampakis I et al (2015) Triggering of RNA interference with RNA-RNA, RNA-DNA, and DNA-RNA nanoparticles. ACS Nano 9(1):251–259
23. Afonin KA, Viard M, Koyfman AY et al (2014) Multifunctional RNA nanoparticles. Nano Lett 14(10):5662–5671
24. Rogers TA, Andrews GE, Jaeger L et al (2015) Fluorescent monitoring of RNA assembly and processing using the split-spinach aptamer. ACS Synth Biol 4(2):162–166
25. Groves B, Chen YJ, Zurla C et al (2016) Computing in mammalian cells with nucleic acid strand exchange. Nat Nanotechnol 11 (3):287–294
26. Low JT, Knoepfel SA, Watts JM et al (2012) SHAPE-directed discovery of potent shRNA inhibitors of HIV-1. Mol Ther 20(4):820–828
27. Adachi A, Gendelman HE, Koenig S et al (1986) Production of acquired immunodeficiency syndrome-associated retrovirus in human and nonhuman cells transfected with an infectious molecular clone. J Virol 59 (2):284–291
28. Waheed AA, Ono A, Freed EO (2009) Methods for the study of HIV-1 assembly. Methods Mol Biol 485:163–184

Chapter 19

Construction and In Vivo Testing of Prokaryotic Riboregulators

Alexander A. Green

Abstract

RNAs that are transcribed and self-assemble within living cells are valuable tools for regulating and organizing cellular activities. Riboregulators, in particular, have been widely used for modulating translation and transcription in response to cognate transactivating or trigger RNAs, enabling cells to evaluate logic operations and to respond to environmental cues. Herein we detail a set of methods for the rapid construction and testing of prokaryotic riboregulators in *Escherichia coli*. These methods enable construction of dozens of riboregulator plasmids at the same time without the use of restriction enzymes. Furthermore, they facilitate rapid screening of devices and can be applied to a variety of other self-assembling in vivo RNA systems.

Key words Riboregulator, Riboswitch, Transformation, Plasmid, Regulation, Toehold switch, Sensor, Cloning, Diagnostic, Self-assembly

1 Introduction

The ability of cells to readily synthesize single-stranded RNAs has enabled artificial RNA structures with useful functions to be assembled within living cells. RNA nanostructures constructed from repeating assemblies of one or two carefully designed RNAs have been used as enzyme scaffolds for channeling of substrates through metabolic pathways. These scaffolds have increased in vivo hydrogen production by nearly 50-fold [1] and enabled precise control of enzyme positioning to enhance the rate of pentadecane synthesis [2]. Self-assembled RNA complexes have also been widely used for regulatory purposes in living cells. Taking inspiration from the natural *hok/sok* system used for plasmid segregation [3, 4], Isaacs et al. carried out seminal work, reporting in 2004 a set of engineered riboregulators capable of translational regulation [5]. These riboregulators activated expression of an output gene in response to a cognate transactivating or trigger RNA. Engineering of natural RNA-based components in subsequent years led to riboregulators

Eckart Bindewald and Bruce A. Shapiro (eds.), *RNA Nanostructures: Methods and Protocols*, Methods in Molecular Biology, vol. 1632, DOI 10.1007/978-1-4939-7138-1_19,

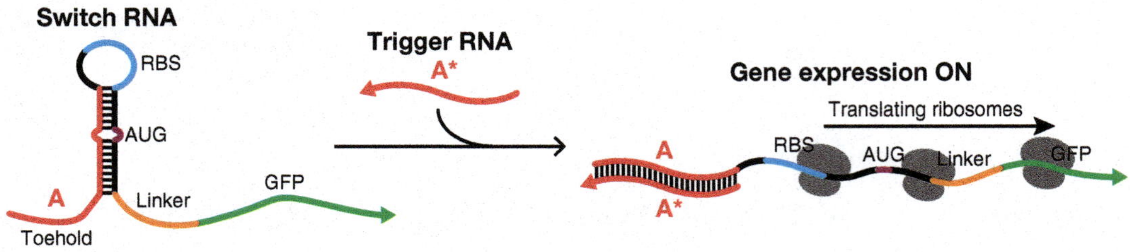

Fig. 1 The secondary structure and mechanism of a toehold switch synthetic riboregulator. Toehold switches consist of two cognate RNAs: a switch RNA regulating an output protein such as GFP and a trigger RNA that activates expression from the switch RNA. In the absence of the trigger RNA, the switch RNA ribosomal binding site (RBS) and start codon (AUG) are sequestered preventing translation. Upon hybridization with the trigger RNA through complementary A and A* sequences, the RBS and start codon are exposed enabling ribosomes to bind and translate GFP

with improved dynamic range, transcriptional-level regulation, and the ability to repress expression in response to a trigger RNA [6–9]. Despite substantial advances, the overall performance of riboregulators engineered from natural systems failed to match that of widely used protein-based regulators [10].

An alternative approach to developing new riboregulators takes advantage of RNA sequence design tools that solve the so-called inverse folding problem, generating de novo RNA sequences predicted to fold into a user-specified secondary structure [11–13]. We have recently reported a new class of completely synthetic prokaryotic riboregulators that employ RNA sequence design tools and design principles from DNA nanotechnology [14, 15] to yield substantial improvements in overall device performance. These riboregulators, termed toehold switches, provide a dynamic range over 400-fold and are sufficiently orthogonal to enable a dozen devices to operate independently within a cell at the same time [16]. These figures of merit meet or exceed those of the best protein-based regulators [10, 17], while using a rational design approach amenable to rapid library generation and iterative device optimization.

The design of the toehold switch riboregulators is shown in Fig. 1. These self-assembling systems are composed of a switch and a trigger RNA. The switch RNA regulates at the translational level an output gene, in this case green fluorescent protein (GFP). In the absence of the trigger RNA, the hairpin secondary structure of the switch sequesters the ribosomal binding site (RBS) in a loop and start codon (AUG) within a bulge, thus preventing translational initiation by the ribosome. In the presence of the trigger RNA, a single-stranded RNA region termed a toehold in the switch binds to the single-stranded trigger RNA. This initial binding enables the trigger and switch to undergo a branch migration process that fully unwinds the stem of the switch RNA. Once unwound, the RBS and start codon of the switch RNA are now fully exposed, enabling recognition and translation by the ribosomes. The design of the toehold switches enables them to detect trigger RNAs with virtually arbitrary sequences [16], thus enabling the devices to sense endogenous RNAs and those associated with pathogens. We have shown toehold switches can be optimized to detect small RNAs in *E. coli* [16] and RNAs from the Ebola and Zika viruses in low-cost paper-based diagnostic systems [18, 19].

The high performance of toehold switches and other recently reported synthetic riboregulators employing small transcription activating RNAs (STARs) [20] and toehold-mediated interactions [21] suggests that riboregulators will play an increasingly important role as regulatory devices in living cells. Herein we describe the methods we employ for constructing and evaluating synthetic riboregulators in *E. coli*. In particular, we detail the procedures used for constructing plasmids for riboregulator expression, including PCR primer design and vector assembly through Gibson assembly. We describe methods for preparing competent *E. coli* cells, transforming these cells with riboregulator plasmids, and protocols for culturing and inducing cells to evaluate riboregulator performance.

Although the discussion is focused on construction and evaluation of riboregulators and toehold switches in particular, these methods are generalizable and can be applied to enable rapid construction and in vivo testing of any system that exploits the cellular machinery to express one or more synthetic RNAs. Such RNA systems include riboswitches that sense small molecules and self-assembled RNA nanostructures for metabolic engineering.

2 Materials

2.1 Plasmid Construction Reagents

1. DNase-free, RNase-free, ultrapure, 18.2 MΩ water is used for all DNA amplification and ligation reactions.
2. DNA oligonucleotides are purchased as either long strands up to 200 nt in length (e.g., Ultramers), or as short strands up to 60 nt in length and consolidated into a longer double-stranded DNA using assembly PCR. No purification procedures are used for the strands prior to amplification via PCR. Strands are initially dissolved in 10 mM Tris, 0.1 mM EDTA, pH 8.0 at concentrations of 2–100 μM.
3. PCR reactions employ either the Phusion or Q5 DNA polymerases, which have error rates 50- and 100-fold lower than Taq DNA polymerase, respectively. 2× PCR master mixes with high-fidelity (HF) buffer can be obtained directly from commercial vendors.
4. Gibson assembly enzyme mix (2×): 8 U/μL *Taq* DNA ligase, 0.05 U/μL Phusion DNA polymerase, 0.008 U/μL T5

exonuclease, 10% PEG-8000, 200 mM Tris–HCl pH 7.5, 20 mM $MgCl_2$, 20 mM DTT, 0.4 mM each of the four dNTPs, and 2 mM NAD in water. Gibson assembly enzyme mix (2×) can also be purchased from commercial vendors. The concentration of T5 exonuclease in the reaction mix specified above is recommended for overlap domains shorter than 150 bp [22].

5. DpnI restriction enzyme: 20,000 U/mL.
6. pET15b: Expression vector available from EMD Millipore.
7. pCOLADuet: Expression vector available from EMD Millipore. GFP can be cloned into pCOLADuet using the methods described below to generate a template vector for switch plasmid construction.

2.2 Cell Culture Media, Antibiotics, Inducers, and Associated Materials

1. IPTG (1000×): 0.1 M isopropyl β-D-1-thiogalactopyranoside (IPTG) in ultrapure, 18.2 MΩ water is sterilized using 0.22-μm pore size filters. This 1000× IPTG stock solution is stored at −20 °C.
2. Ampicillin (1000×): 100 mg/mL ampicillin in ultrapure, 18.2 MΩ water is sterilized using 0.22-μm pore size filters. This 1000× ampicillin stock solution is stored at −20 °C.
3. Kanamycin (1000×): 50 mg/mL kanamycin in ultrapure, 18.2 MΩ water is sterilized using 0.22-μm pore size filters. This 1000× kanamycin stock solution is stored at −20 °C.
4. Lysogeny broth (LB) medium: 10 g tryptone, 5 g yeast extract, and 10 g NaCl in 1 L of ultrapure, 18.2 MΩ water. The 1 L of LB medium is autoclaved at 121 °C for 40 min to sterilize.
5. Super optimal broth (SOB) medium: 10 g tryptone, 5 g yeast extract, 0.5 g NaCl, 0.186 KCl, and 2.47 g $MgSO_4$ heptahydrate in 1 L of ultrapure, 18.2 MΩ water. The 1 L of SOB medium is autoclaved at 121 °C for 40 min to sterilize.
6. Inoue Transformation Buffer (ITB): 10 mM HEPES (4-(2-hydroxyethyl)-1-piperazineethanesulfonic acid), 55 mM $MnCl_2$, 15 mM $CaCl_2$, and 250 mM KCl. Add all buffer components except $MnCl_2$ to 80% of the final intended volume of ultrapure 18.2 MΩ water, dissolving the components by shaking. Carefully adjust the pH of the solution to 6.7 by adding 1 M KOH. Add the appropriate amount of $MnCl_2$ and dissolve by shaking. Next, add ultrapure, 18.2 MΩ water to the solution to reach the intended final volume. Filter-sterilize the ITB solution by running it through a sterile 0.22-μm pore size filter. Store the ITB at 4 °C.
7. LB/agar plates: Plates are prepared using the same formulation as LB medium with the addition of 14 g/L of agar. The mixture is autoclaved at 121 °C for 40 min to sterilize. Use a

container that is at least 25% larger than the volume of the LB/agar in order to avoid having the solution boil over. After removal from the autoclave, the molten LB/agar is left to cool to 50–55 °C whereupon 50 μg/mL of kanamycin, 100 μg/mL, and 0.1 mM IPTG are added from 1000× stock solutions as required. 20–25 mL of the LB/agar is added to each 100 mm diameter petri dish. After agar solidification, the plates are stored at 4 °C.

8. Glass beads for spreading bacterial cultures onto LB/agar plates are 4 mm in diameter and made out of soda-lime glass. Beads are sterilized by autoclaving at 121 °C for 30 min and stored at room temperature.
9. Deep 96-well plates.
10. 14-mL cell culture tubes with round bottoms and snap-on caps.
11. 250- to 500-mL centrifuge tubes.
12. Dimethyl sulfoxide (DMSO).
13. Liquid nitrogen is used for flash-freezing chemically competent cells.
14. Gas permeable adhesive plate seals are used to prevent contamination of evaporation from 96-wellplates with cell cultures.

2.3 *E. coli* Strains

1. DH5α: recommended cloning strain for construction of riboregulator plasmids.
2. BL21 Star DE3: a strain containing a chromosomal copy of T7 RNA polymerase that is inducible with IPTG; it also features a mutation in RNase E that reduces RNase levels and increases mRNA stability.
3. BL21 DE3: an alternative strain to BL21 Star DE3; it has the same chromosomal T7 RNA polymerase cassette, but has the full complement of RNases.

2.4 Equipment

1. Spectrophotometer: enables measurement of optical density of cell cultures.
2. Microplate reader: enables multimode measurements of cellular fluorescence and optical density as cell cultures are being incubated.
3. Flow cytometer: enables measurements of fluorescence and light scattering properties of individual cells for populations of thousands of cells.
4. Microbiological incubator: used for growth and maintenance of bacterial cultures.
5. Incubation shaker: used for growth of liquid cultures under shaking.

6. Gel imaging system: used for imaging DNA in agarose gels and observing fluorescence from *E. coli* colonies expressing GFP or other fluorescent proteins.

3 Methods

All procedures are carried out at room temperature unless otherwise specified. All DNA sequences are listed from the 5′ end to the 3′ end. Although GFP is used as the reporter protein in the procedures below, a variety of other potential output proteins can be used instead [16].

3.1 Construction of Riboregulator Plasmids

We test the synthetic riboregulators using a two-plasmid system in which switch RNAs regulating a reporter protein (e.g., GFP) are expressed from a medium copy number plasmid (pCOLADuet vector with ColA origin and kanamycin resistance), and trigger RNAs are expressed from a high copy number plasmid (pET15b vector with a ColE1 origin and ampicillin resistance). Both RNA components are expressed using T7 RNA polymerase and transcription is terminated using a T7 terminator present on the source pCOLADuet and pET15b vectors. The methods and primers described herein will be focused on construction of these switch and trigger plasmids with T7 RNA polymerase driven expression. However, the same experimental protocols can be used for constructing other riboregulators or plasmids in general with other promoters, terminators, and source vectors as long as the sequences of the PCR primers are modified accordingly.

We construct our riboregulator plasmids using Gibson assembly [22]. This method employs two different linear DNAs produced using PCR: a vector backbone containing at minimum the desired antibiotic resistance cassette and an origin of replication for the final plasmid, and a vector insert containing the sequence of the synthetic riboregulator. PCR primers are used to generate homologous regions of 30 bp on both the backbone and insert linear DNAs. During the Gibson assembly reaction, these homologous regions enable ligation of the two DNAs and formation of circular plasmid DNA (Fig. 2). More than two DNAs can also be assembled into a circular plasmid DNA using Gibson assembly, if required [22].

3.1.1 Design of PCR Primers and Insert Sequences

The sequences to be amplified for plasmid construction are shown schematically in Fig. 2. For switch plasmids, the backbone is generated from a medium copy plasmid that contains a kanamycin resistance marker, a ColA origin of replication, and the GFP open reading frame followed by a T7 terminator (Fig. 2a). An existing switch plasmid can be used as the template for the backbone DNA. For trigger plasmids, the backbone is generated from a high copy

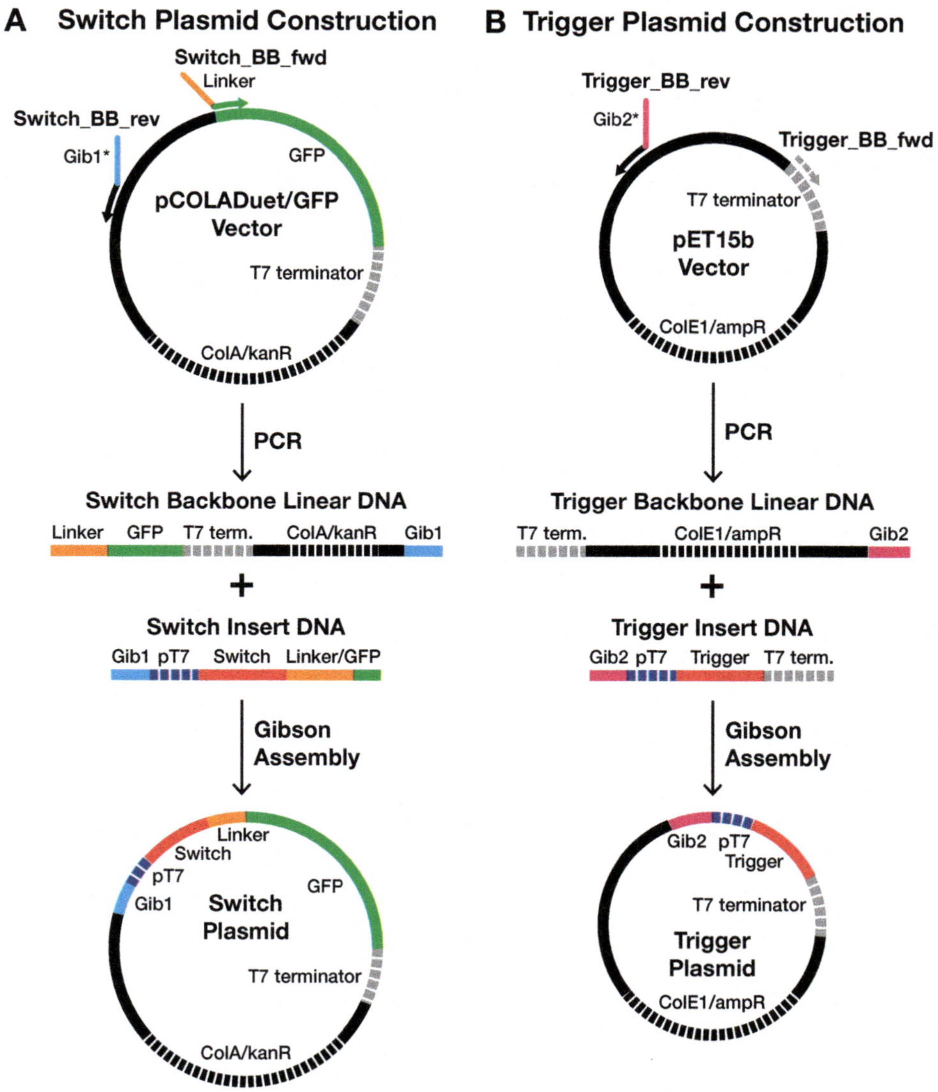

Fig. 2 Procedure for constructing riboregulator switch and trigger plasmids. (**a**) Scheme used for switch plasmid construction. A template pCOLADuet vector containing a GFP cassette is amplified using primers that add a 30-bp Gibson homology domain (Gib1*) and a 21-nt linker sequence to the product DNA. The resulting backbone linear DNA is combined with a switch insert DNA that features homologous Gib1 and linker/GFP domains. The homologous domains from the two linear DNAs are ligated using Gibson assembly to form the final circular plasmid DNA. (**b**) Scheme used for trigger plasmid construction. A template pET15b vector is amplified using a forward primer that overlaps with the T7 terminator sequence and a reverse primer that adds a 30-bp Gibson homology domain (Gib2*). The resulting backbone linear DNA is combined with a linear trigger insert DNA with flanking homology domains Gib2 and the first 30 bp of the T7 terminator. A Gibson assembly reaction ligates the two linear DNAs to form the circular trigger plasmid. The "*" symbol on DNA domains is used to denote reverse complement sequences. *Thicker lines* represent double-stranded DNAs, while *thinner lines* designate single-stranded DNA primers. Domains for the double-stranded DNAs are written in the direction of transcription of the riboregulator RNAs

plasmid that contains an ampicillin resistance marker, a ColE1 origin of replication, and a T7 terminator (Fig. 2b). The switch plasmid insert DNA sequence contains a T7 promoter and the switch RNA hairpin sequence (Fig. 2a). This central regulatory module is flanked by an upstream 30-bp domain for homology with the backbone and a downstream 30-bp domain containing a 21-bp linker sequence (AACCTGGCGGCAGCGCAAAAG) featuring low molecular weight amino acids and the first 9 bp of the GFP reporter. Similarly, the trigger plasmid insert DNA sequence contains a T7 promoter and the trigger sequence with an upstream 30-bp homology domain and a downstream region containing the first 30 bp of the T7 terminator (Fig. 2b).

To produce linear backbones with suitable homology domains, we employ backbone reverse primers that hybridize to a region upstream of the desired insertion site and have at their 5′ ends a 30-bp domain that is complementary to the one used for the desired DNA inserts. For switch backbones, a forward primer that overlaps with the start of the GFP sequence is used and has an additional 21-bp domain at the 5′ end that encodes the linker sequence (Fig. 2a). For trigger backbones, a forward primer that overlaps with the start of the T7 terminator sequence is used (Fig. 2b).

The linear DNA inserts are generated using synthetic oligonucleotide templates purchased as either single strands with lengths up to 200 nt or through assembly of multiple strands of overlapping DNA segments following published protocols [23]. These template DNAs contain from the 5′ end to the 3′ end a conserved T7 promoter primer sequence (GCGCTAATACGACTCACTATAGGG), the sequence of the riboregulator switch or trigger, and another conserved sequence overlapping with the desired homology domain. All the flanking conserved sequences have melting temperatures of at least 57 °C to enable amplification with pairs of universal switch or trigger primers. For switch RNAs, the conserved 3′ sequence contains a region of the 21-bp linker with a >57 °C melting temperature (AACCTGGCGGCAGCGCAA). For trigger RNAs, the conserved 3′ sequence contains the start of the T7 terminator (TAGCATAACCCCTTGGGGC). Thus, the forward primers for the inserts have the general sequence 5′-[30-bp homology domain]-GCGCTAATACGACTCACTATAGGG-3′. The reverse primer for the switch inserts is a 30-nt sequence that is complementary to the 21-bp linker and the first 9 bp of GFP. The reverse primer for the trigger inserts is the reverse complement of the first 30 bp of the T7 terminator.

The four pairs of primers used for generating the vector backbones and inserts for the switch and trigger plasmids are listed in Table 1, along with several additional biologically conserved sequences relevant for primer design. All primers are designed to

Table 1
PCR primers used for construction of riboregulator plasmids and other relevant DNA sequences

Primer name	Sequence (5′–3′)	Template
Switch_BB_fwd	AACCTGGCGGCAGCGCAAAAG ATGCGTAAAGGAGAAGAACTTTTCACT	pCOLADuet/GFP vector or existing switch plasmid
Switch_BB_rev	TGTTGGGGTTCTCTTAGCTTTGTTTCGCCG CATAAGGGAGAGCGTCGAGATC	
Trigger_BB_fwd	TAGCATAACCCCTTGGGGC	pET15b vector or existing trigger plasmid
Trigger_BB_rev	GAGCTATATCGCGAACCACTGGCAGACTAC CGAGATCTCGATCCTCTACGC	
Switch_insert_fwd	CGGCGAAACAAAGCTAAGAGAACCCCAACA GCGCTAATACGACTCACTATAGGG	Synthetic switch oligonucleotide
Switch_insert_rev	TTTACGCATCTTTTGCGCTGCCGCCAGGTT	
Trigger_insert_fwd	GTAGTCTGCCAGTGGTTCGCGATATAGCTC GCGCTAATACGACTCACTATAGGG	Synthetic trigger oligonucleotide
Trigger_insert_rev	CCCGTTTAGAGGCCCCAAGGGGTTATGCTA	
T7 promoter	TAATACGACTCACTATAGGG	N/A
T7 terminator	TAGCATAACCCCTTGGGGCCTCTAAAC GGGTCTTGAGGGGTTTTTTG	N/A
GFP (First 30 bases of GFPmut3b)	ATGCGTAAAGGAGAAGAACTTTTCACTGGA	N/A

PCR primers are generated as described in Subheading 3.1.1. Each of the primers is designed to have an overlap with a melting temperature of at least 57 °C with its corresponding initial DNA template binding site. 30-bp overlapping Gibson assembly domains are included in the primers for plasmid construction. Additional sequences of biologically conserved features are included in the table for reference

have at least 57 °C melting temperatures with their initial templates. We routinely employ a set of eight additional 30-bp homology domains for Gibson assembly to enable modular construction of different riboregulator plasmids (*see* Table 2 for all eight of our commonly used Gibson homology domain sequences), including those that express multiple trigger and/or switch RNAs from the same plasmid [16].

3.1.2 Preparation of Vector Backbones

1. Prepare a backbone primer master mix containing 1 μM of both the forward and reverse primers in nuclease-free water. Separate primer master mixes must be prepared for the switch and trigger amplifications. Switch backbones are amplified using primers Switch_BB_fwd and Switch_BB_rev. Trigger backbones are amplified using primers Trigger_BB_fwd and Trigger_BB_rev. Primer sequences are listed in Table 1.

Table 2
Orthogonal 30-bp Gibson assembly homology domains

Domain name	Sequence (5′–3′)
Gib1	CGGCGAAACAAAGCTAAGAGAACCCCAACA
Gib2	GTAGTCTGCCAGTGGTTCGCGATATAGCTC
Gib3	GCCGGGTTTAGAAATCTGAAGCTCTAGAGA
Gib4	CTAGTAAATTCGCGTTTCTACGGTAGCCGG
Gib5	GGCGAGCTACTCGTTTGCGTAGAAGTAGTT
Gib6	TTGAAAGTGACCACCGCTCACGTGTTGTGT
Gib7	AGTGATAGCCGTTTGTCTGGTGTCTACGCC
Gib8	GAATCTGTGTTCAATCTGGGTGGCGAGACG

The set of eight 30-bp Gibson assembly homology domains were designed to satisfy the following requirements: GC content between 45% and 55%, lack of mononucleotide runs of five bases or more, melting temperatures of at least 53 °C, free energies above −1 kcal/mol at 50 °C, and a dimerization probability of under ~1% with any other of the primers in the set at 50 °C

2. Prepare a diluted template plasmid mix by diluting the source vector to a concentration of approximately 1–4 ng/μL using nuclease-free water. For most minipreps, the plasmid DNA concentrations are in the range of 100–400 ng/μL, so that plasmids are diluted by 100-fold.
3. Inside a PCR tube, combine 24 μL of the primer master mix, 1 μL of the diluted template plasmid, and 25 μL of 2× PCR master mix. Pipette the mixture up and down briefly and place in thermal cycler.
4. Promptly run the PCR reaction using the following series of steps:
   (a) Incubate at 98 °C for 30 s.
   (b) Incubate at 98 °C for 5 s, followed by incubation at 57 °C for 15 s, and then hold at 72 °C for 2 min. Carry out this three-temperature procedure a total of 35 times. The 72 °C polymerization extension time can be adjusted depending on the length of the final DNA product and the polymerization rate specified by the manufacturer.
   (c) Incubate at 72 °C for 8 min and then return to room temperature.
5. To eliminate any residual template DNA plasmid, add 1 μL of DpnI enzyme (20 U) to the PCR reaction.
6. Mix by pipetting up and down.
7. Incubate the reaction at 37 °C for at least 30 min (*see* **Note 1**).

8. Heat-inactivate the DpnI enzyme by incubating at 80 °C for 20 min.
9. Confirm successful PCR by measuring the DNA product on a 0.8% agarose gel. The switch plasmid backbone DNA should be ~4.5 kb and the trigger plasmid backbone DNA should be ~5.5 kb.

3.1.3 Preparation of Riboregulator Vector Inserts

1. Prepare a primer master mix containing 1 μM of each of the forward and reverse primers in nuclease-free water. Separate primer mixes should be prepared for the switch and trigger insert amplifications. Switch inserts are amplified using primers Switch_insert _fwd and Switch_insert_rev. Trigger inserts are amplified using primers Trigger_insert_fwd and Trigger_insert_rev. Primer sequences are listed in Table 1.
2. Prepare a template mix with a concentration of between 10 nM and 1 μM of the template DNA in nuclease-free water.
3. Inside a PCR tube, combine 9.8 μL of the primer master mix, 0.2 μL of diluted template, and 10 μL of 2× PCR master mix. Pipette the mixture up and down briefly and place in thermal cycler.
4. Promptly run the PCR reaction using the following series of steps:
   (a) Incubate at 98 °C for 30 s.
   (b) Incubate at 98 °C for 5 s, then incubate at 57 °C for 15 s, and then incubate at 72 °C for 15 s. Carry out this three-temperature procedure a total of 35 times. The 72 °C polymerization extension time can be adjusted depending on the length of the final DNA product and the polymerization rate specified by the manufacturer.
   (c) Incubate at 72 °C for 5 min and then return to room temperature.
5. Confirm successful amplification of the switch and trigger insert DNAs by measuring the PCR products on a 2% or 2.5% agarose gel. Inserts should be approximately 100 bp depending on the riboregulator design.

3.1.4 Gibson Assembly and Transformation of E. coli

1. Inside a 200 μL capacity PCR tube, combine 2.5 μL of the backbone PCR product, 2.5 μL of the insert PCR product, and 5 μL of 2× Gibson assembly mix.
2. Pipette up and down to mix.
3. Incubate at 50 °C for 30 min. PCR products do not need to be purified prior to use in Gibson assembly reaction. More than two DNA fragments can be ligated during the Gibson reaction, provided they have suitable homology domains (*see* **Note 2**).

4. Remove chemically competent *E. coli* cells from the −80 °C freezer and leave to thaw on ice for 10 min. A cloning strain such as DH5α is recommended for plasmid construction. *See* protocol Subheading 3.4 for preparation of chemically competetent *E. coli.*
5. Distribute 50 μL aliquots of competent cells into chilled plastic tubes.
6. After the 30-min Gibson reaction, place the products on ice for ~2 min.
7. Pipette 2 μL of the Gibson reaction into the aliquots of chemically competent cells (*see* **Note 3**).
8. Mix by pipetting up and down.
9. Leave cells on ice for 10–15 min. Gibson products do not need to be purified prior to addition to the competent cells.
10. Complete transformation of the cells using protocol Subheading 3.5 **steps 6–12**.
11. Spread transformed cells onto LB/agar plates with 50 μg/mL kanamycin on LB/agar plates for the switch plasmids and 100 μg/mL ampicillin on LB/agar plates for the trigger plasmids.
12. Sequences of the plasmids can be verified from colonies by using commercial DNA sequencing services. Miniprep colonies with validated sequences to obtain switch and trigger DNA plasmids for later studies on the synthetic riboregulators.

3.2 Double Transformation of Expression Strains for Riboregulator Testing

This protocol is used to transform cells with a pair of switch and trigger plasmids for evaluating the performance of different synthetic riboregulators. Since switch and trigger RNAs are expressed using T7 RNA polymerase, we typically employ BL21 Star DE3 and BL21 DE3 strains of *E. coli* that possess an IPTG-inducible chromosomal copy of the phage polymerase for riboregulator testing. BL21 Star DE3 is an RNase-deficient strain that is known to have better mRNA stability and could thus lead to better riboregulator performance. However, testing of the riboregulators in BL21 Star DE3 and BL21 DE3 has revealed similar performance in both strains [16]. This protocol is generalizable and can also be applied to other *E. coli* strains with riboregulators, riboswitches, or other transcripts that are expressed using non-T7 RNA polymerases.

1. Prepare plasmid minipreps of the riboregulator switches and triggers of interest. Ensure that the concentrations of plasmid DNA are at least 100 ng/μL.
2. Remove chemically competent BL21 Star DE3 or BL21 DE3 cells from the −80 °C freezer and leave to thaw on ice for 10 min.

3. Distribute 50 μL aliquots of competent cells into chilled plastic tubes.
4. Add 1 μL of switch plasmid and 1 μL of trigger plasmid to the competent cells (*see* **Note 4**).
5. Mix by pipetting up and down. Leave mixture on ice for 10–15 min.
6. Complete transformation of the cells using the protocol in Subheading 3.5 starting from **step 3**.
7. Spread the transformed cells onto LB/agar plates with 50 μg/mL kanamycin and 100 μg/mL ampicillin. For faster screening of the riboregulators, 0.1 mM IPTG can be added to the LB/agar plates in order to induce expression of the riboregulator RNAs. GFP output from colonies can be detected under blue light illumination with a suitable emission filter.

3.3 Induction and Evaluation of Riboregulators

This protocol is performed on BL21 Star DE3 or BL21 DE3 cells that have been co-transformed with riboregulator switch and trigger plasmids. T7 RNA polymerase expression is induced with the addition of IPTG.

1. Prepare a deep 96-well plate with 500 μL aliquots of LB containing 30 μg/mL kanamycin and 50 μg/mL ampicillin (*see* **Note 5**).
2. Using a sterile pipette tip or toothpick, inoculate each well with a single colony of *E. coli* double transformant. Use at least three colonies from each batch of transformed cells for biological replicates. Use a gas permeable adhesive plate seal to ensure the liquid media does not evaporate overnight and to prevent contamination.
3. Incubate the cells overnight at 37 °C with shaking at 800–1000 rpm.
4. Prepare a deep 96-well plate with 247.5 μL aliquots of LB containing 30 μg/mL kanamycin and 50 μg/mL ampicillin (*see* **Note 5**).
5. Dilute the overnight cultures 100-fold by adding 2.5 μL of the cultures into the fresh LB media.
6. Incubate at 37 °C with shaking at 800–1000 rpm for 80 min.
7. Induce expression of T7 RNA polymerase and transcription of the switch and trigger RNAs by adding IPTG to a concentration of 0.1 mM in each well.
8. Return plate to shaking at 800–1000 rpm and 37 °C.
9. For flow cytometry measurements, harvest cells at time points of interest by diluting into PBS to determine GFP expression levels. The plate can also be shaken in the plate reader at 37 °C for kinetic measurements of cell growth and GFP expression.

3.4 Preparation of Chemically Competent E. coli

This method for preparing chemically competent cells is based on the protocol described by Inoue et al. in 1990 [24]. The procedure uses a 1 L cell culture but can be scaled down depending on the number of transformations required. A 1 L preparation of competent cells yields 800 aliquots of 50 μL suitable for large scale screening of different riboregulators and large scale plasmid construction. Competent cells prepared using this procedure should yield dozens of doubly transformed colonies when cotransformed with switch and trigger plasmids as described in Subheading 3.2.

1. Plate the desired strain of *E. coli* onto LB/agar. If the *E. coli* strain harbors plasmids, the LB/agar can contain 50 μg/mL kanamycin, 100 μg/mL ampicillin, or other antibiotics as appropriate.
2. Incubate overnight at the recommended growth temperature, typically 37 °C.
3. After 12–16 h of incubation, pick a single colony of *E. coli* from the plate and inoculate 6 mL of LB with the appropriate antibiotics in a 14-mL culture tube.
4. Incubate cells overnight at 37 °C with shaking at 250–300 rpm.
5. Refrigerate the centrifuge tubes and ITB solution required for the next day of experiments. Do not refrigerate the DMSO as it could freeze overnight.
6. Transfer 5 mL of the overnight culture into 1 L of SOB medium with the appropriate antibiotics.
7. Shake cells at 250–300 rpm at 30 °C. Use a flask of at least 4 L volume to ensure the culture receives adequate aeration (*see* **Note 6**).
8. Monitor cell growth at 30-min intervals by extracting aliquots of the culture and measuring the optical density at a wavelength of 600 nm (OD600). Once cells reach an OD600 between 0.4 and 0.6, remove them promptly from the incubator and leave them to chill on ice for 10 min. Cells typically require 3–4 h to reach this OD600 level at 30 °C, depending on the *E. coli* strain.
9. Transfer the cell culture to chilled (~4 °C) centrifuge tubes (250–500 mL). Distribute the culture evenly between tubes. Centrifuge for 10 min at 2500 × *g* at 4 °C.
10. Carefully pour out the supernatant making sure to preserve the integrity of the cell pellet.
11. Take 200 mL of the refrigerated ITB solution and distribute it evenly between the centrifuge tubes. Resuspend the cells in ITB by shaking. Leave the centrifuge tubes on ice for 10 min.
12. Centrifuge the tubes for 10 min at 2500 × *g* at 4 °C.

13. Prepare a mixture of 37.2 mL ITB and 2.8 mL of DMSO. Chill on ice.
14. Carefully pour out the supernatant, again making sure to not disturb the pellet. Transfer the 40 mL ITB/DMSO mixture to each centrifuge tube, using it to resuspend each pellet. Leave this highly concentrated mixture of cells in ITB/DMSO to chill on ice for 10 min.
15. Prepare a bath of liquid nitrogen in a thermally insulated container for flash freezing of the competent cells.
16. Prepare recipient tubes for the competent cells by cooling them on ice (*see* **Note 7**).
17. After the recipient tubes are cooled to ~4 °C, transfer competent cell mixture to the tubes at the desired volume (typically 50–500 μL). Promptly freeze the cells using the liquid nitrogen.
18. Extract the tubes of frozen competent cells from the liquid nitrogen and store at −80 °C. Cells should remain competent for at least 1 year at −80 °C, although transformation efficiency is highest when the cells are freshly prepared.

3.5 Transformation of Chemically Competent Cells

1. Prior to adding DNA, remove chemically competent cells from −80 °C freezer and leave them to thaw on ice for 10 min.
2. Distribute 50 μL aliquots of competent cells into chilled plastic tubes (*see* **Note 8**).
3. Add the DNA of interest to the competent cells.
4. Mix by pipetting up and down.
5. Leave the mixture on ice for 10–15 min.
6. Incubate mixture of chemically competent cells and DNA for 30–60 s in a 42 °C water bath.
7. Return cells to ice for 1–5 min.
8. Prepare a deep 96-well plate or 14-mL culture tubes for cell growth by adding 1 mL of antibiotic-free LB to wells or tubes to be used for cultures.
9. Transfer the transformed cells to the wells or culture tubes with LB.
10. Place cultures in a shaker at 37 °C for at least 1 h. For 96-well plates, cover the wells with a gas permeable adhesive plate seal and shake the cells at 800–1000 rpm for best growth. For culture tubes, cells can be shaken at 250–300 rpm.
11. After at least an hour of shaking, cells can be harvested by centrifugation. Spin down the 96-well plates by centrifuging at 1750 × *g* for 10 min at 4 °C. Remove the adhesive seal and quickly pour out the supernatant into biological waste being

careful to not cross-contaminate the wells. Use a stack of paper towels to absorb any supernatant remaining at the top of the plate. Spin down the culture tubes by centrifuging at 3000 × *g* for 10 min at 4 °C. Pour supernatant into biological waste.

12. Resuspend the cell pellets in 200 μL of antibiotic-free LB by pipetting them up and down carefully.
13. Transfer the resuspended cells to their recipient LB/agar petri dishes containing the appropriate antibiotics. Use glass beads to spread cells onto the plates. Let the plates stand with lid side up at room temperature for 10–15 min.
14. Discard glass beads and leave plates with lid side down at 37 °C overnight. Colonies should be visible after 12–16 h of incubation.

4 Notes

1. Residual template DNA plasmid can lead to a high background of unwanted colonies without the desired DNA insert. DpnI digests methylated DNA present in the miniprepped template plasmids produced in *E. coli*. Linear DNA produced during PCR reactions is not methylated and is thus unaffected by the DpnI treatment.
2. As an example of Gibson plasmid assembly with three different linear DNAs A, B, and C, one would use the Gibson homology domains listed in Table 2 with DNAs having the following general sequences:
   (a) DNA A: 5′-Gib1-[Main DNA A Sequence]-Gib2-3′.
   (b) DNA B: 5′-Gib2-[Main DNA B Sequence]-Gib3-3′.
   (c) DNA C: 5′-Gib3-[Main DNA C Sequence]-Gib1-3′.

   These three DNAs can be combined in equimolar concentrations in a Gibson assembly reaction to yield circular DNAs containing all three sequences. Assembly yields will decrease as additional DNA species are ligated. Intermediate PCR steps can be used to ligate different DNAs together prior to the Gibson reaction and increase the ultimate yield of the Gibson plasmid assembly step.
3. Larger volumes of the Gibson reaction can be added if plasmid assembly yields are low.
4. Volumes of plasmid added can be increased or decreased depending on miniprep concentration and transformation efficiency of the competent cells.
5. When culturing cells transformed with more than one plasmid in liquid media, we typically use lower concentrations of antibiotics than with cells having only a single plasmid to reduce

stresses on the cells. In practice, this means that antibiotic levels are reduced to 50–60% of their typical levels, yielding 30 μg/mL for kanamycin and 50 μg/mL for ampicillin.

6. For convenience, the culture can also be divided into two 500 mL volumes in 2 L flasks, instead of a single large 4 L flask.
7. Do not cool the recipient tubes too aggressively, for instance in a freezer. Overly cool tubes will cause the competent cell mixture to freeze rapidly and could lead to lower transformation efficiencies if they thaw before being transferred to liquid nitrogen. Tubes can be prechilled in a refrigerator or by leaving them on ice.
8. Aliquots smaller than 50 μL can also be used for transformations. In particular, smaller volumes are helpful to conserve competent cells when transforming with only a single plasmid.

Acknowledgment

This work was supported by start-up funds from Arizona State University.

References

1. Delebecque CJ, Lindner AB, Silver PA, Aldaye FA (2011) Organization of intracellular reactions with rationally designed RNA assemblies. Science (New York, NY) 333(6041):470–474
2. Sachdeva G, Garg A, Godding D, Way JC, Silver PA (2014) In vivo co-localization of enzymes on RNA scaffolds increases metabolic production in a geometrically dependent manner. Nucleic Acids Res 42(14):9493–9503
3. Thisted T, Gerdes K (1992) Mechanism of post-segregational killing by the hok/sok system of plasmid R1. Sok antisense RNA regulates hok gene expression indirectly through the overlapping mok gene. J Mol Biol 223 (1):41–54
4. Franch T, Petersen M, Wagner EG, Jacobsen JP, Gerdes K (1999) Antisense RNA regulation in prokaryotes: rapid RNA/RNA interaction facilitated by a general U-turn loop structure. J Mol Biol 294(5):1115–1125
5. Isaacs FJ, Dwyer DJ, Ding C, Pervouchine DD, Cantor CR, Collins JJ (2004) Engineered riboregulators enable post-transcriptional control of gene expression. Nat Biotechnol 22 (7):841–847
6. Lucks JB, Qi L, Mutalik VK, Wang D, Arkin AP (2011) Versatile RNA-sensing transcriptional regulators for engineering genetic networks. Proc Natl Acad Sci 108(21):8617–8622
7. Callura JM, Cantor CR, Collins JJ (2012) Genetic switchboard for synthetic biology applications. Proc Natl Acad Sci 109 (15):5850–5855
8. Mutalik VK, Qi L, Guimaraes JC, Lucks JB, Arkin AP (2012) Rationally designed families of orthogonal RNA regulators of translation. Nat Chem Biol 8(5):447–454
9. Takahashi MK, Lucks JB (2013) A modular strategy for engineering orthogonal chimeric RNA transcription regulators. Nucleic Acids Res 41(15):7577–7588
10. Lutz R, Bujard H (1997) Independent and tight regulation of transcriptional units in *Escherichia coli* via the LacR/O, the TetR/O and AraC/I1-I2 regulatory elements. Nucleic Acids Res 25(6):1203–1210
11. Zadeh JN, Steenberg CD, Bois JS, Wolfe BR, Pierce MB, Khan AR et al (2011) NUPACK: analysis and design of nucleic acid systems. J Comput Chem 32(1):170–173
12. Zadeh JN, Wolfe BR, Pierce NA (2011) Nucleic acid sequence design via efficient ensemble defect optimization. J Comput Chem 32(3):439–452
13. Wolfe BR, Pierce NA (2015) Sequence design for a test tube of interacting nucleic acid strands. ACS Synth Biol 4(10):1086–1100

14. Yin P, Choi HMT, Calvert CR, Pierce NA (2008) Programming biomolecular self-assembly pathways. Nature 451 (7176):318–322
15. Qian L, Winfree E (2011) Scaling up digital circuit computation with DNA strand displacement cascades. Science (New York, NY) 332 (6034):1196–1201
16. Green AA, Silver PA, Collins JJ, Yin P (2014) Toehold switches: de-novo-designed regulators of gene expression. Cell 159(4):925–939
17. Rhodius VA, Segall-Shapiro TH, Sharon BD, Ghodasara A, Orlova E, Tabakh H et al (2013) Design of orthogonal genetic switches based on a crosstalk map of sigmas, anti-sigmas, and promoters. Mol Syst Biol 9702
18. Pardee K, Green AA, Ferrante T, Cameron DE, DaleyKeyser A, Yin P et al (2014) Paper-based synthetic gene networks. Cell 159(4):940–954
19. Pardee K, Green AA, Takahashi MK, Braff D, Lambert G, Lee JW et al (2016) Rapid, low-cost detection of Zika virus using programmable biomolecular components. Cell 165(5):1255–1266
20. Chappell J, Takahashi MK, Lucks JB (2015) Creating small transcription activating RNAs. Nat Chem Biol 11(3):214–220
21. Krishnamurthy M, Hennelly SP, Dale T, Starkenburg SR, Martí-Arbona R, Fox DT et al (2015) Tunable riboregulator switches for post-transcriptional control of gene expression. ACS Synth Biol 4(12):1326–1334
22. Gibson DG, Young L, Chuang R-Y, Venter JC, Hutchison CA, Smith HO (2009) Enzymatic assembly of DNA molecules up to several hundred kilobases. Nat Methods 6(5):343–345
23. Rouillard J-M, Lee W, Truan G, Gao X, Zhou X, Gulari E (2004) Gene2Oligo: oligonucleotide design for in vitro gene synthesis. Nucleic Acids Res 32(Web Server):W176–W180
24. Inoue H, Nojima H, Okayama H (1990) High efficiency transformation of *Escherichia coli* with plasmids. Gene 96(1):23–28

Chapter 20

Preparation of a Conditional RNA Switch

Paul Zakrevsky, Lorena Parlea, Mathias Viard, Eckart Bindewald, Kirill A. Afonin, and Bruce A. Shapiro

Abstract

RNA has gained great interest for use in biomedical and therapeutic applications. This is due in part to RNA's ability to perform multiple functions, including the regulation of endogenously expressed genes. However, the ability of RNA based drugs to distinguish target diseased cells from healthy tissue remains challenging. Here we present methods for the production of a recently developed conditional RNA switch that releases a Dicer substrate RNA in response to interaction with a specific RNA biomarker.

Key words RNA, Toehold, Nanotechnology, Nanomedicine, RNase H, RNA switch, Conditional RNA

1 Introduction

Nucleic acids have become one of the preferred materials for construction of objects and scaffolds at the nanoscale [1, 2]. RNA in particular has gained significant interest in the fields of nanomedicine and nanotechnology as a material for fabrication of nanoconstructs for therapeutic application due to several characteristics inherent to this biopolymer. Natural RNAs possess an assortment of recurrent structural motifs composed of classic and noncanonical base pair interactions [3, 4] providing a repository of building blocks that can be used to define the geometry, positioning, and assembly of monomeric strands within a rationally designed RNA construct [5–9]. In addition, the wide array of known functions performed by different RNA elements offers the potential to encode diverse functionalities within an RNA nanoconstruct. Such functions can include cell-specific targeting, knockdown of target gene expression, or organization of cellular machinery [10–12].

Critical to the design and function of RNA nanoconstructs is the specificity provided by Watson–Crick base pairing interactions (WC bps). From a structural standpoint, the specificity of WC bps

Eckart Bindewald and Bruce A. Shapiro (eds.), *RNA Nanostructures: Methods and Protocols*, Methods in Molecular Biology, vol. 1632, DOI 10.1007/978-1-4939-7138-1_20, © Springer Science+Business Media LLC 2017

provides a means to program the formation of helical regions which act as the scaffolding backbone that ultimately defines construct geometry [13, 14]. From a function standpoint, specific base pairing allows RNA to convey information in the form of the nucleotide sequence. This ability to store, search for, and recognize specific sequence information is central to many RNA-based regulatory functions, such as RNA interference (RNAi) and the CRISPR-Cas system, and RNA-based nanoconstructs that regulate gene expression have been designed utilizing such functions [11, 15].

However, the ability of RNA to store and read information at the level of its nucleotide sequence can also be used to regulate the function of the RNA nanoconstruct itself. Single-stranded nucleic acid toeholds can be used to produce conditional structural changes in response to the presence of a cognate "trigger" sequence that is complementary to the toehold region. Such a method can be used to conditionally knock down target gene expression if the induced conformational change either generates or liberates an RNA element capable of inducing an RNAi response [16, 17].

We have recently developed a two-stranded RNA switch (Fig. 1) that is able to knock down target gene expression in response to the presence of a specific mRNA biomarker and verified its function by gel electrophoresis and in a cultured human breast cancer cell line [18]. The switch is assembled from two individual RNA strands, named the "anti-trigger" and "anti-target" strands. On its own, the anti-target strand is designed to fold into an shRNA-like structure, which is able to be processed by Dicer and results in knockdown of a target gene. However, assembly of the anti-target with the anti-trigger strand sequesters the anti-target into a more thermodynamically favorable, intermolecular assembly and prevents formation of the active shRNA-like fold. The anti-trigger strand contains a single-stranded toehold within the assembly, which is used to perform the diagnostic function of the switch. Upon recognition of the biomarker trigger RNA through its single-stranded toehold, the anti-trigger strand unzips from the anti-target strand and results in the disassembly of the switch. This allows the anti-target strand to undergo a conformational rearrangement to form its active shRNA-like structure. As the switch design contains no overlap between the sequence regions that encode the siRNA component and the region complementary to the RNA trigger, virtually any combination of trigger and target pairs can be incorporated within the switch design. Additionally, as the 3′ end of the anti-trigger sequence contains no design constraints, it allows for further potential functionalization. Conjugation to a targeting aptamer or attachment to a multivalent RNA scaffold could be utilized to create a more robust, conditional drug delivery vehicle [19].

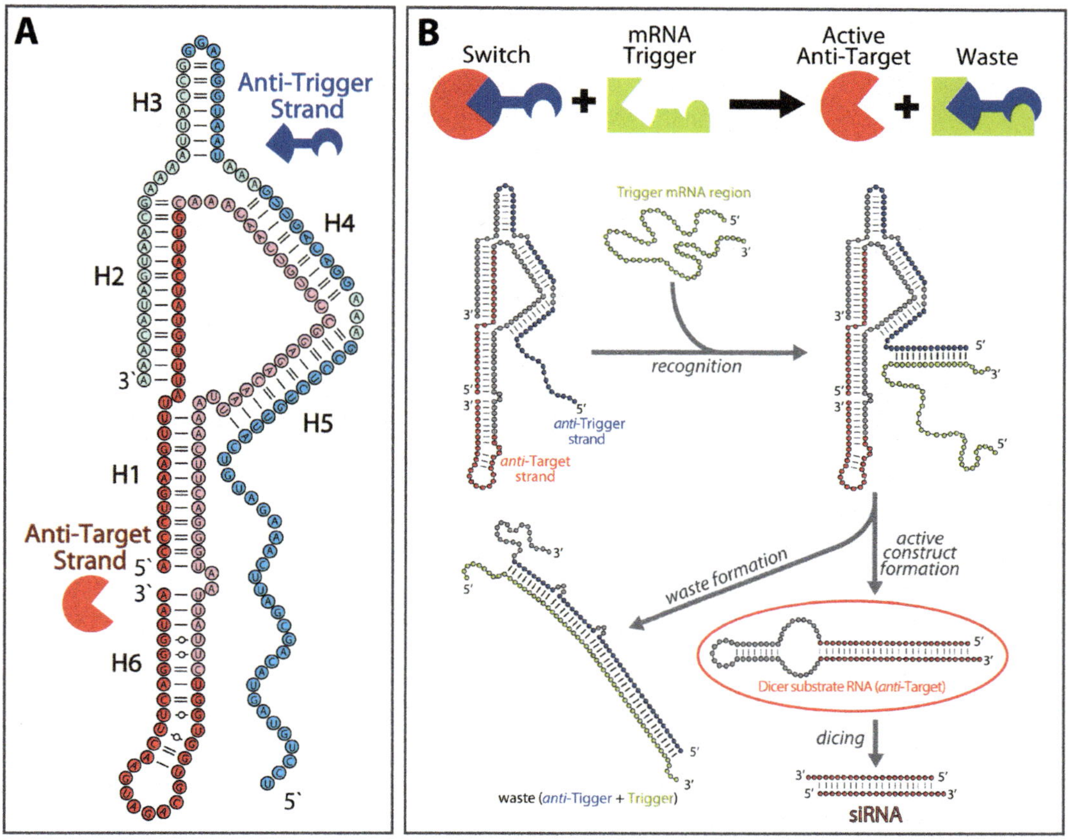

Fig. 1 Design of a two-stranded RNA switch. (**a**) The two-stranded switch is assembled from an anti-trigger strand (*blue*) and an anti-target strand (*red*). The majority of the anti-trigger strand (*dark blue nodes*) is designed to be complementary to the trigger RNA sequence, while the 5′ and 3′ end regions of the anti-target strand (*dark red nodes*) encode siRNA sense and antisense sequences, respectively. (**b**) The RNA switch is initially in an inert conformation. Interaction of the anti-trigger's single-stranded toehold with a trigger RNA induces a conformational change that releases the anti-target strand as an active shRNA-like Dicer substrate. Adapted and reprinted with permission from [18], copyright 2016 American Chemical Society

In the protocols that follow we describe detailed methods for the synthesis, assembly, purification, and functional validation of the two-stranded RNA switch designed to silence eGFP expression in response to the presence of connective tissue growth factor mRNA (CTGF/CCN2). It is possible to incorporate other target/trigger pairs within the switch design, and a brief discussion of sequence design is included in Subheading 3.1. In vitro transcription and PAGE purification of transcripts are discussed in Subheadings 3.2 and 3.3, respectively. As the two strands, and particularly the anti-target strand, require specific 5′ end sequences that deviate from the typical purine-rich 5′ sequence preferred by T7 RNA polymerase [20], in Subheading 3.4 we provide a method for the removal of a 5′ end leader sequence by addition of RNase H [18, 21]. Assembly and native purification of the RNA switch are described in Subheading 3.5 and 3.6, respectively. Methods for functional validation by nondenaturing PAGE and in cultured human cells are described in Subheading 3.7 and 3.8, respectively.

2 Materials and Equipment

All buffers and solutions should be made with ultrapure 18.2 MΩ water. The stock buffers and solutions should be filtered using a 0.22-μm filter. Ensure that reagents and enzymes used are RNase- and DNase-free. Any solutions and experiments prepared for cell culture studies should be made with water that is certified RNase-, DNase-, and endotoxin-free.

2.1 *In Vitro* Transcription Components

1. DNA templates encoding desired RNA sequences (*see* **Note 1**).
2. Transcription buffer, final concentrations (1×): 5 mM $MgCl_2$, 2 mM spermidine, 40 mM Tris pH 7.5 (*see* **Note 2**). Typically prepared as a 1 mL stock solution at 10× concentration.
3. NTP mixture: 25 mM each ATP, GTP, CTP, UTP.
4. $MnCl_2$, 5.0 mM solution.
5. Inorganic pyrophosphatase (IPP), 0.1 U/μL solution.
6. Dithiothreitol (DTT), 100 mM solution.
7. T7 RNA polymerase enzyme.
8. DNase 1 enzyme.

2.2 RNA Strand Purification Components

1. Urea, 8 M.
2. EDTA, 0.5 M pH 8.0.
3. Urea blue loading dye: 6 M urea, 20 mM EDTA, 10% (v/v) glycerol, 0.005% (w/v) xylene cyanol, 0.005% (w/v) bromophenol blue. Typically prepared in a 25 mL volume.
4. Tris–borate–EDTA (TBE) buffer, final concentrations (1×): 89 mM Tris, 89 mM boric acid, 2 mM EDTA, pH 8.3. Prepare 1 L as a 10× stock solution.
5. Denaturing PAGE gel: 8% acrylamide–bis-acrylamide (19:1), 1× TBE, 5.5 M urea; 1.6 mm thickness.
6. Ammonium persulfate (APS), 10% solution w/v. Prepare in a 10 mL volume.
7. Tetramethylethylenediamine (TEMED).
8. Tx elution buffer: 10 mM Tris pH 7.5, 200 mM NaCl, 0.5 mM EDTA. Prepare in a 25 mL volume.
9. Ethanol, 100% and 90% solution, chilled (−20 °C).

10. Vertical PAGE system (16.5 cm × 28 cm or 33 cm × 39 cm gel).
11. Plastic film.
12. Handheld UV lamp.
13. Scalpel or razor blade.
14. Shaker/mixer for 1.5-mL microcentrifuge tubes (temperature controlled or in refrigerated environment, 4 °C).
15. Temperature controlled benchtop centrifuge.
16. Centrifugal vacuum concentrator.
17. UV spectrophotometer.

2.3 RNase H Processing Components

1. Purified RNA transcripts containing a 5′ leader sequence.
2. DNA oligonucleotide complementary to 5′ RNA leader sequence.
3. RNase H enzyme.
4. RNase H buffer, final concentration (1×): 50 mM Tris pH 8.3, 75 mM KCl, 3 mM $MgCl_2$, 10 mM dithiothreitol (DTT) (*see* **Note 3**). Prepare 1 mL as a 10× stock solution.

2.4 RNA Assembly and Nondenaturing PAGE Components

1. Purified RNA transcripts.
2. Tris–borate (TB) buffer, final concentrations (1×): 89 mM Tris, 89 mM boric acid, pH 8.3. Prepare 1 L as a 10× stock solution.
3. Assembly buffer, final concentrations (1×): 1× TB buffer, 2 mM $Mg(OAc)_2$. Prepare 1 mL as a 5× concentrated solution.
4. Non-denaturing PAGE gel: 10% acrylamide–bis-acrylamide (19:1), 1× TB buffer, 2 mM $Mg(OAc)_2$; 0.8 mm thickness.
5. Native gel running buffer: 1× TB buffer, 2 mM $Mg(OAc)_2$. Prepare 1 L.
6. Native gel loading buffer: 1× assembly buffer, 0.005% (w/v) xylene cyanol, 0.005% (w/v) bromophenol blue, 50% (v/v) glycerol (*see* **Note 4**). Prepare in a 1 mL volume.
7. Cold room or refrigerated environment (4 °C).
8. Vertical PAGE system (16.5 cm × 28 cm or 33 cm × 39 cm gel).
9. Nucleic acid stain solution (such as: ethidium bromide solution, 0.5 μg/mL; SYBR Gold, 1×; SYBR Green II, 1×).
10. Fluorescence gel scanner (*see* **Note 5**).
11. Plastic film.
12. Handheld UV lamp.
13. Scalpel or razor blade.
14. Flat-bottom microcentrifuge tubes.

2.5 Cell Culture, Transfection, and Analysis Components

1. MDA-MB 231 cell line stably expressing eGFP (*see* **Note 6**).
2. Dicer substrate RNA designed against the endogenous RNA trigger molecule (for the CTGF/eGFP switch described here, an anti-CTFG Dicer substrate is used).
3. Complete growth medium: DMEM high glucose medium supplemented with 10% FBS, 2 mM L-glutamine, 100 U/mL penicillin, 100 μg/mL streptomycin.
4. Opti-MEM medium (Gibco).
5. Lipofectamine 2000 (Invitrogen) (*see* **Note 7**).
6. Cell dissociation buffer. Available from various commercial sources.
7. Phosphate buffered saline (PBS), 1× concentration: 1.47 mM KH_2PO_4, 4.3 mM Na_2HPO_4, 2.7 mM KCl, 137 mM NaCl, pH 7.4, sterile. Can be purchased from commercial source, or prepare 1 L of 1× concentration.
8. Grid slide.
9. Optical microscope.
10. Flow cytometer.

3 Methods

3.1 Overview of RNA Sequence Design

Both the anti-trigger and the anti-target strand of the switch must adhere to strict sequence constraints across long stretches of each molecule (Fig. 1). Much of the anti-trigger strand sequence is defined by its necessity to be complementary to the RNA biomarker trigger. Within the anti-target strand, the 5′ and 3′ regions must encode the sense and anti-sense sequences of the siRNA against the gene targeted for knockdown, respectively. These two constraints establish the starting point for RNA sequence design.

The anti-trigger sequence is designed to contain the single-stranded toehold (approximately 25–30 nt) at its 5′ end before hybridizing with the anti-target strand to form helices H5 and H4, and then forms the intramolecular hairpin containing helix H3. All three of these helical sequences are defined by the RNA trigger sequence. The 3′ end of the anti-trigger strand is designed to be complementary to a portion of the siRNA region within the anti-target strand, forming helix H2.

The anti-target strand begins at its 5′ end by encoding the sense sequence for the target siRNA. The central portion of the molecule is designed to be complementary to the anti-trigger strand, forming helices H4 and H5, before forming intrastrand pairings with its own 5′ end (helix H1). Additional nucleotides are then incorporated to ensure that the 3′ end of the molecule, that encodes the siRNA antisense sequence, forms a stable hairpin

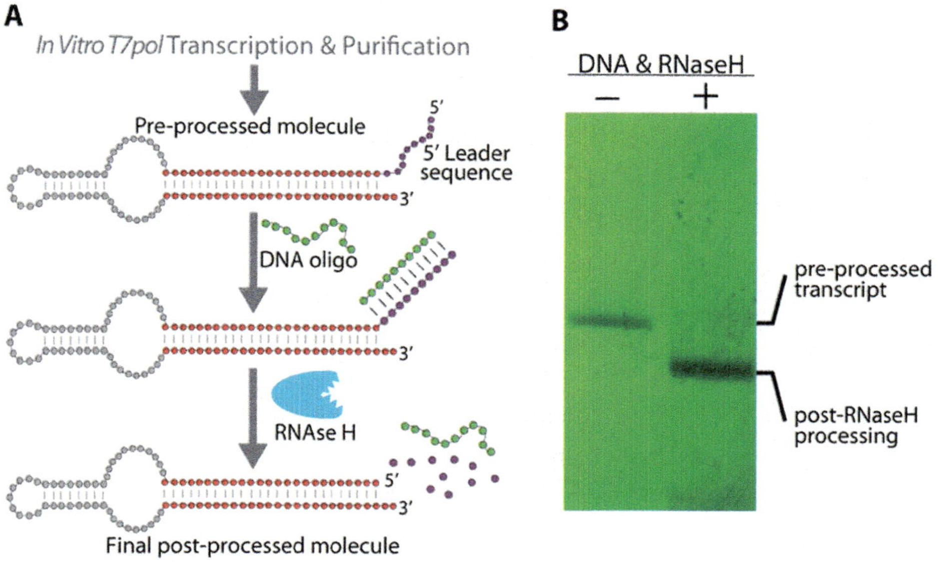

Fig. 2 5′-End processing by RNase H. (**a**) RNA molecules are transcribed containing a purine rich 5′ leader sequence which increases transcription yield by T7 RNA polymerase. The 5′ leader sequence can be removed by processing the transcript with RNase H. A short DNA oligonucleotide is added, complementary to the 5′ leader sequence of the RNA, forming an RNA–DNA hybrid duplex. The RNA strand of the hybrid duplex is degraded by addition of RNase H. (**b**) Transcripts processed by RNase H can be purified by denaturing PAGE. Molecules not subjected to the RNase H processing can be used as a full-length negative control. Adapted and reprinted with permission from [18], copyright 2016 American Chemical Society

denoted helix H6. Single-stranded poly-A regions are incorporated into the anti-trigger and anti-target strands at various places to provide structural flexibility.

Importantly, both anti-trigger and anti-target RNAs are initially transcribed containing a purine rich 5′ leader sequence that helps increase synthesis yield by T7 RNA polymerase. The leader sequence is later removed by processing with RNase H (Fig. 2). The DNA templates for each switch strand should be designed to contain the T7 promoter sequence, followed by a 14 nucleotide RNA leader sequence at positions +1 to +14, relative to the start of transcribed regions. The sequence of the final desired RNA should begin at position +15 on the template. Typically, we use the promoter sequence 5′TTCTAATACGACTCACTATA at positions −20 to −1, followed by the leader sequence 5′GGGAAAGGAAGAGC at positions +1 to +14, after which the desired switch strand sequence is encoded. The process of removing the leader sequence is outlined in Subheading 3.4.

3.2 In Vitro Transcription Using T7 RNA Polymerase

1. Prepare in vitro transcription reactions in a 1.5-mL microcentrifuge tube using approximately 20 pmol DNA template.
2. Transcription reactions can be prepared to a final volume of up to 200 μL. Based on your desired reaction volume, add the

appropriate volume of 10× transcription buffer to achieve a 1× concentration.

3. Add NTPs to a final concentration of 2.5 mM each.
4. Add IPP to a final concentration of 0.01 U/μL.
5. Add DTT to a final concentration of 2 mM.
6. Add $MnCl_2$ to a final concentration of 0.5 mM.
7. Bring to the final reaction volume by addition of water.
8. Initiate transcription by adding T7 RNA polymerase to the transcription mix, to a final concentration of 2.5 U/μL. Pipet to mix; avoid vortexing.
9. Incubate the reaction at 37 °C for 4 h.
10. Terminate the transcription reaction by adding 5 U of DNase 1 per 100 μL of transcription volume.
11. After addition of DNase 1, incubate the transcription reaction for 30 min at 37 °C to ensure digestion of the DNA template.
12. It is recommended that transcribed RNA be purified immediately following transcription termination and DNA template digestion. However, if necessary, the transcription reaction can be stored at −20 °C overnight and purified the following day (*see* **Note 8**).

3.3 Purification of RNA by Denaturing PAGE

1. Prepare a 1.6 mm thick denaturing PAGE gel. For a 16.5 cm × 28 cm gel, prepare 100 mL of denaturing gel solution and initiate polymerization by adding 600 μL of 10% APS solution and 80 μL TEMED. For a 33 cm × 39 cm gel, increase the volume of gel solution to 300 mL and polymerize by adding 1800 μL of 10% APS solution and 240 μL TEMED (*see* **Note 9**). Insert a comb that produces wells with a volume of 300–400 μL.
2. One polymerized, set up the PAGE system at room temperature using 1× TBE as the running buffer.
3. Using a syringe or plastic transfer pipet, rinse the wells of the gel with the running buffer to remove any debris or air bubbles.
4. Prerun the gel for 20–30 min. For a 16.5 × 28 cm gel, run at a constant power of 30 W. For a 33 × 39 cm gel, run at a 60 W.
5. Add 100 μL urea blue loading dye to the RNA sample(s) to be purified.
6. Heat to 94 °C for 2 min to denature RNA molecules, then remove from heat.
7. Rerinse the wells of PAGE gel and load RNA in the cleaned wells. For small gels (16.5 cm × 28 cm), perform electrophoresis at a constant 30 W. For large gels (33 cm × 39 cm), electrophoresis should be performed using a constant power

of 60–70 W. Gels should be run for approximately 2–3 h (*see* **Note 10**).

8. Following electrophoresis, carefully open the glass plates used to cast the gel. The gel should adhere to one of the two plates.
9. With the gel adhered to one of the plates, cover the exposed gel surface with plastic film. Flip the gel over so the plastic film is down on the benchtop.
10. Carefully peel the gel off the second plate and onto the plastic film. Cover the newly exposed side of the gel with plastic film.
11. Locate the RNA within the gel by examining the gel under UV light. For greater contrast a UV reflective surface can be placed under the gel, but a white surface should be sufficient. Circle the location of RNA in the gel using a fine-tipped marker.
12. Use a scalpel blade to cut the band of RNA from the rest of the gel and place the excised gel piece in a 1.5-mL microcentrifuge tube.
13. Add 600 μL of Tx elution buffer to the tube and place on a shaker at 900 rpm and 4 °C overnight (~12–14 h).
14. The following morning, transfer the elution buffer to a new 1.5-mL microcentrifuge tube.
15. Add 900 μL of cold 100% ethanol to the elution buffer. Invert the tube several times to mix.
16. Cool the RNA to −20 °C by placing in a − 20 °C freezer for 1 h. Alternatively, the sample can be placed in dry ice for 3–5 min, but should not be allowed to freeze.
17. Centrifuge the RNA at 14,000 × *g*, at 4 °C for 30 min to precipitate RNA.
18. Remove the supernatant (leaving about 100 μL of liquid) and wash the pelleted RNA by adding 900 μL of cold 90% ethanol.
19. Centrifuge the RNA at 14,000 × *g* at 4 °C for 10 min.
20. Repeat the wash step described in **steps 18** and **19**.
21. Following the second wash with 90% ethanol, remove the supernatant, leaving ≤50 μL of liquid in the tube. Be careful not to disturb the pelleted RNA.
22. Dry the RNA by centrifuging under vacuum to remove any remaining solvent.
23. Reconstitute the purified RNA in 30 μL of ultrapure water. RNA concentration can be determined by OD_{260} using the molecule's molar extinction coefficient.
24. RNA should be stored at −20 °C. For long-term storage, store RNA at −70 °C.

3.4 5′-End Processing of RNA Transcripts Using RNase H

1. The reaction should be conducted at a RNA concentration of 10 μM. Based on the concentration of purified RNA obtained from in vitro transcription, adjust the final reaction volume to achieve a 10 μM solution (*see* **Note 11**).
2. Add the appropriate volume of 10× RNase H buffer to produce a final 1× solution.
3. Add 500 pmol of the DNA oligonucleotide that is complementary to the 5′ RNA leader sequence you wish to remove from the RNA transcript.
4. Bring to the final volume with water.
5. Heat the reaction to 94 °C for 2 min.
6. Remove from the heat block and allow 5 min for the tube to cool to room temperature.
7. Add 6 U of RNase H per 1 nmol RNA. Mix by pipetting.
8. Incubate the reaction at 37 °C for 1 h.
9. Follow Subheading 3.3 to purify the RNase H processed molecules by denaturing PAGE (*see* **Note 12** and Fig. 2).

3.5 Two-Stranded Switch Assembly from Individual Component Strands

1. For a 10 μL final assembly volume, combine stoichiometric quantities of RNA monomer stands in a 1.5-mL microcentrifuge tube to achieve the desired nanoconstruct concentration (*see* **Note 13**). Add water to bring the total volume to 8 μL (*see* **Note 14**).
2. Heat the assembly mixture to 94 °C for 2 min to denature the monomer strands.
3. Quickly transfer the assembly mixture from 94 °C to a 55 °C heat block. Allow the mixture to incubate at 55 °C for 10 min (*see* **Note 15**).
4. Add 5× assembly buffer to the mixture to achieve a final concentration of 1× assembly buffer with respect to the final volume. Incubate the assembly mixture for an additional 10–15 min.
5. Remove the assembly from the heat block and allow to cool to room temperature.
6. Spin the assembly mixture in a tabletop centrifuge for 30 s at room temperature to allow any condensation on the walls of the tube to collect in the bottom of the tube.
7. Assembled RNA nanoconstructs can be stored short term (hours) at 4 °C. For longer term storage, store assembled constructs at −20 °C (*see* **Note 16**).
8. *See* Subheading 3.7 for verification of switch assembly and function by nondenaturing PAGE.

3.6 Native Purification of RNA Switch by Nondenaturing PAGE

Before purifying a large quantity of assembled RNA switch for the first time, it is recommended to first verify that strand assembly is occurring. For a detailed description and protocol for analytical PAGE analysis, *see* Subheading 3.7.

1. Prepare a nondenaturing PAGE gel with a thickness of 0.8 mm. For a 16.5 cm × 28 cm gel prepare 50 mL of nondenaturing gel solution. Initiate polymerization by adding 300 μL of 10% APS solution and 40 μL TEMED (*see* **Note 9**). Insert a comb that produces wells with a volume of 80–100 μL. This should result in 13–15 wells across a 16.5 cm wide gel.
2. Once polymerized, set up the PAGE system in a refrigerated environment (4–6 °C) using 1× native gel running buffer.
3. Using a syringe or transfer pipet, rinse the wells of the gel with 1× native gel running buffer to remove any gel debris or air bubbles.
4. Prerun the gel at 4 W for 10 min while preparing RNA samples for electrophoresis.
5. Prepare approximately 0.5 nmol of assembled RNA switch in a 50 μL volume according to the protocol presented in Subheading 3.5 (*see* **Note 17**).
6. Add 0.2 volumes of native gel loading buffer to each assembled sample to be purified (for 50 μL of assembled construct, add 10 μL of loading buffer).
7. Load the nondenaturing PAGE gel with the constructs to be purified. If running a 16.5 cm × 28 cm gel, perform electrophoresis at a constant power of 6–8 W. This can be increased to 14–16 W if using a larger 33 cm × 39 cm gel. Run the electrophoresis until sufficient separation has been achieved, approximately 4 h (*see* **Notes 18** and **19**).
8. Following electrophoresis remove the gel from the glass plates and visualize the location of the assembled construct using a UV lamp. Refer to Subheading 3.3, **steps 8–11** for details.
9. Cut out the area of gel containing the assembled construct using a scalpel blade.
10. Carefully transfer the excised gel piece to a flat-bottomed 2-mL microcentrifuge tube. Place the piece of gel in tube so that the gel lies flat on the bottom of the tube (*see* **Note 20** and Fig. 3).
11. Add 50–60 μL of 1× assembly buffer directly on top of the gel piece(s) in the flat-bottom tube. This volume should be sufficient to completely cover the gel piece if lying flat on the bottom. Store overnight at 4 °C to allow the construct to elute from the gel.

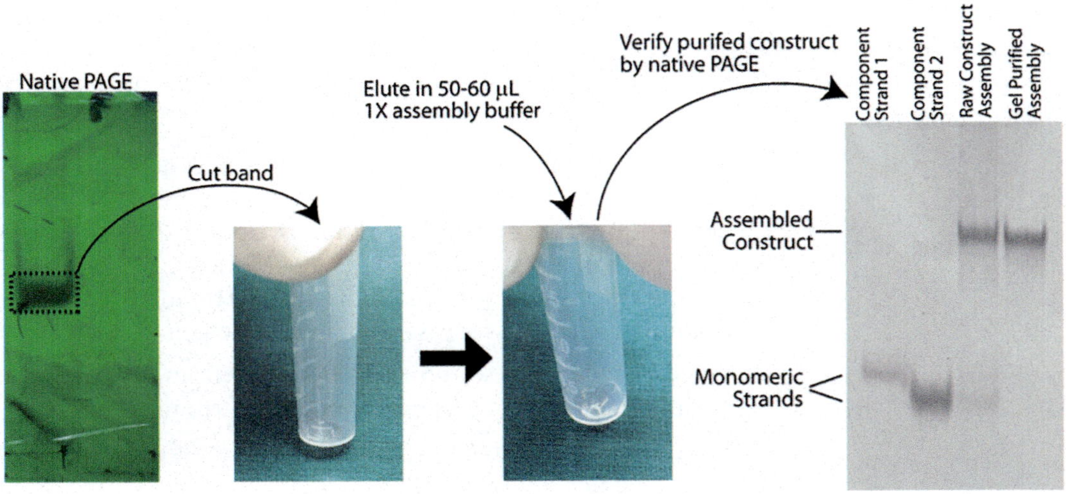

Fig. 3 RNA nanoconstructs purified by nondenaturing PAGE. After performing electrophoresis in nondenaturing conditions, the RNA can be located by visualizing the gel under UV light. The RNA is cut form the gel and eluted in 1× assembly buffer overnight. It is important the gel lies flat on the bottom of a centrifuge tube to ensure maximal interface between the gel and buffer. The purity of the eluted nanoconstruct can be qualitatively assessed by performing analytical nondenaturing PAGE with monomeric and unpurified assembly controls. Assembly of the switch from component anti-trigger and anti-target strands is not 100% efficient and some monomeric strands remain. Native PAGE purification removes any unassembled monomer strands, yielding only the assembled switch

12. The following day, remove the assembly buffer used to elute the gel-purified RNA construct and transfer it to a new tube. The concentration of the assembled nanoconstruct can be determined by OD_{260} (*see* **Note 21**). The assembled construct can be used immediately or stored at −20 °C (*see* **Note 16**).
13. Verify the purity of the gel-purified construct by running on a nondenaturing PAGE gel along with an unpurified assembly and monomeric strand controls (Fig. 3). Refer to Subheading 3.7 for details regarding analytical nondenaturing PAGE analysis.

3.7 Verification of Switch Assembly and/or Function by Nondenaturing PAGE Analysis

1. Prepare a nondenaturing PAGE gel as described in Subheading 3.6, **steps 1–4**.
2. Prepare RNA samples in 10 μL volumes containing a 1× concentration of assembly buffer. The concentration of each RNA should be approximately 1 μM. If assessing switch assembly, prepare one tube containing the assembled switch and control tubes containing individual component strands (anti-trigger and anti-target). If assessing switch function, prepare one tube containing the gel-purified switch incubated with the trigger RNA, as well as appropriate molecular controls. These controls may include the gel-purified switch, individual RNA molecules (anti-trigger, anti-target, trigger RNA), and

combinations of the RNA molecules (anti-trigger with trigger, anti-target with trigger) (*see* **Note 22** and Fig. 4).

3. Add 0.2 volumes of native gel loading buffer to each sample tube (for a 10 μL sample add 2 μL of loading buffer).
4. Rinse the wells after prerunning the nondenaturing PAGE gel and load the samples.
5. Perform electrophoresis by applying a constant power of 8–10 W for a 16.5 cm × 28 cm gel. Increase this to 16–20 W for 33 cm × 39 cm gel. Run the gel at constant power for approximately 3 h (*see* **Note 18**).
6. Following electrophoresis, remove the gel from the stand and carefully remove one of the glass plates. Place the plate that retained the gel in a secondary container, gel side up.
7. Pour a nucleic acid stain over the gel; enough to cover the entire gel with a thin layer of stain solution (*see* **Note 23**). Allow the stain to sit and permeate the gel for 10–15 min. If the gel appears dry, add additional stain solution.
8. Remove the stain solution by tilting the gel and allowing the stain to flow off into the secondary container (*see* **Note 24**).
9. Rinse the gel with deionized water to remove any excess stain. Use paper towels to blot excess water around the gel and remove any gel or salt residue from the bottom of the plate, as these can decrease the quality of the image taken.
10. Use a fluorescence gel scanner to image the gel. Use appropriate excitation and emission filters based on the nucleic acid stain used.

3.8 Verification of Switch Function in Cultured Human Cells

It is important to use a cell line that that expresses both the trigger and target gene. In the case of the CTGF/eGFP switch, we use a human breast cancer cell line, MDA-MB 231, that stably expresses eGFP. If other trigger–target pairs are to be used, consult the literature to find an appropriate cell line. All cell culture experiments should be performed in an appropriate biosafety cabinet. All cell culture experiments should be prepared using RNase-, DNase-, and endotoxin-free water and solutions. Cells are cultured at 37 °C with 5% CO_2 and 85–90% humidity. All cell culture media should be pre-equilibrated to 37 °C before use.

3.8.1 Preparing Cells in a 24-Well Plate for RNA Transfection

1. Culture cells in a flask with complete growth medium until 90% confluence is reached.
2. Add cell dissociation buffer to lift cells. Incubate the flask at 37 °C until cells have completely detached, but no longer than 20 min (*see* **Note 25**).
3. Add an equal volume of complete growth medium to the flask.

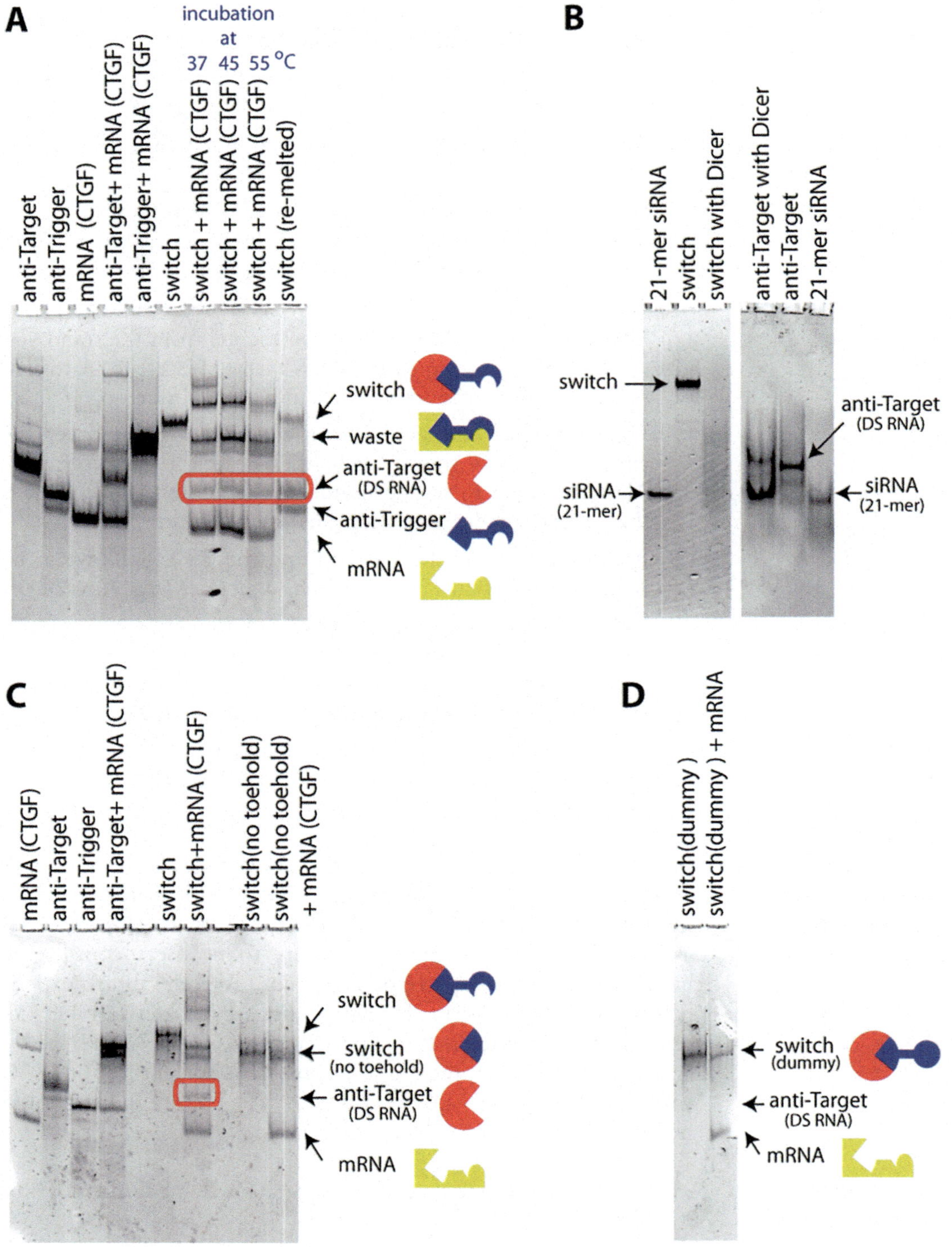

Fig. 4 Total staining RNA native PAGE results demonstrating the proper assembly and function of the RNA switch. (**a**) *Lanes* 1–5 contain molecular controls, including various combinations of anti-target strand, anti-trigger strand and mRNA fragment. The lane labeled "switch" corresponds to the gel-purified switch, assembled from anti-target and anti-trigger strands. *Lanes 7–9* demonstrate for several temperatures that the RNA switch releases the functional anti-target RNA in the presence of the trigger mRNA fragment. (**b**) Dicer processing of the RNA switch. *Lanes 1 & 6*: control of 21-mer siRNA. *Lanes 2–3* show that Dicer can degrade

4. Transfer dissociation buffer/medium/cell solution to a 15-mL tube.
5. Spin down the cells in a centrifuge at 500 × *g* for 5 min.
6. Aspirate medium from the 15-mL tube, leaving pelleted cells in the bottom.
7. Add 1 mL of complete growth medium to cells. Mix well by pipetting.
8. Count cells using a grid slide and an optical microscope.
9. Add between 5000 and 30,000 cells in complete growth medium to each well of a 24-well plate, depending on the length of the experiment. For longer transfection experiment, fewer cells should initially be plated, so they do not become overconfluent. The final well volume should be 500 μL.
10. Allow cells to grow in incubator for 24 h.

3.8.2 Verification of Gene Silencing by Two-Stranded RNA Switch

1. Follow Subheading 3.7, **step 1** to prepare a 24-well plate seeded with 30,000 cells per well.
2. In sterile microcentrifuge tubes, prepare RNA switch and individual strand controls at 50× concentration in a 10 μL volume. The final 1× RNA concentration should be between 1 and 50 nM. Each sample should be prepared in 1× assembly buffer. A 10 μL sample of only 1× assembly buffer should be prepared as a negative control.
3. Prepare a second set of sterile tubes containing 2 μL Lipofectamine 2000 (L2K) and 20 μL Opti-MEM medium.
4. Add 10 μL of the prepared 50X RNA solution to the tubes containing L2K–Opti-MEM. Mix by pipetting.
5. Incubate RNA–L2K–Opti-MEM mixture at room temperate for 20 min.
6. Add 470 μL Opti-MEM to the RNA–L2K–Opti-MEM mixture. Mix by pipetting.
7. Aspirate growth media from a well of the 24-well plate and quickly add 500 μL of the corresponding RNA–L2K–Opti-MEM mixture.
8. After addition of all samples to the plated cells, incubate the plate at 37 °C for 4 h.

Fig. 4 (continued) the assembled 2-stranded switch. The degradation products do not correspond, however, to distinct bands (*lane 3*). *Lanes 4–5* demonstrate that the Dicer enzyme processes the anti-target strand into a 21-mer siRNA strand. (**c**) A version of the RNA switch where the anti-trigger strand has no toehold does not lead to the release of the functional anti-target strand. (**d**) Similarly, a version of the RNA switch where the anti-trigger strand has a toehold sequence with scrambled (i.e., randomized) nucleotide content does not lead to the release of the functional anti-target strand. Adapted and reprinted with permission from [18], copyright 2016 American Chemical Society

9. After 4 h, aspirate the transfection mixture from each well of the 24-well plate and replace with 500 μL of complete growth medium.
10. Place the 24-well plate in the incubator for additional 68 h (3 days total).
11. After 3 days, aspirate the growth medium and wash wells with 1× PSB.
12. Add 325 μL of cell dissociation buffer to each well and place in incubator for 15–20 min.
13. Rinse the bottom of each well with the added dissociation buffer and transfer to a new vessel appropriate for use in flow cytometry analysis. Place the vessels in ice.
14. Perform flow cytometry to analyze the extent of switch function by measuring the level of eGFP expression in each sample (Fig. 5a).

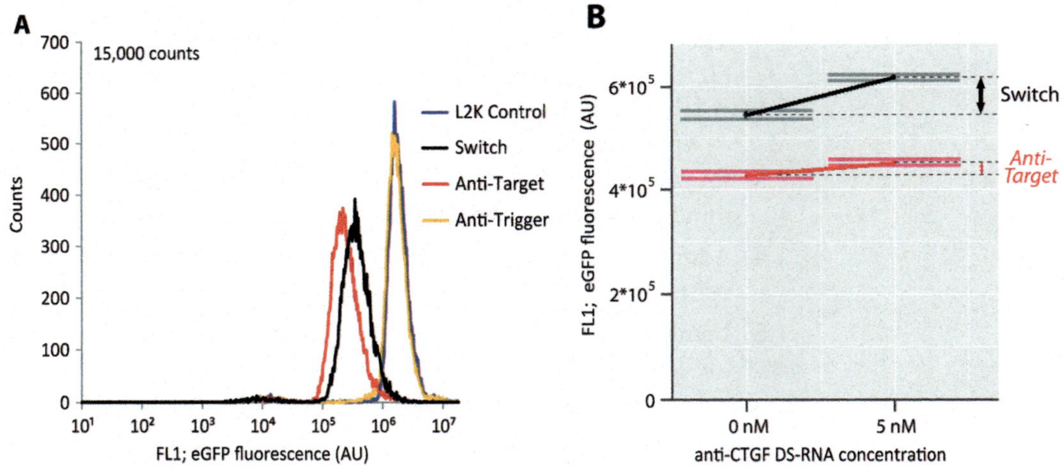

Fig. 5 Analysis of switch function using flow cytometry. As the RNA switch described throughout the protocol targets the knockdown of eGFP, flow cytometry can be used to measure switch activity. (**a**) Analysis of eGFP expression reveals that MDA-MB 231 cells transfected with the CTGF/eGFP RNA switch show a significant reduction in eGFP levels 3 days post transfection. (**b**) To assess how trigger RNA (CTGF mRNA) expression levels affect switch function, MDA-MB 231 cells were initially transfected with a Dicer substrate RNA (DS-RNA) against CTGF. After 2 days, cells were transfected with the RNA switch or anti-target control, targeting eGFP knockdown. Three days later eGFP expression level was analyzed. Cells that were initially transfected with the anti-CTGF DS-RNA saw a large increase in eGFP expression compared to cells were no anti-CTGF DS-RNA was transfected, indicating reduced switch activation. Cells transfected with the anti-target control displayed minimal changes in eGFP expression as a function of anti-CTGF DS-RNA. Adapted and reprinted with permission from [18], copyright 2016 American Chemical Society

3.8.3 Sequential Transfection to Knock Down RNA Trigger Expression

This protocol is use to knock down expression of the endogenous RNA trigger molecule prior to transfection with the RNA switch. It allows for examination of switch activity as a function of trigger expression level (Fig. 5b).

1. Follow Subheading 3.8.1 to prepare a 24-well plate seeded with 5000 cells per well.
2. In sterile microcentrifuge tubes prepare a Dicer-substrate RNA, targeting the knockdown of the RNA trigger, at 50× concentration in a 10 μL volume. The final 1× RNA concentration should be between 1 and 50 nM. Each sample should be prepared in 1× assembly buffer. A 10 μL sample of only 1× assembly buffer should be prepared as a negative control.
3. Follow Subheading 3.8.2, **steps 3** through **9**.
4. Place the 24-well plate in the incubator for additional 44 h (2 days total).
5. After 2 days, prepare individual samples of the RNA switch and the anti-target strand at 50× concentration in a 10 μL volume in sterile microcentrifuge tubes (*see* **Note 26**). The final 1× RNA concentration should be between 1 and 50 nM. Each sample should be prepared in 1× assembly buffer. A 10 μL sample of only 1× assembly buffer should be prepared as a negative control.
6. Follow Subheading 3.8.2, **steps 3** through **14**.

4 Notes

1. DNA templates for transcription can be purchased from a commercial source or generated by PCR.
2. The conditions of this buffer have been optimized for use with an in-house produced T7 RNA polymerase enzyme, but also work well with many commercially available T7 polymerases. However, buffer conditions may need to be optimized for different T7pol enzymes. If commercial T7 polymerase is used for transcription, we recommend initially using the buffer provided or suggested by the commercial source.
3. This buffer may need to be optimized based on the source of the enzyme. Generally, commercial vendors of RNase H will provide or specify a particular reaction buffer for the enzyme. If obtaining RNase H from a commercial source, we recommend initially using the buffer conditions given by the supplier.
4. Depending on the excitation and emission filters used to image the gel following electrophoresis, xylene cyanol and bromophenol blue may produce a fluorescence signal when imaging.

If this is problematic, either or both dyes can be omitted from the gel loading buffer.

5. If a fluorescence gel scanner is not available, imaging can be performed by staining the gel with ethidium bromide, visualizing the gel on a UV-transilluminator, and capturing the image with a digital camera. Due to the reduced sensitivity of this technique, the concentration of ethidium bromide solution used for staining may need to be increased above the 0.5 μg/mL suggested in the main text.
6. Choose an appropriate cell line based on the trigger and target of the RNA switch. Based on the choice of cell line used, the complete growth medium may need to be adjusted.
7. We find that Lipofectamine 2000 works very well for transfection of cultured cells and the transfection protocol outlined in Subheading 3.8 is based on the use of Lipofectamine 2000. Other transfection agents could be used in conjunction with the RNA switch; however, transfection protocols would likely need to be tailored and optimized for different transfection agents.
8. The transcription buffer contains a relatively high concentration of Mg^{2+} ions, which may promote backbone hydrolysis of the RNA and reduce full-length transcript yield if left for an extended duration. It is advised to purify RNA transcripts immediately following transcription.
9. TEMED should always be added immediately prior to pouring the gel, as polymerization will begin as soon as TEMED is added.
10. For denaturing PAGE gels, the power supplied should be sufficient to heat the gel approximately 45–50 °C. If the gel feels extremely hot to the touch while running, reduce the applied power. Many power supplies support the use of an electronic temperature probe that can be attached to the outside of the glass plate during electrophoresis, if desired. Adhesive temperature indicator strips can also be purchased and placed on the glass plate. The duration of electrophoresis should be optimized based on the length of molecules and the conditions of the gel to ensure adequate separation of the full-length and incomplete transcripts. Marker dyes within the urea blue loading buffer (xylene cyanol and bromophenol blue) can be used to estimate the migration distance of transcripts.
11. We suggest a final reaction volume in the range of 50–200 μL. A 50 μL reaction should contain enough RNA to visualize within a PAGE gel by shadowing with a UV lamp. However, we do not suggest larger reaction volumes than 200 μL to

ensure the entire reaction contents can be loaded into a single well of a polyacrylamide gel for the purpose of purification.

12. To ensure that cleavage of the 5′ RNA leader sequence by RNase H has occurred, prepare a negative control sample of the transcript that has not be subjected to the RNase H protocol. Run this control side by side with the RNase H processed transcript on the denaturing PAGE gel used for purification as a means to visually confirm the reduction in length of the RNase H processed molecules.
13. For the purpose of verifying switch assembly by nondenaturing PAGE, preparing a 10 μL sample at a final construct concentration of 1 μM is generally satisfactory. If the assembly is to be gel-purified, we suggest preparing a 50 μL sample at a final assembled concentration of approximately 10–12 μM.
14. Because a concentrated 5× assembly buffer will be added at a later step, the initial preparation of monomer strands in water should be prepared to 4/5 of the final desired volume.
15. The incubation step at 55 °C is designed to allow the thermodynamically most-favorable intermolecular interactions to form, while preventing formation of less stable alternative folds. We find that incubation at 55 °C works best for the CTGF/eGFP two-stranded switch. However, if variant switches are being assembled, we suggest trying assembly temperatures ranging between 45 and 65 °C to find the optimal condition to promote proper nanoconstruct assembly.
16. If assembled RNA constructs have been stored for an extended duration of time, or repeatedly thawed and frozen, we recommend checking the composition of the construct by both native and denaturing PAGE to verify that the assembly is intact and no strand degradation has occurred.
17. Due to several factors including the efficiency of nanoconstruct assembly, the need to identify the nanoconstruct's location in the gel by UV shadowing, and the efficiency of RNA elution from polyacrylamide gels, the starting quantity of the RNA nanoconstruct must be relatively large to purify by nondenaturing PAGE. Generally, a 50 μL volume of assembled construct at a concentration of 10 μM is sufficient to perform purification.
18. The power supplied to the gel should be low enough that the gel does not warm above ~10 °C. Keeping the gel cool will ensure that the assembled nanoconstructs do not denature during electrophoresis.
19. Due to the large quantity of RNA loaded in an individual well during nondenaturing PAGE purification, running the gel slowly at fairly low power helps to reduce the extent of

streaking tails as the band of RNA migrates through the gel. The duration of electrophoresis should be sufficient to clearly separate the assembled nanoconstruct from any partial assemblies or monomeric strands. This can be optimized by performing nondenaturing PAGE electrophoresis using relatively small amounts of assembled RNA and visualizing the migration using fluorescent nucleic acid stains (described in Subheading 3.7)

20. It may be necessary to cut the gel piece into multiple smaller sections so that the pieces fit flat on the bottom of the tube. If this is the case, place each gel section from a single band into a single tube to maximize the final concentration of RNA construct eluted. Try to position the gel pieces to minimize gel overlap and maximize exposure to the elution buffer. Use of a pipette tip may be helpful to position the sections of gel flat on the bottom of the tube.
21. Due to the extensive base pairing formed within an assembled RNA nanoconstruct, the RNA should be denatured to obtain an accurate measure of concentration by OD_{260}. This can be accomplished by measuring absorbance at a temperature above the melting temperature of the RNA, or by adding a denaturant and briefly heating the RNA to 94 °C. We find that adding a sample of the RNA assembly to a 6 M urea, 20 mM EDTA solution (final concentrations), heating for 1 min to 94 °C, and allow the solution to cool to room temperature is sufficient to denature most RNA constructs.
22. The trigger RNA can be synthesized using in vitro transcription and PAGE purification as described in Subheadings 3.2 and 3.3. If the switch is designed to trigger upon interaction with an endogenous mRNA, the entire mRNA does not need to be synthesized for use in nondenaturing PAGE analysis, but only a portion of the mRNA sequence that contains the region complementary to the anti-trigger molecule.
23. Nucleic acid stains pose potential health hazards. These stains should be handled and disposed of in accordance with local environmental health and safety practices.
24. Nucleic acid stains such as ethidium bromide, SYBR Gold, and SYBR Green II can be collected post-staining and reused multiple times until the extent of staining is no longer satisfactory. Dispose of the stain in accordance with local environmental health and safety practices.
25. If cells have not completely detached after 20 min, try washing the bottom of the flask with the cell dissociation buffer that has already been added to the flask.

26. The anti-target molecule should be used as an internal control and be subjected to identical conditions as the switch samples. The ability of the anti-target molecule to silence eGFP should not be affected by the knockdown of RNA trigger levels. As such, transfection of the anti-target molecule provides a control measurement indicating the maximal extent of possible target knockdown, regardless of endogenous RNA trigger expression levels.

Acknowledgments

This work has been funded in whole or in part with Federal funds from the Frederick National Laboratory for Cancer Research, National Institutes of Health, under Contract No. HHSN261200800001E. This research was supported [in part] by the Intramural Research Program of the NIH, National Cancer Institute, Center for Cancer Research. The content of this publication does not necessarily reflect the views or policies of the Department of Health and Human Services, nor does the mention of trade names, commercial products, or organizations imply endorsement by the US Government.

References

1. Grabow WW, Jaeger L (2014) RNA self-assembly and RNA nanotechnology. Acc Chem Res 47(6):1871–1880
2. Guo P (2010) The emerging field of RNA nanotechnology. Nat Nanotechnol 5 (12):833–842
3. Batey RT, Rambo RP, Doudna JA (1999) Tertiary motifs in RNA structure and folding. Angew Chem Int Ed Engl 38(16):2326–2343
4. Butcher SE, Pyle AM (2011) The molecular interactions that stabilize RNA tertiary structure: RNA motifs, patterns, and networks. Acc Chem Res 44(12):1302–1311
5. Parlea L et al (2016) Ring Catalog: a resource for designing self-assembling RNA nanostructures. Methods 103:128–137
6. Bindewald E, Hayes R, Yingling YG, Kasprzak W, Shapiro BA (2008) RNAJunction: a database of RNA junctions and kissing loops for three-dimensional structural analysis and nanodesign. Nucleic Acids Res 36(Database issue): D392–D397
7. Parlea LG, Sweeney BA, Hosseini-Asanjan M, Zirbel CL, Leontis NB (2016) The RNA 3D motif atlas: computational methods for extraction, organization and evaluation of RNA motifs. Methods 103:99–119
8. Grabow WW et al (2011) Self-assembling RNA nanorings based on RNAI/II inverse kissing complexes. Nano Lett 11(2):878–887
9. Severcan I, Geary C, Verzemnieks E, Chworos A, Jaeger L (2009) Square-shaped RNA particles from different RNA folds. Nano Lett 9 (3):1270–1277
10. Zhou J et al (2013) Functional in vivo delivery of multiplexed anti-HIV-1 siRNAs via a chemically synthesized aptamer with a sticky bridge. Mol Ther 21(1):192–200
11. Afonin KA et al (2015) Triggering of RNA interference with RNA-RNA, RNA-DNA, and DNA-RNA nanoparticles. ACS Nano 9 (1):251–259
12. Delebecque CJ, Lindner AB, Silver PA, Aldaye FA (2011) Organization of intracellular reactions with rationally designed RNA assemblies. Science 333(6041):470–474
13. Afonin KA et al (2010) In vitro assembly of cubic RNA-based scaffolds designed in silico. Nat Nanotechnol 5(9):676–682

14. Yu J, Liu Z, Jiang W, Wang G, Mao C (2015) De novo design of an RNA tile that self-assembles into a homo-octameric nanoprism. Nat Commun 6:5724
15. Zalatan JG et al (2015) Engineering complex synthetic transcriptional programs with CRISPR RNA scaffolds. Cell 160 (1–2):339–350
16. Afonin KA et al (2016) The use of minimal RNA toeholds to trigger the activation of multiple functionalities. Nano Lett 16 (3):1746–1753
17. Hochrein LM, Schwarzkopf M, Shahgholi M, Yin P, Pierce NA (2013) Conditional Dicer substrate formation via shape and sequence transduction with small conditional RNAs. J Am Chem Soc 135(46):17322–17330
18. Bindewald E et al (2016) Multistrand structure prediction of nucleic acid assemblies and design of RNA switches. Nano Lett 16(3):1726–1735
19. Zakrevsky P, Bindewald E, Shapiro BA (2016) RNA toehold interactions initiate conditional gene silencing. DNA RNA Nanotechnol 3 (1):11–13
20. Milligan JF, Groebe DR, Witherell GW, Uhlenbeck OC (1987) Oligoribonucleotide synthesis using T7 RNA polymerase and synthetic DNA templates. Nucleic Acids Res 15 (21):8783–8798
21. Zhu X, Li T, Dang Y, Feng Y, Huang P (2005) A novel in vitro transcription method for producing siRNAs without specific sequence requirements. Mol Biotechnol 31(3):187–192

Chapter 21

Rational Engineering of a Modular Group I Ribozyme to Control Its Activity by Self-Dimerization

Takahiro Tanaka, Yoshiya Ikawa, and Shigeyoshi Matsumura

Abstract

We designed and constructed a dimer of the *Tetrahymena* group I ribozyme the activity of which is regulated by self-dimerization. This dimer was rationally designed by utilizing the P5abc and ΔP5abc domains as large RNA motifs. This strategy enabled us to install large ribozyme functions into an RNA structure. This is a step toward expanding the field of RNA nanotechnology beyond the limitation of using only relatively small functional motifs. Self-dimerization can also be rationally programmed by modular engineering of RNA interaction motifs. In this chapter, we present the procedure for the rational design and construction of large ribozyme domains based on RNA tertiary structures. We also describe the electrophoresis mobility shift assay (EMS) and several ribozyme activity assays to confirm the ribozyme function and its regulation. We have succeeded in construction of tecto-GIRz.

Key words Large ribozyme, *Tetrahymena* group I ribozyme, Self-splicing, Rational design, Modular structure, RNA motif

1 Introduction

RNA is an attractive biomolecule for construction of self-assembled nanostructures in a programmable manner. Several two- and three-dimensional (2D and 3D, respectively) RNA assemblies have been reported, including filaments, polygons, tiles, and polyhedrons [1]. Naturally occurring and artificial RNA functions, e.g., binding and catalysis, have been integrated into the nanostructures using relatively small and simple RNA units, such as aptamers and small ribozymes [2, 3]. In contrast, large ribozymes, e.g., group I and II self-splicing introns, and ribonuclease P (RNase P), have been rarely utilized for functionalization. As these large ribozymes have more sophisticated functions and play essential roles in cellular events, it is expected that their introduction into the RNA supramolecular structures would expand the potential and applications in the field of RNA nanotechnology. For example, the stabilization and modification of RNA nanostructure for more complex 3D structure or

Eckart Bindewald and Bruce A. Shapiro (eds.), *RNA Nanostructures: Methods and Protocols*, Methods in Molecular Biology, vol. 1632, DOI 10.1007/978-1-4939-7138-1_21,

molecular system might be accomplished by the introduction of a large ribozyme. Besides, RNA processing such as splicing reactions might be able to be regulated directly by group I and II ribozymes based RNA nanostructure.

Naturally occurring RNAs are composed of several modular units, as proposed by Westhof et al. [4] as "RNA tectonics." Artificial RNA nanostructures can be rationally designed by assembling these units. Recent development and knowledge accumulation of structural biology and biochemistry revealed modular structures and folding properties of functional RNAs, allowing us to design desired RNA molecules. For example, Jaeger et al. constructed a self-assembling RNA, named tecto-RNA, using tetraloop-receptor interaction motifs [5].

Natural large ribozymes also have modular structures. For example, the *Tetrahymena* group I ribozyme can be divided into two domains: P5abc (activator domain) and ΔP5abc (catalytic domain). The RNase P RNA also can be split into two domains: specificity domain and catalytic domain. These domains are catalytically inactive alone, but they assemble together in solution and recover the activity [6, 7]. They are also composed of small RNA motifs, e.g., loop-receptor and kissing loop. Hence, the modular engineering of these motif interactions can modify the structural formation and the ribozyme activity.

We have rationally designed and constructed "tecto-GIRz," which is a *Tetrahymena* group I dimer based on the modularity of P5abc and ΔP5abc domains [8]. In this design, we regarded the domain structure as a "large RNA motif." Initially, a homodimer structure is designed by connecting the wild-type P5abc and ΔP5abc domains using a linker dsRNA. This homodimer could self-assemble and activate itself. To regulate dimer formation, the interaction motifs (tetraloop-receptor and kissing loop) are replaced with other orthogonal motifs to heterodimerize tecto-GIRz (Fig. 1).

In this chapter, we present the procedure for rational design and construction of the large ribozyme domains based on the RNA tertiary structures. We also describe the electrophoresis mobility shift assay (EMS) and several ribozyme activity assays to confirm the ribozyme function and its regulation. We have succeeded in construction of the group I ribozyme dimer the self-splicing activity of which is regulated by its dimerization.

2 Materials

2.1 Molecular Design of tecto-GIRz

1. Molecular modeling software, Discovery Studio (Accelrys Software/BIOVIA, San Diego, CA).

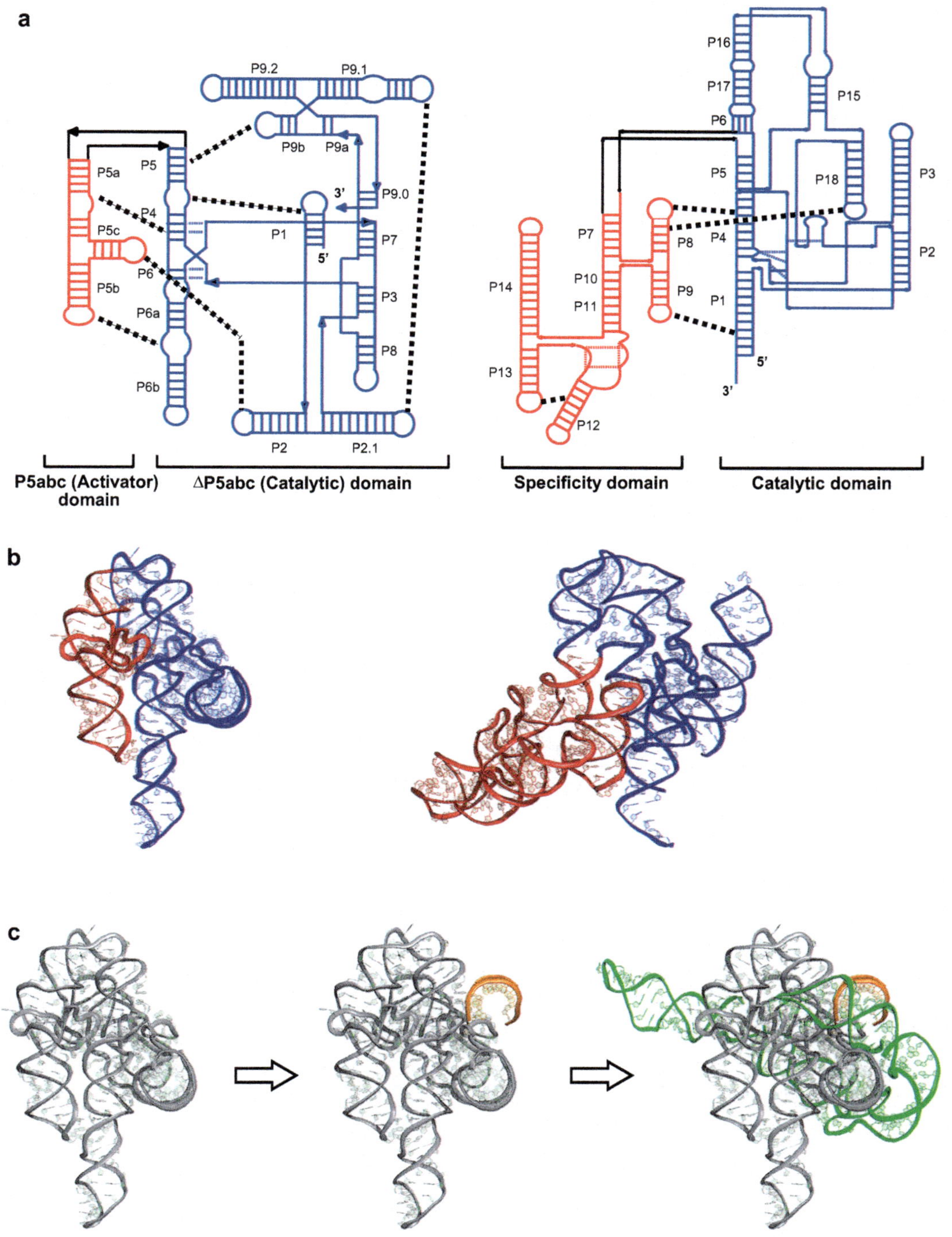

Fig. 1 Modularity of the *Tetrahymena* group I ribozyme and the *T. maritima* ribonuclease P RNA. (**a**) Simplified secondary structures of the *Tetrahymena* group I ribozyme and the *T. maritima* ribonuclease P RNA. The activator domain and specificity domain are shown in *red*. The catalytic domains are shown in *blue*. (**b**) Crystal structures of the *Tetrahymena* group I ribozyme (PDB ID: 1X8W) and the *T. maritima* ribonuclease P RNA (PDB ID: 3Q1Q). *Colored regions* correspond to those of the secondary structures. The P1–P2–P2.1 and P9.1–P9.2 regions are deleted in the group I ribozyme structure. (**c**) Schematic representation of the full-length 3D molecular design. *Left*: The crystal structure of the *Tetrahymena* group I ribozyme (PDB ID: 1X8W). *Center*: Addition of the P1 stem from the *Azoarcus* group I ribozyme (PDB ID: 1ZZN). *Right*: Addition of the P2–P2.1 and P9.1–P9.2 stem from the predicted 3D model. A 3D structure of the full-length *Tetrahymena* group I ribozyme was obtained

2.2 Plasmid Construction

1. Synthetic DNA oligonucleotides.

   kyi-767: 5′-ACAGCCC*GGTCTC*CTACGTCCGTCAGCTT-ATTACCATACCCT-3′, where the *Bsa*I site is shown in italics.
   kyi-768: 5′-ACAGCCC*GGTCTC*GTCCGTCAGTACCAAGT CTCAGGG-3′,
   where the *Bsa*I site is shown in italics.
   kyi-772: 5′-ACAGCCC*GGTCTC*ACGGACGTAGCATGG CCCCTTTCCCGCAA-3′, where the *Bsa*I site is shown in italics.
   kyi-773: 5′-ACAGCCC*GGTCTC*ACGTAGCATGGTCCTA-ACCACGCAGCCAAGT-3′,
   where *Bsa*I site is shown in italics.
   M13 Fw: 5′-CGCCAGGGTTTTCCCAGTCACGAC-3′.
   M13 Rev.: 5′-CGGATAACAATTTCACACAGGAAAC-3′.
2. pTZ-IVSU and pTZ18U plasmid.
3. dNTPs solution (2.5 mM each). Store at −20 °C.
4. Recombinant *Taq* (r*Taq*) DNA polymerase (Takara Bio, Otsu, Japan).
5. PCR thermal cycler Dice (Takara Bio).
6. Phenol saturated with TE (10 mM Tris, pH 8.0, 1 mM EDTA). Store at 4 °C protected from light.
7. Diethylether.
8. 3 M sodium acetate pH 7.0.
9. 100% (v/v) ethanol.
10. 70% (v/v) ethanol.
11. Restriction enzymes *Bsa*I-HF (New England Biolabs, Ipswich, MA), *Eco*RI and *Hin*dIII (Takara Bio).
12. Low melting point agarose gel.
13. Rapid DNA Dephos & Ligation Kit (Roche, Penzberg, Germany).
14. Competent *E. coli* JM109 cells.
15. LB medium.
16. LB agar plate.
17. TB medium.
18. FastGene Plasmid Mini Kit (Nippon Genetics Co., Ltd., Tokyo, Japan).
19. 4300 DNA analyzer (Li-COR, Lincoln, NE).

2.3 In Vitro Transcription (IVT)

1. Synthetic DNA oligonucleotides.

   L-21 sense-b: 5′-C*TAATACGACTCACTATA*GGAGGGAA-AAGTTATCAGGCA-3′, where the T7 promoter is shown in italics.
   Anti-ScaI: 5′-ACTCCAAAACTAATCAATAT-3′

TIL-29: 5′-GTCCCACTCTTTCGAATGCGCCCAAGCTTG TGACGAGGCA-3′.
TIL-30: 5′-*CAATTAACCCTCACTAAA*GGCGTACTCAC GCGTATACAGTCTCCGGCTCGTATGTTGTGTGG-3′, where the T3 promoter is shown in italics.
TIL-31: 5′-GTCCCACTCTTTCGAATGCGCGGTAACGC-CAGGGTTTTCCC-3′.

2. Template plasmid (constructed in this study).
3. 10× T7 transcription buffer: 400 mM Tris–HCl (pH 7.5), 150 mM $MgCl_2$, 50 mM DTT, 10 mM spermidine.

   10× T3 transcription buffer: 400 mM Tris–HCl (pH 7.5), 30 mM $MgCl_2$, 50 mM DTT, 10 mM spermidine.
4. 10 mM each NTP.
5. T7 RNA polymerase.
6. T3 RNA polymerase.
7. Recombinant RNase inhibitor (Takara Bio).
8. RQ I RNase-free DNase (Promega, Madison, WI).
9. 10 M ammonium acetate.
10. Denaturing dye: 70% formamide, 0.15% XC, 0.2% BPB.

2.4 Denaturing Polyacrylamide Gel Electrophoresis

1. Glass plate.
2. Spacer (1 mm thickness).
3. Comb (8 wells, 1.5 cm each).
4. 5× TBE buffer.
5. 30% acrylamide–bis solution (29:1).
6. Urea (Wako, Osaka, Japan).
7. 20% (w/v) ammonium persulfate (APS).
8. *N*,*N*,*N*′,*N*′-tetramethylethylenediamine (TEMED).
9. Gel electrophoresis apparatus.
10. Electrophoresis power supply.
11. Thin layer chromatography plates F254 (20 × 20 cm; Merck, Billerica, MA).
12. UV lamp (UVP, Upland, CA).
13. RNA elution solution: 0.3 M NaOAc, 0.1% SDS.
14. FluoroImager Pharos FX (Bio-Rad, Hercules, CA).

2.5 EMS (Electrophoretic Mobility Shift) Assay

1. Glass plate.
2. Comb (12 wells, 0.5 cm each).
3. 10× Tris–borate (pH 8.3) buffer.
4. 1 M $Mg(OAc)_2$.

5. 10× folding buffer: 712 mM Tris–Borate (pH 8.3); 0, 30, 50 mM $Mg(OAc)_2$.
6. 6× loading buffer: 71.2 mM Tris–Borate (pH 8.3); 0, 3, 5 mM $Mg(OAc)_2$; glycerol; xylene cyanol (XC; 0.1%).
7. 30% acrylamide–bis solution (29:1).
8. 20% (w/v) ammonium persulfate (APS).
9. *N*,*N*,*N*′,*N*′-tetramethylethylenediamine (TEMED).
10. Gel electrophoresis apparatus.
11. Power supply.
12. FluoroImager Pharos FX (Bio-Rad).

2.6 Ribozyme Activity Assay

1. Precise heat block or PCR thermal cycler.
2. 10× reaction buffer: 300 mM Tris–HCl (pH 7.5); 10, 30, or 50 mM $MgCl_2$.
3. 2 mM GTP.
4. Stop solution: 75% formamide, 100 mM EDTA, 0.1% XC.5. Fluorophore-labeled synthetic RNA oligonucleotides.
5. Fluorophore-labeled synthetic RNA olig onucleot ides.

   5'-FAM-GGCCCUCUAAAAA-3' (FAM: carboxyfluorescein)
   5'-AAAAAGCCAUAAAAA-TAMRA-3' (TAMRA: carboxytetramet hylrhodamine)

2.7 Splicing Assay

1. Precise heat block or PCR thermal cycler.
2. 5× splicing buffer: 150 mM Tris–HCl (pH 7.5), 25 mM $MgCl_2$, 1 mM GTP, 10 mM spermidine.
3. Stop solution: 75% formamide, 100 mM EDTA, 0.1% XC.

2.8 Cotranscriptional Splicing Assay

1. Plate reader; Infinite F200 Pro (Tecan, Männedorf, Switzerland).
2. Microtiter plate.
3. 10× T7 cotranscription buffer: 400 mM Tris–HCl (pH 7.5), 90 mM $MgCl_2$, 50 mM DTT, 10 mM spermidine; 100 μM DFHBI (3,5-difluoro-4-hydroxybenzylidene imidazolinone).
4. 10 mM each NTP.
5. T7 RNA polymerase.
6. Recombinant RNase inhibitor (Takara Bio).
7. Mineral oil (Sigma, St. Louis, MO).

3 Methods

3.1 Molecular Design of tecto-GIRz

The crystal structure of the full-length *Tetrahymena* group I ribozyme has not been solved. A full-length 3D model structure is necessary to grasp the configuration of unsolved regions and their interactions (P1–P2–P2.1, P9.1–P9.2) for this ribozyme-based molecular design (*see* **Note 1**). Therefore, it was first necessary to design the full-length 3D model of this ribozyme.

1. Download atom coordinate files in PDB format from the RCSB Protein Data Bank database (http://www.rcsb.org). For this work, we obtained PDB files corresponding to the *Tetrahymena* group I ribozyme (PDB ID: 1X8W) and the *Azoarcus* group I ribozyme (PDB ID: 1ZZN). We also used the predicted 3D model of the full-length *Tetrahymena* group I ribozyme [9]. Import these files individually into the Discovery Studio software.
2. Using the molecular modeling software, extract the L4, P1, and P7 regions of 1ZZN as well as the P2–P2.1, and P9.1–P9.2 regions of the predicted model and attach them to the 1X8W structure based on the previous report [9]. Divide the molecule into the P5abc region and the ΔP5abc region. Remove the J5/5a region and part of the P5–P5a region based on a previous biochemical report [10].
3. Generate a 15- to 20-bp A-form dsRNA using the "Build Action: Create/grow 3′" function of the "Build and Edit Nucleic Acid" tool. For this work, a 17 bp dsRNA is chosen as a linker sequence. Change the display style of the 2–4 nucleotides at the end of RNA strands to ball-and-stick style. Superimpose these nucleotides for aligning the dsRNA, P5abc, and ΔP5abc regions in the software (*see* **Note 2**).
4. Remove one of the overlapping nucleotides. Select the phosphorus and oxygen atoms, and create a covalent bond from "bond" and "single" functions of the "Chemistry" tool (Fig. 2).
5. Based on the 3D modeling results, determine the sequence of the linker region. The linker sequence should not form undesired secondary structures.

3.2 Plasmid Construction

It is difficult to mutate the P5abc or receptor domain (P6-P6a) of the *Tetrahymena* group I ribozyme by PCR-based mutagenesis because it has 414 nucleotides. We divided the DNA sequence of the ribozyme into three fragments (5′ region, P5abc region, and 3′ region), and then introduced mutations separately by PCR (*see* **Note 3**). The three mutated fragments are ligated, followed by insertion into the plasmid (Fig. 3). In this procedure, the wild-type *Tetrahymena* group I ribozyme is converted to homodimeric tecto-GIRz by replacing the J5/5a with a double strand. Orthogonalization of the interaction motif for a heterodimer can be also performed by the same procedure.

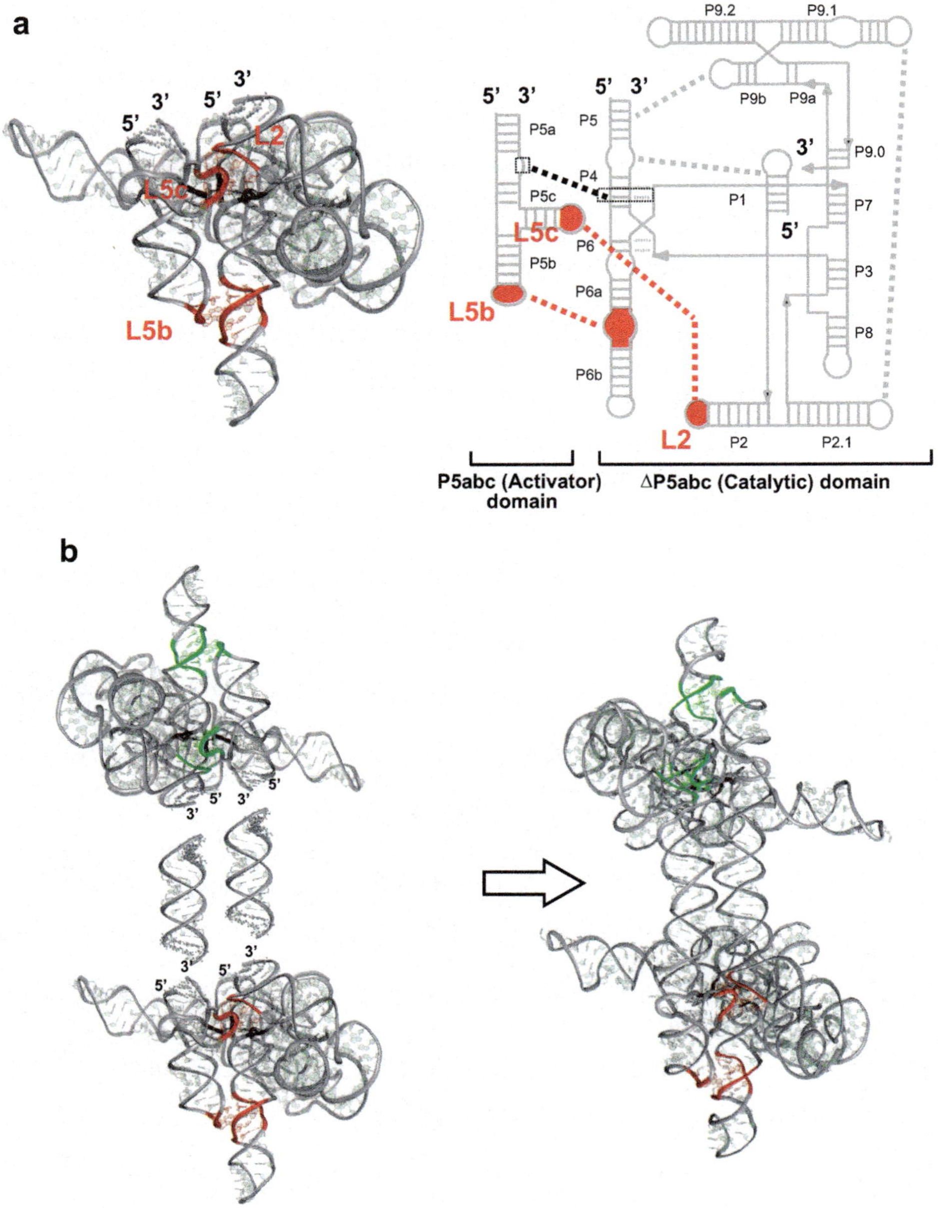

Fig. 2 (**a**) Interactions between the P5abc and ΔP5abc domains in the secondary and tertiary structures. The RNA motifs that participate in the interactions are shown in *red*. The L2/L5c and L5b/P6a-P6b represent kissing loop and loop-receptor interactions, respectively. (**b**) Schematic representation of the 3D molecular design. P5abc and ΔP5abc should be moved together. Terminus two nucleotides are shown in the "ball-and-stick" display style. Each nucleotide is superimposed, and then the dimer structure is obtained

1. Prepare PCR mixes (total volume 100 μL) containing the following components: 10 ng PCR template (plasmid: pTZ-IVSU), 0.5 μM primers (5′ region: M13 Rev. and kyi-772, P5abc region: kyi-767 and kyi-768, 3′ region kyi-773 and

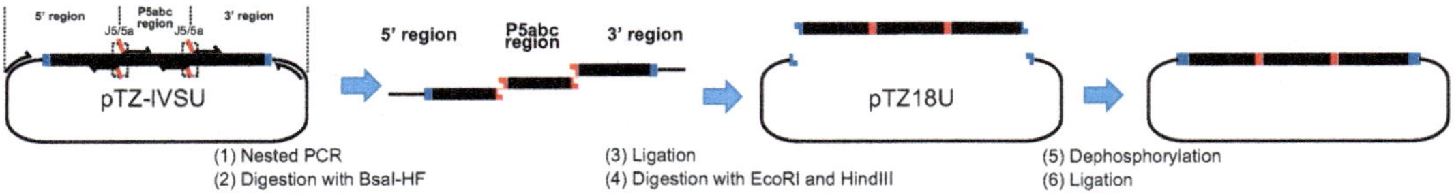

Fig. 3 Schematic representation of the plasmid construction procedure. (*1*) Mutations were introduced by nested PCR. (*2*) The PCR products were digested with *Bsa*I-HF at the sites indicated in *red*. (*3*) These three DNA fragments were ligated with T4 DNA ligase. (*4*) The ligated product and pTZ18U plasmid were digested with *Eco*RI and *Hin*dIII. (*5*) Dephosphorylation was performed for pTZ18U. (*6*) The digested DNA and pTZ18U were ligated to form a circular plasmid

M13 Fw), 0.2 mM dNTP, 1× r*Taq* reaction buffer, r*Taq* DNA polymerase 10 U.

2. Perform three PCR cycles according to the following profile: initial denaturation at 94 °C for 3 min; 25 cycles of 94 °C for 1 min, 50 °C for 1 min, 72 °C for 1 min; final extension at 72 °C for 5 min.
3. Collect the PCR products into a 1.5-mL tube, purified by adding the same volume TE-saturated phenol, vortexing, and centrifuging for 5 min at 15,000 × *g*. Transfer the aqueous phase into a new tube, followed by addition of 1.5× volume of diethylether, vortexing, and centrifuging as above. Discard the diethylether phase and add 1/9 volume 3 M sodium acetate and 1 mL of 100% ethanol. Incubate the mixture for 30 min at −20 °C and centrifuge for 5 min at 15,000 × *g* at 4 °C. Discard the supernatant, add 70% ethanol, and centrifuge for 1 min at 15,000 × *g* at 4 °C. After discarding the supernatant, the dry the pellets, and dissolve them in water.
4. Digest the PCR products with 5 U of *Bsa*I-HF restriction enzyme in 1× CutSmart Buffer by incubation at 37 °C for 3 h.
5. Separate the reaction mixtures in 1% low melting point agarose gels. If the enzyme digestion proceeds completely, the desired band appears as a largest band. Collect the desired bands in a tube, and add 300 μL of TE buffer followed by incubation at 65 °C for 30 min to dissolve the gel completely.
6. Extract the reaction mixtures three times with 300 μL of TE-saturated phenol and 450 μL diethyl ether. Then add 33 μL of 3 M sodium acetate (pH 7.5) and 1 mL of 100% ethanol, incubate for 30 min at −20 °C and centrifuge for 5 min at 15,000 × *g* at 4 °C. Discard the supernatant, add 70% ethanol, and centrifuge for 1 min at 15,000 × *g* at 4 °C. After discarding the supernatant, allow pellets to dry, and dissolve them in water.
7. Ligate three DNA fragments using the Rapid DNA Dephos and Ligation Kit. Mix the three fragments with 1× dilution

buffer and T4 DNA ligation buffer with addition of 5 U of T4 DNA ligase and incubate at room temperature for 30 min.

8. Digest the ligation product (insert DNA) and pTZ18U (vector) with *Eco*RI and *Hin*dIII restriction enzyme in 1× M buffer (Takara Bio). Incubate at 37 °C for 1 h.
9. Separate the reaction mixtures by 1% (ligation product) or 0.5% (pTZ18U) low melting point agarose gel electrophoresis. Collect the desired band and extract the DNA as described above.
10. Dephosphorylate the digested pTZ18U with the Rapid DNA Dephos and Ligation Kit. Mix the plasmid solution with the 1× rAPid alkaline phosphatase buffer and 1 U rAPid alkaline phosphatase, followed by incubation at 37 °C for 10 min, and inactivation of the enzyme at 75 °C for 2 min.
11. Ligate the inserted DNA and vector with the Rapid DNA Dephos and Ligation Kit, and incubate at room temperature for 30 min.
12. Then, add 100 μL of competent *E. coli* JM109 cells and incubate on ice for 10 min. Next, heat-shock at 42 °C for 30 s in water bath and incubate on ice for 10 min.
13. Then, add 1 mL of LB medium and incubate on a shaker at 37 °C for 1 h.
14. Spread aliquots of 100 μL on LB agar plates containing ampicillin, X-gal (5-bromo-4-chloro-3-indolyl β-D-galactopyranoside) and IPTG (isopropyl β-D-1-thiogalactopyranoside). Centrifuge the remaining mixture at 15,000 × *g* for 3 min. Discard the 900-μL supernatant, and resuspend the pellet and residual supernatant. Aliquots of 100 μL are spread on other plates.
15. Incubate these plates at 37 °C for 12–15 h.
16. Pick colonies, transfer into 3 mL of ampicillin-containing TB medium, and incubate on a shaker at 180 rpm, 37 °C for 12–15 h.
17. Collect the cultures into tubes, centrifuge, and discard the supernatant. The plasmid is extracted with a FastGene Plasmid Mini Kit.
18. Confirm the insert sequence using a 4300 DNA analyzer (Li-COR, Lincoln, NE).

3.3 RNA Preparation

3.3.1 In Vitro Transcription (IVT)

The ΔP5abc domain is the catalytic domain of the *Tetrahymena* group I ribozyme and is activated by the P5abc domain. However, the ΔP5abc domain is also modestly activated under high-salt conditions (>15 mM Mg^{2+}). Therefore, we use two conditions for RNA synthesis: standard conditions (15 mM Mg^{2+}) for ribozyme synthesis and low-salt (3 mM Mg^{2+}) conditions for precursor RNA synthesis (*see* **Note 4**).

1. Synthesize IVT templates by PCR in a mix (total volume: 100 μL) containing the following components: 10 ng of PCR template plasmid (constructed as described above), 0.5 μM primers (L-21 sense-b and Anti-ScaI, TIL-29 and TIL-30 (for the precursor ribozyme), TIL-30 and TIL-31 (for the precursor ribozyme)), 0.2 mM dNTP, 1× r*Taq* reaction buffer, and 10 U of r*Taq* DNA polymerase. Perform PCR according to the following profile: initial denaturation, 94 °C for 3 min; 25 cycles of 94 °C for 1 min, 50 °C for 1 min, and 72 °C for 1 min; final extension, 72 °C for 5 min. Extract the template with phenol and diethylether, ethanol precipitation, and rinsing. Dry the pellet and dissolve in water.
2. Prepare IVT mix (total volume 200 μL) containing the following components:

   Standard conditions: 20 μL of 10× T7 reaction buffer; 20 μL of 10 mM NTPs mix; 100 pmol of DNA template; 2 μL of recombinant RNase inhibitor; and 4 μL of T7 RNA polymerase. Incubation for 4.5 h at 37 °C.
   Low-salt conditions: 20 μL of 10× T3 reaction buffer; 10 μL of 10 mM NTPs mix; 100 pmol of DNA template; 2 μL of recombinant RNase inhibitor; and 4 μL of T3 RNA polymerase. Incubation for 4.5 h at 37 °C.
3. Degrade the DNA template by addition of RQ I DNase followed by incubation for 30 min at 37 °C.
4. Then add 50 μL of 10 M ammonium acetate and 250 μL of TE-saturated phenol, followed by vortexing, and centrifugation for 5 min at 15,000 × *g*. Next, add 750 μL of 100% ethanol followed by vortexing, and centrifugation at 15000 × *g* for 20 min at 4 °C. Remove the supernatant, open the test tube, and dry the pellet for 10 min at 37 °C before resuspending the pellet in 20 μL of denaturing dye.
5. Prepare 6% (ribozyme) or 4% (precursor RNA) polyacrylamide solution by mixing 5 mL of 5× TBE, 10 mL or 6.7 mL of 30% acrylamide–bis solution, 25 g of urea, and water to a final volume of 50 mL. After dissolution of urea, start polymerization by addition of 250 μL of 20% APS and 25 μL of TEMED. Cast a gel immediately. Remove air bubbles if necessary by knocking gently on the glass plates. Put a comb in place until the gel has polymerized (30 min to 1 h). Then remove the comb and polyacrylamide rests.
6. Assemble the gel on the electrophoresis apparatus with an aluminum plate for heat dispersion and reservoirs filled with 0.5× TBE.

7. Connect the gel to the power supply and prerun for 30 min at 500 V. After stopping the power supply, rinse the wells with 0.5× TBE using a syringe with a needle to remove urea. Immediately, load RNA into the wells and run the gel for approximately 2 h at 750 V until xylene cyanol (XC) has migrated out of the gel.
8. Remove the gel from the apparatus and carefully separate the glass plates. Leave the gel on one plate, which is placed on wrapping film and remove the second glass plate. Place the wrapping film on the other side of the gel.
9. Place the gel on a thin layer chromatography plate containing a fluorescence indicator F254 and illuminate with UV light (254 nm) to reveal the RNA as a dark shadow band. Cut the band corresponding to the full-length RNA out of the gel and place into a 1.5-mL tube. Next, add 500 μL of RNA elution solution and incubate overnight at 16 °C.
10. Extract the RNA with TE-saturated phenol and diethylether, ethanol precipitation, and rinsing. Remove the supernatant and leave the pellets to dry. Dissolve the RNA in water.

3.4 EMS Assay

1. Before RNA folding, prepare electrophoresis buffer and nondenaturing polyacrylamide gel. The electrophoresis buffer (800 mL) contains the following components: 71.2 mM Tris-acetate (pH 7.0); 0, 3, 5 mM $Mg(OAc)_2$.

   The nondenaturing polyacrylamide gel contains the following components: 71.2 mM Tris–acetate (pH 7.0); 3, 5 mM $Mg(OAc)_2$; 6.7 mL 30% acrylamide–bis solution; 0.1 mM EDTA (for 0 mM $Mg(OAc)_2$ conditions).

   Start polymerization by addition of 250 μL of 20% APS and 25 μL of TEMED. Cast a gel immediately. Put a comb in place until the gel has polymerized (30 min to 1 h). Assemble the gel on the electrophoresis apparatus, but do not remove the comb (*see* **Note 5**). Cool the buffer and gel to 4 °C in a refrigerator.
2. Incubate the fluorophore-labeled ribozyme (250 nM) and its partner (250 nM) or unlabeled ribozyme (250 nM) separately at 80 °C. After 5 min, mix the ribozymes at 37 °C followed by addition of 1 μL of 10× folding buffer and incubation at 37 °C for 30 min and then at 4 °C for 30 min. Then, add 2 μL of 6× loading buffer. Keep the samples on ice until loading.
3. Remove the comb and wash all wells using a syringe. Load samples onto the gel immediately and electrophorese at 4 °C, 200 V for 5 min, followed by 75 V for 4.5 h.
4. After electrophoresis, carefully separate the glass plates with the gel remaining on one plate. Analyze the gels with FluoroImager Pharos FX or equivalent.

3.5 Ribozyme Activity Assay

1. Incubate the tecto-GIRz variant (10 pmol: final 500 nM) and its partner (10 pmol: final 500 nM) separately at 80 °C for 5 min. Next, mix ribozymes at 37 °C followed by addition of 10× buffer and GTP, and incubation at 37 °C for 30 min.
2. Before starting the reaction, take a 4 μL aliquot and mix with 1.5 volumes of stop solution, followed by addition of 2 pmol of the fluorophore-labeled substrate as the 0 min reaction sample.
3. Start the reaction by adding 8 pmol of the fluorophore-labeled substrate (final 500 nM), and incubate at 37 °C for 20 min.
4. Take aliquots of 4 μL at given time points, and treat with 1.5 volumes of stop solution.
5. Separate the mixtures on 15% denaturing polyacrylamide gels, and analyze with FluoroImager Pharos FX or equivalent.

3.6 Splicing Assay

To confirm the activity of each ribozyme in the heterodimer, a different length of exon is attached to each ribozyme. Ligated exons from the ribozymes can be distinguished by denaturing polyacrylamide gel electrophoresis.

1. Label precursors with the BODIPY fluorophore as described in the report [11].
2. Incubate the precursor (10 pmol: final 500 nM) and its partner (10 pmol: final 500 nM) separately at 80 °C for 5 min. Mix the RNAs at 37 °C and incubate at 37 °C for 30 min.
3. Before starting the reaction, take a 4 μL aliquot and mixed with 1.5 volumes of stop solution as the 0 min reaction sample.
4. Start the reaction by adding 10× buffer, and incubate at 37 °C for 2 h.
5. Stop the reaction by adding 1.5 volumes of the stop solution.
6. Separate the mixture on a 4% denaturing polyacrylamide gel, and analyze with the gel imaging system.

3.7 Cotranscriptional Splicing Assay

Previously, our group reported *Tetrahymena* group I ribozyme variants that have the split spinach aptamer as 5′ and 3′ exons [12]. The activity of this variant can be easily monitored as fluorescence of the spinach aptamer, since ligation of the exons leads to the formation of functional spinach. Therefore, the ribozyme activity and folding properties could be determined under cotranscriptional conditions without any gel-purified RNAs. Here, we describe the procedure for cotranscriptional splicing assay.

1. Construct a plasmid in which the *Tetrahymena* group I ribozyme is replaced by tecto-GIRz.
2. Prepare template DNA for transcription by PCR as described above.

3. After workup, perform agarose gel electrophoresis to determine the DNA concentration (*see* **Note 6**). Load a DNA ladder marker of known concentration in the same agarose gel. After ethidium bromide staining, analyze the gel with the gel imaging system.
4. Warm the plate reader to 37 °C for more than 30 min before starting the measurement.
5. Mix the following components: 10× T7 cotranscription buffer; 100 μM DFHBI; 10 mM each NTP; T7 RNA polymerase; recombinant RNase inhibitor (each sample contained 2 μL of 10× T7 cotranscription buffer, 1 μL of 10 mM each NTP, 0.5 μL of T7 RNA polymerase, 0.5 μL of recombinant RNase inhibitor). Keep the mixture on ice (*see* **Note** 7).
6. Then, add 4 μL of the mixture to 16 μL of template DNA solution (containing 2 μg of template DNA). Keep on ice.
7. Immediately add aliquots of 20 μL to the wells of a microtiter plate, and cover with 10 μL of mineral oil to prevent evaporation.
8. Immediately set the microtiter plate into the plate reader. Perform the reactions at 37 °C for 4.5 h, with fluorescence measured every 10 min.

4 Notes

1. Peripheral domains are sometimes not included in crystal structures. These domains stabilize the catalytic core, and do not participate in the catalytic reaction. It is possible to design RNA nanostructures with the non-full-length reported structures. However, the peripheral domains protrude and would cause steric hindrance. In addition, they are useful as platforms for insertion of other functional domains. Therefore, we recommend using the full-length structure if it has been solved or predicted.
2. Do not separate the interacting domain (P5abc-ΔP5abc) when domains are used as a large RNA motif as changing the relative orientation of the interacting nucleotides strongly affects tertiary interactions. The joint of linker and domains also needs to connect naturally, as gaps and deformation also affect ternary interactions. We recommend that domains should be connected on duplex regions, and more than four nucleotides (2 bp) should be superimposed to minimize gaps and deformation.
3. In this procedure, we prepared the insert DNA and the vector individually. Invert PCR is also available to construct the vector DNA. However, some plasmids contain a *Bsa*I site and digestion produces several fragments. Then, the ligation step became complex. In addition, if the plasmid is used for in vivo expression

assay, it is important to verify the sequence of the reporter gene, as a PCR-generated long DNA fragment sometimes contains mutations. Thus, use of a long plasmid increases the risk of mutation.

4. It is well known that nucleotides chelate magnesium ions. This means that the concentration of magnesium ions must be more than the total nucleotide concentration. The ΔP5abc domain is slightly activated under conditions of at least 5 mM Mg^{2+}. We confirmed that ΔP5abc is inactive at 3 mM Mg^{2+}. Therefore, we use 3 mM Mg^{2+} and 2 mM total NTPs in this study. Besides, we use a T3 RNA polymerase for the precursor synthesis since pTZ-IVSU contains T7 promoter at upstream of EcoRI site. Hence, T3 promoter is attached by PCR mutagenesis.
5. A nondenaturing polyacrylamide gel sometimes produces small gel scraps in its wells, which make bands broad or unclear. To obtain clear results, remove the comb and add buffer to the upper tank immediately before performing electrophoresis.
6. The PCR products recovered as described above contain large amount of dNTPs. Therefore, we perform agarose gel electrophoresis to determine DNA concentrations from the intensities of the bands. Spin columns are also useful to clean up the PCR products. In this case, DNA concentration can be calculated from the optical density.
7. The fluorescence intensity is affected by the amounts of DFHBI and T7 RNA polymerase. To minimize this effect, we recommend that the components other than the template DNA should be premixed.

Acknowledgments

This work was supported by Grants-in-Aid for Scientific Research on Innovative Area "Synthetic Biology" (No. 23119005 to Y.I.), Scientific Research (C) (No. 15K05561 to Y.I.) from the Ministry of Education, Culture, Sports, Science, and Technology (MEXT), Grant for Basic Science Research Projects from The Sumitomo Foundation (No. 151473 to S.M.), and also partly supported by University of Toyama Discretionary Funds of the President "Toyama RNA Research Alliance" (to Y.I. and S.M.), and "Evolutionary Microdroplets" (to S.M.), Japan.

References

1. Guo P (2010) The emerging field of RNA. Nat Nanotechnol 5:833–842
2. Delebecque CJ, Lindner AB, Silver PA, Aldaye FA (2011) Organization of intracellular reactions with rationally designed RNA assemblies. Science 333:470–474
3. Jasinski DL, Khisamutdinov EF, Lyubchenko Y, Guo P (2014) Physicochemically tunable

polyfunctionalized RNA square architecture with fluorogenic and ribozymatic properties. ACS Nano 8:7620–7629

4. Westhof E, Masquida B, Jaeger L (1996) RNA tectonics: towards RNA design. Fold Des 1: R78–R88
5. Jaeger L, Leontis NB (2000) Tecto-RNA: one-dimensional self-assembly through tertiary interaction. Angew Chem Int Ed 39: 2521–2524
6. Horst van der G, Christian A, Inoue T (1991) Reconstitution of a group I intron self-splicing reaction with an activator RNA. Proc Natl Acad Sci U S A 88:184–188
7. Loria A, Pan T (1996) Domain structure of the ribozyme from eubacterial ribonuclease P. RNA 2:551–563
8. Tanaka T, Matsumura S, Furuta H, Ikawa Y (2016) Tecto-GIRz: engineered group I ribozymes the catalytic ability of which can be controlled by self-dimerization. ChemBioChem 17:1448–1455
9. Lehnert V, Jaeger L, Michele F, Westhof E (1996) New loop-loop tertiary interactions in self-splicing introns of subgroup IC and ID: a complete 3D model of the Tetrahymena thermophile ribozyme. Chem Biol 3:993–1009
10. Doherty EA, Herschlag D, Doudna JA (1999) Assembly of an exceptionally stable RNA tertiary interface in a group I ribozyme. Biochemistry 38:2982–2990
11. Ikawa Y, Furuta H (2008) Facile syntheses of BODIPY derivatives for fluorescent labeling of the 3′ and 5′ ends of RNAs. Anal Biochem 378:166–170
12. Furukawa A, Maejima T, Matsumura S, Ikawa Y (2016) Characterization of an RNA receptor motif that recognizes a GCGA tetraloop. Biosci Biotech Bioch 80:1386–1389

Chapter 22

CRISPR-Cas RNA Scaffolds for Transcriptional Programming in Yeast

Jesse G. Zalatan

Abstract

CRISPR scaffold RNAs (scRNAs) provide a modular system for locus-specific transcriptional programming. scRNAs are generated by extending CRISPR guide RNA sequences with domains that recruit RNA-binding proteins, thus physically linking DNA binding and protein recruitment activities. A single scRNA molecule encodes information about the target locus and instructions about what regulatory function to execute at that locus. Sets of scRNA constructs can be used to generate synthetic multigene transcriptional programs in which some genes are activated and others are repressed. Such programs can be executed by inducing expression of the dCas9 protein, which acts as a single master regulatory control point, and this approach has been recently applied to flexibly redirect flux through a complex branched metabolic pathway in yeast. This protocol describes how to construct multi-scRNA transcriptional programs in yeast, including target site selection, cloning strategies, and yeast engineering.

Key words CRISPR-Cas, Scaffold RNA, Transcriptional programming, Guide RNA

1 Introduction

The recently developed CRISPR-Cas system enables a wide range of applications from genome editing to transcriptional reprogramming [1–3]. For transcriptional regulation, a catalytically inactive Cas9 (dCas9) protein can be used together with a single guide RNA (sgRNA) to recognize specific DNA target sequences. Gene repression results when this complex is targeted downstream of a transcription start site (TSS) to physically block RNA polymerase, a phenomenon referred to as CRISPR interference (CRISPRi) [4]. The CRISPR-Cas system can also enable more sophisticated transcriptional regulation by fusing activators or repressors to dCas9 [5, 6]. A limitation of this approach is that all genes targeted by the dCas9 fusion protein are subject to the same regulatory instructions—genes will be activated or repressed depending on whether an activator or repressor is fused to dCas9. To circumvent this limitation, we have redesigned the guide RNA to encode both

Eckart Bindewald and Bruce A. Shapiro (eds.), *RNA Nanostructures: Methods and Protocols*, Methods in Molecular Biology, vol. 1632, DOI 10.1007/978-1-4939-7138-1_22, © Springer Science+Business Media LLC 2017

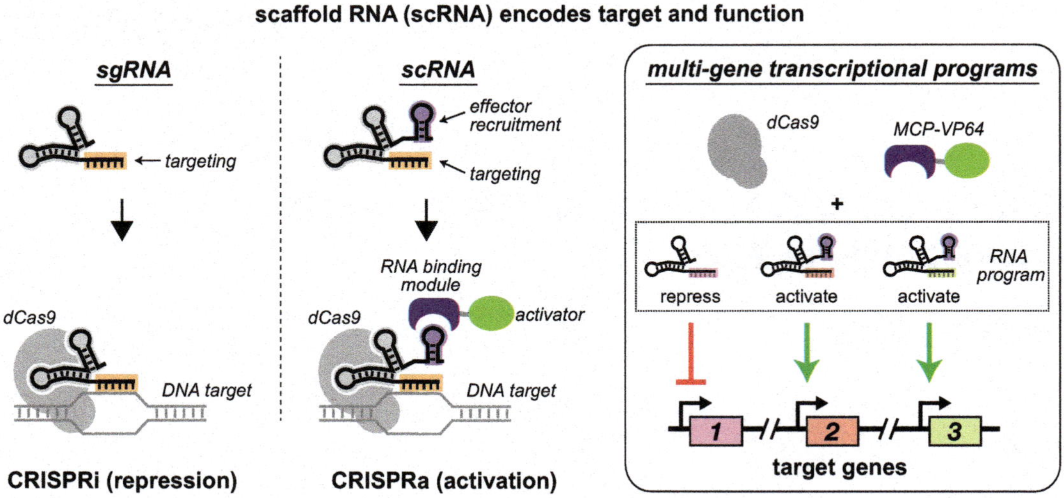

Fig. 1 Single guide RNAs (sgRNAs) contain a 20 base DNA targeting sequence and hairpins that bind the dCas9 protein. A 3′ extension of the sgRNA produces a scaffold RNA (scRNA) that encodes both gene target and the function to be recruited to that target. Multiple sgRNAs and scRNAs can be expressed to encode a program of simultaneous gene activation and repression in a single cell. To activate genes, scRNAs with a 3′-MS2 hairpin recruit the RNA binding protein MCP (MS2 coat protein), which is fused to the transcriptional activator VP64. To repress genes, sgRNAs targeted downstream of the TSS, or at transcription factor binding sites, physically block transcription [6, 7]

the DNA target and effector protein recruitment [7]. These scaffold RNAs (scRNAs) consist of an sgRNA with an extended 3′ sequence containing a modular RNA–protein interaction domain (Fig. 1). Multiple scRNAs can be expressed simultaneously, with each scRNA encoding a different gene target and a regulatory function to execute at that target. Most simply, this approach enables simultaneous activation (CRISPRa) and repression (CRISPRi) at different gene targets, and these programs have been successfully implemented in yeast (*S. cerevisiae*) and human cells [7].

In practice, implementing multigene expression programs requires selecting appropriate target sites, screening for the desired functional effects on transcription of individual genes, and assembling all of the scRNA sequences in a single expression construct. This protocol will provide instructions for executing these programs in yeast, which can be used for a range of applications from basic research to synthetic biology. In yeast cells, we have not identified any repressor proteins that can be recruited by scRNAs to improve silencing beyond that observed with an sgRNA targeted to physically block gene expression. Thus, we use standard sgRNAs for gene repression and an scRNA that recruits a VP64 activator for gene activation (Fig. 1) [7].

dCas9 can be expressed constitutively or under the control of an inducible promoter, which allows regulation of the transcriptional program. In yeast, a galactose-inducible promoter provides tight control, with minimal leaky expression when yeast cells are grown with glucose and strong induction when cells are grown with galactose in the absence of glucose [7, 8].

Several alternative approaches to modifying guide RNAs for protein effector recruitment and imaging have been reported [9–14]. Once these modified guide RNAs have been cloned in to the appropriate yeast expression vector, the protocols described here can be readily adapted to be used with these alternative RNA designs.

The protocol below describes how to construct and evaluate multi-scRNA transcriptional programs in yeast. Key steps include how to select appropriate target sites for gene activation or repression, how to clone the desired expression constructs, and how to deliver these constructs to yeast. Once the constructs have been delivered, RT-qPCR can be used to directly assess effects on transcription of the target genes.

2 Materials

2.1 Plasmids

1. Plasmids for expression of the protein effectors dCas9 and the MCP-VP64 activator (MCP binds to the MS2 RNA hairpin). These plasmids integrate in single copy at the *leu2* locus in the yeast genome. Constitutive expression of dCas9 and MCP-VP64 (pJZC595, Addgene #62322) or galactose-inducible dCas9 and constitutive MCP-VP64 (pJZC638, Addgene #62283) are available (*see* **Note 1**).
2. sgRNAs and scRNAs are expressed from a pol III SNR52 promoter [15] in a CEN/ARS self-replicating plasmid derived from pRS316 with a *ura3* selection marker [16]. Constitutive expression of sgRNA for CRISPRi (pJZC523, Addgene #62320) and MS2 scRNAs for CRISPRa (pJZC545 1× MS2, Addgene #62313; pJZC583 2× MS2, Addgene #62314; pJZC588 2× MS2 [wt + f6], Addgene #62315) are available (*see* **Note 1**).

2.2 Oligonucleotide Primers

1. Primers for cloning new guide RNA constructs (*oligos* 1–4, Table 1).
2. Colony PCR primers to verify genomic integration of dCas9 cassettes at the yeast *leu2* locus (*oligos* 5–6, Table 1).
3. Primers for RT-qPCR of the yeast *ACT1* reference gene (*oligos* 7–8, Table 1). Additional primers will be necessary for new target genes.
4. Primers to assemble plasmids with multiple guide RNAs (*oligos* 4, 9–11, Table 1).

Table 1
Oligonucleotide primers for cloning, colony PCR, and RT-qPCR

Name	Sequence	Usage
Oligo1	GCCCTCTTGGGCTAGCGG	Cloning
Oligo2	GATCATTTATCTTTCACTGCGGAGAA	Cloning
Oligo3	GAAAGATAAATGATCnnnnnnnnnnnnnnnnnnnnGT TTTAGAGCTAGAAATAGCAAGTT	Cloning[a]
Oligo4	GCTGGTACCGAACTCTGCAG	Cloning
Oligo5	CCCCCTTTGCTTATAATTGTGTGG	Colony PCR
Oligo6	GTACTGAAGAGGAGGTCGACTACG	Colony PCR
Oligo7	GAGGTTGCTGCTTTGGTTATTG	RT-qPCR
Oligo8	ACCGACGATAGATGGGAAGA	RT-qPCR
Oligo9	TGTTTTTTATGTCTCTCGATCTTTGAAAAGATAATGTATGATTATGC	Cloning
Oligo10	AGCGTGCAGTAGTCCGAACTCTGCCGAGACATAAAAAAC	Cloning
Oligo11	TACTGCACGCTACGCTCTTTGAAAAGATAATGTATGATTATGC	Cloning

[a] *Oligo3* is used to generate new guide RNA target sites. The 20 × *n* bases (underlined) should be replaced by new target sites selected as described in Subheading 3.1

2.3 Cloning

1. NheI and PstI restriction enzymes (New England Biolabs or Thermo Scientific FastDigest), supplied with appropriate buffers.
2. *Escherichia coli* NEB Turbo competent cells (New England Biolabs), or any standard cloning strain.
3. LB media and agar plates.
4. Carbenicillin.
5. Phusion DNA polymerase (New England Biolabs) or equivalent PCR polymerase.
6. dNTPs.
7. In-Fusion cloning kit (Clontech).
8. DNA gel purification and miniprep kits.

2.4 Yeast Growth and Manipulation

1. A haploid yeast strain with the *leu2* and *ura3* marker slots available. Derivatives of S288C and W303 are commonly used [17]. A wide range of yeast strains are available from ATCC and Euroscarf. The online *Saccharomyces* Genome Database (http://www.yeastgenome.org/) maintains a list of commonly used strains with descriptions and links to suppliers.
2. Yeast Peptone Dextrose (YPD) media and agar plates (YPD: 1% yeast extract, 2% peptone, 2% dextrose) (*see* **Note 2**). For plates, use 2% agar. Sterilize by autoclaving.

3. For yeast synthetic dropout media, use dropout mixes (US Biological D9525 (−Leu) and D9535 (−Ura)) and yeast nitrogen base (US Biological Y2025). To prepare SD −Leu and SD −Ura media, follow manufacturer's instructions for quantities of dropout mix and yeast nitrogen base, and supplement with a final concentration of 2% dextrose (SD = synthetic dextrose). For galactose inductions, prepare SSG (synthetic sucrose galactose) using 2% sucrose and 2% galactose. For plates, use 2% agar (*see* **Note 3**).
4. PmeI (NEB) or MssI (Thermo FD) restriction enzymes.
5. 10× TE pH 7.5, sterile filtered (10× TE is 100 mM Tris–HCl pH 7.5, 10 mM EDTA).
6. 1 M lithium acetate (LiOAc), sterile filtered, pH should be ~8.
7. 50% PEG 3350, sterile filtered, store at 4 °C protected from light, use within 3 months.
8. DMSO (Fisher BioReagent grade).
9. Salmon sperm DNA (ssDNA, Sigma D-1626). To prepare 10 mg/mL solution, dissolve 250 mg ssDNA in 25 mL of sterile 1× TE pH 8.0 at 4 °C using a sterile beaker and magnetic stir bar and store 1 mL aliquots at −20 °C. It can take ~6 h for this solution to dissolve.
10. DreamTaq Green DNA polymerase (Fisher Scientific), supplied with DreamTaq Green Buffer, or equivalent low-cost, low-fidelity PCR polymerase for colony PCR.
11. Absorbance spectrometer. NanoDrop 2000c (ThermoFisher Scientific) or equivalent.

2.5 RT-qPCR

1. cDNA synthesis kit. iScript RT Supermix (Bio-Rad) or equivalent.
2. RNA extraction kit. Aurum Total RNA mini kit (Bio-Rad) or equivalent.
3. Fluorescent DNA probes. SsoAdvanced SYBR Green (Bio-Rad) or equivalent.
4. PCR Tubes and Caps: 0.2 mL 8-Tube PCR Strips, 0.2 mL Flat PCR Tube 8-Cap Strips (Bio-Rad) or equivalent.
5. Real-time thermal cycler. Stratagene Mx3005P or equivalent.

3 Methods

3.1 Guide RNA Target Site Selection

Extensive research and genome-wide screens in eukaryotes suggest general rules for selecting target sites for CRISPRi and CRISPRa [18–20]. For *S. pyogenes* Cas9, each 20 base target site must be followed by the 5′-NGG-3′ protospacer adjacent motif (PAM) sequence, henceforth referred to as NGG. Key additional criteria to evaluate include the presence of off-target sites elsewhere in the genome (*see* **Note 4**), the distance from the transcriptional start site (TSS), and the absence of poly-T sequences that can act as terminators for pol III-mediated expression of short RNAs. In eukaryotic cells, targeting the template or nontemplate strand can be effective for either activation or repression [6, 7, 18]. There are automated computational design tools available on the web (e.g., http://crispr-era.stanford.edu/) [19], but the outputs of these programs should be manually inspected and evaluated, particularly because TSSs may be poorly annotated or ambiguous. Because TSSs typically reside within 100 bases of the ATG start codon [21], a general guideline is to screen potential activating sites from position −300 to −50 upstream of the ATG for activation, and from −100 upstream to +300 downstream of the ATG for repression.

1. Obtain the target gene sequence from the Saccharomyces Genome Database (http://www.yeastgenome.org/). Retrieve the open reading frame (ORF) along with 500 bases upstream (5′ of the start codon) (*see* **Note 5**).
2. Identify all potential NGG PAM sites by searching for all GG motifs on both strands.
3. Identify candidate sequences (20 bases upstream of the NGG PAM site) according to the criteria described above: for gene activation identify sites from −300 to −50 upstream of the ATG start site, and for gene repression identify sites from −100 upstream to +300 downstream of the ATG site.
4. In parallel, obtain candidate sites from an automated design tool such as http://crispr-era.stanford.edu/ which will evaluate off-target specificity across the genome.
5. Select 6–8 sites distributed across the region of interest for experimental validation. Choose sites that are appropriately positioned by manual inspection and that are ranked for minimal off-target binding by the design tool. Typically expect 2–4 target sites to be effective in functional assays.
6. The new 20-base targeting sequences (the 20 bases upstream of the NGG PAM site) should be incorporated into *oligo3* (Subheading 2.2) for cloning as described below in Subheading 3.2. The guide RNA will base-pair to the DNA strand that is complementary to the strand containing the NGG PAM [22].

3.2 Cloning Alternative Target Sites into Guide RNA Vectors

Cloning new 20-base targeting sites into guide RNA expression plasmids can usually be accomplished with standard PCR-based cloning methods. Yeast guide RNA plasmids present a minor complication in that the SNR52 promoter used for pol III expression of short RNAs [6, 15] lacks a convenient restriction site adjacent to

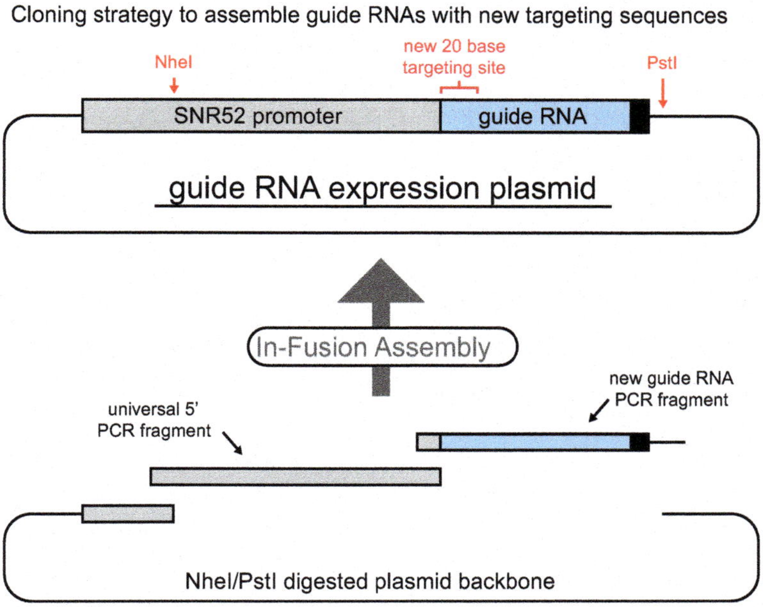

Fig. 2 Guide RNA expression constructs with new targeting sequences can be cloned using a three-component In-Fusion assembly strategy. The necessary components are a plasmid backbone (NheI/PstI-digested) and two PCR fragments with 15 base overlaps at the assembly junctions. The 5′ primer for the guide RNA PCR fragment contains a new 20 base targeting sequence. Assembly with two PCR fragments is necessary because there is no convenient restriction site at the junction between the SNR52 promoter and the guide RNA. The *black box* at the end of the guide RNA represents the SUP4 terminator sequence

the 20-base targeting site. To modify the targeting site, an inverse PCR method has been described [23, 24] involving a single PCR step to amplify the entire plasmid with a primer containing the new target site, followed by a blunt-end ligation. While reasonably effective, this approach is often prone to small deletions, substantial plasmid rearrangements, and frequent recovery of the parent plasmid template. We have devised an alternative approach using In-Fusion fragment assembly (Fig. 2). In our hands, this approach is much more likely to generate the desired target plasmid. The inverse PCR method is advantageous when the goal is to construct a large library of guide RNA constructs in parallel because it involves fewer manipulations and shorter PCR oligonucleotides, but when a small set of specific constructs is desired, the In-Fusion assembly method is more robust. Three fragments are necessary: a restriction-digested plasmid backbone, a universal 5′ PCR fragment, and a guide RNA PCR product containing the new targeting sequence. The protocol below generates the necessary fragments with 15-base overhangs for In-Fusion assembly.

1. Digest a parent guide RNA plasmid (any of the guide RNA plasmids described in Subheading 2.1) with NheI and PstI. Use 1 μg of plasmid DNA.
2. Run the digested product on an agarose gel.
3. Perform a gel extraction and purification of the 4.9 kb product.
4. PCR amplify the universal 5′ fragment using *oligo1* and *oligo2* (Subheading 2.2). Any guide RNA plasmid can be used as a template for this reaction.
5. Gel purify the 146 bp PCR product.
6. PCR amplify guide RNA constructs with new 20 base targeting sites using *oligo3* and *oligo4* (Subheading 2.2). For *oligo3*, replace the 20 × *n* sequence with new 20 base target sites designed in Subheading 3.1. Use the appropriate guide RNA plasmid as a template. For repression, use the sgRNA plasmid (pJZC523) and for activation use the 1× or 2× MS2 scRNA plasmids (pJZC545, pJZC583, or pJZC588).
7. Gel purify the ~150–200 bp PCR product (exact size depends on choice of template).
8. Prepare an In-Fusion assembly reaction. Use a total volume of 5 μL with ~50 ng of plasmid backbone and ~100 ng of each PCR fragment (*see* **Note 6**).

5× In-Fusion premix	1 μL
Digested plasmid backbone (100 ng/μL)	0.5 μL
Universal 5′ PCR fragment (100 ng/μL)	1 μL
Guide RNA PCR fragment (100 ng/μL)	1 μL
Water	1.5 μL
Total	5 μL

9. Incubate the reaction at 50 °C for 15 min. Place the reaction on ice.
10. Using the entire volume of the In-Fusion reaction, transform to competent *E. coli* cells following the manufacturer's instructions.
11. Plate cells on LB + carbenicillin (50 μg/mL carbenicillin).
12. Grow at 37 °C overnight. Expect ~100 colonies.
13. Pick colonies and inoculate in 5 mL LB + 50 μg/mL carbenicillin (*see* **Note 7**).
14. Grow at 37 °C overnight (for standard *E. coli* strains) or 6 h (for fast-growing strains like NEB Turbo *E. coli*).
15. Purify plasmids using any standard miniprep kit.
16. Verify DNA sequence using the M13 reverse sequencing primer.

3.3 Yeast Cloning and Manipulation

Standard methods for yeast cloning and manipulation [17] are effective with yeast engineered to express CRISPR-Cas transcriptional regulatory programs. The following yeast transformation protocol closely follows previously published versions [25, 26] with minor modifications. For the plasmids described in Subheading 2 above, start with a haploid yeast strain with the *leu2* and *ura3* marker slots available. If these markers are not available or already in use, the CRISPR-Cas constructs can be easily moved to other standard yeast vectors with alternative auxotrophic markers or antibiotic resistance cassettes. The protocol below describes how to generate a yeast strain with dCas9 and the MCP-VP64 activator integrated into the genome at the *leu2* locus. The final step explains how to apply this protocol to deliver guide RNA plasmids to this new yeast strain.

1. Grow target yeast strain in 5–10 mL YPD overnight at 30 °C.
2. The next day, dilute overnight culture ~1:20 into fresh YPD. Choose a culture volume that will provide a final 2.5 mL yeast culture for each transformation reaction. Include a few extra mL for OD measurements.
3. Grow 3–5 h at 30 °C to $OD_{600} = 0.5–0.8$ (doubling time ~90 min, dilute culture starts at ~OD 0.05–0.10). While cells are growing, complete **steps 4** and **5** below.
4. Obtain the plasmid vector containing dCas9 and MCP-VP64 with a *leu2* selection marker (either pJZC595 for constitutive dCas9 or pJZC638 for galactose-inducible dCas9). Linearize the integrating vector using PmeI or MssI following the manufacturer's instructions for appropriate reaction conditions. Use 0.5–1 μg plasmid DNA in a total volume ≤10 μL. The digested DNA can be used as is without further purification for the transformation reaction.
5. Prepare ssDNA: denature the DNA in a boiling water bath or a heat block at >95 °C for 10 min, then cool on ice for at least 10 min (*see* **Note 8**). Use this ssDNA in **step 11**.
6. Harvest yeast cells in 15 mL conical tubes by centrifugation at ~2000 × *g*, 2 min.
7. Wash cells with sterile water. Use a volume equal to the original culture volume.
8. Centrifuge again at ~2000 × *g*, 2 min.
9. Resuspend the pellet in sterile water. The final volume should correspond to 100 μL for each 2.5 mL of culture (i.e., for four transformations from 10 mL of yeast cell culture, resuspend in 400 μL).
10. Split the resuspended cells into 100 μL aliquots.
11. To 100 μL resuspended cells, add 100 μg ssDNA (10 μL of 10 mg/mL stock) followed by the linearized target DNA for integration (typically 10 μL).

12. Add 480 μL 50% PEG 3350, 60 μL 10× TE pH 7.5, 60 μL 1 M LiOAc. Vortex to mix.
13. Add 75 μL DMSO. Vortex to mix, ensuring that PEG and DMSO are thoroughly mixed with the cells.
14. Incubate at 42 °C for 30 min.
15. Pellet cells at 3500 × *g* for 2 min in a microcentrifuge.
16. Discard supernatant, taking care to remove the PEG solution completely with a micropipette (*see* **Note 9**).
17. Resuspend the cells in 500 μL YPD or sterile water.
18. Pellet the cells at 6000 rpm for 2 min in a microcentrifuge.
19. Discard the supernatant.
20. Resuspend the cells in 50 μL YPD or sterile water. Plate the entire volume on SD −Leu agar plates.
21. Incubate at 30 °C for 2–3 days. Expect to obtain >20–50 colonies from this procedure (*see* **Note 10**).
22. Pick 3–5 single colonies and use a sterile toothpick or pipette tip to make small patches (~0.5 cm × 1.5 cm) on a fresh SD −Leu plate. Patching the yeast cells enables repeated use of cells generated from the same single colony.
23. Grow overnight at 30 °C.
24. Using the patched yeast, confirm successful genomic integration by yeast colony PCR with *oligo5* (internal integrating cassette primer) and *oligo6* (Subheading 2.2). *Oligo5* binds inside the integrating cassette and *oligo6* binds outside of the integration cassette in the endogenous *leu2* locus. Amplification of a 573 bp PCR product confirms successful genomic integration. Using a sterile toothpick or pipette tip, collect a small amount of yeast cells (should be visible on the tip) and smear the cells into a PCR tube containing 3 μL of 20 mM NaOH. Lyse the yeast cells at 95 °C for 20 min. Add the PCR reagents below, run the PCR reaction using the manufacturer's recommended thermocycling conditions, and analyze the reactions by agarose gel electrophoresis.

	per rxn (μL)
10× DreamTaq Green Buffer	2.5
2 mM dNTPs	2.5
50 μM F primer	0.25
50 μM R primer	0.25
DreamTaq	0.125
Water	19.4
Total	25

25. Yeast patch plates can be stored at 4 °C for 2–3 months. Once genomic integration has been confirmed by colony PCR, it is no longer necessary to maintain selection on SD −Leu. Subsequent propagation steps with these strains can be performed in YPD media.
26. To transform CEN/ARS self-replicating guide RNA plasmids into the newly constructed yeast strain, follow the same transformation protocol as above from **steps 1** to **23** with the following modifications: (a) use 100 ng of intact (undigested) plasmid in the transformation reaction and (b) use SD −Ura as the selection media. Expect >>100 colonies. If yields are high, the total transformation volume can be scaled down by 4×.

3.4 Evaluate Guide RNA Function

Once yeast strains have been constructed with dCas9, MCP-VP64, and guide RNA plasmids, it is necessary to evaluate if gene expression has been modified as expected. Depending on the desired target, fluorescent reporter genes or functional assays can allow rapid screening [7]. If these methods are unavailable, RT-qPCR provides a general method to evaluate whether gene expression has been successfully perturbed.

1. Use any primer design software to generate primers that will amplify a 100–200 bp fragment from the target gene (*see* **Note 11**). Primers should have a $T_m > 55$ °C.
2. Choose a reference gene to use as a relative standard. *ACT1* can be used as a reference gene with *oligo7* and *oligo8* (Subheading 2.2) to amplify a 99 bp product.
3. Validate qPCR primers in a regular PCR reaction following the yeast colony PCR protocol described in Subheading 3.3. Make sure that only a single band of the expected size is amplified.
4. For yeast strains with constitutive dCas9 expression, grow cells in SD −Ura selection media. For yeast cells with galactose-inducible dCas9 expression, grow cells in SD −Ura (noninducing) or SSG −Ura (inducing). Be sure to include a control strain for comparison. Typically, this will be the parent strain transformed with an empty *ura3*-selectable plasmid or an off-target guide RNA plasmid.
5. Grow yeast, harvest cultures, and extract RNA following manufacturer's instructions in Aurum RNA extraction kit or equivalent. Make sure pipettes are clean, remove the tip ejector (if possible), and use filter tips to minimize RNase contamination. Use sterile, autoclaved water for all solutions after cells have been harvested.

6. Check RNA quality on a 1% agarose gel. Expect two sharp bands corresponding to 25S and 18S rRNA (3.4 and 1.8 kb, respectively). These bands will run at roughly 1.8 and 1.0 kb relative to a dsDNA ladder. Smeared or blurry bands are indicative of RNase contamination and a poor quality preparation.
7. Use an absorbance spectrometer to measure the absorbance at 260 nm (A_{260}) for the purified RNA solution.
8. Calculate RNA concentration using the Beer–Lambert law ($A_{260} = \varepsilon_{260} \cdot C \cdot l$) where $\varepsilon_{260} = 0.025\ \text{cm}^{-1}\ (\text{ng}/\mu\text{L})^{-1}$, C is the concentration, and l is the path length in cm. Expect ~200 ng/μL.
9. Convert RNA to cDNA using a cDNA synthesis kit (containing oligo-dT primers). Also perform a reaction without reverse transcriptase (no-RT control reaction), which will be used to assess genomic DNA contamination. Assume 1:1 conversion of RNA to cDNA and dilute this stock to ~2 ng/μL.
10. Set up a qPCR reaction. Include at least three biological replicates and one no-RT control reaction.

SsoAdvanced SYBR Green Supermix	10 μL
10 μM F primer	1 μL
10 μM R primer	1 μL
cDNA (2 ng/μL)	3 μL
Water	5 μL
Total	20 μL

11. Perform the RT-qPCR reaction on a real-time thermal cycler with the appropriate fluorescent DNA probe (*see* Subheading 2). Follow fluorescent probe manufacturer's instructions for instrument-specific thermocycling parameters.
12. Calculate relative gene expression levels using the $\Delta\Delta C_T$ method [27]. This method gives a gene expression level in a modified strain relative to a control strain. Each RT-qPCR reaction produces a threshold cycle (C_T) value, which is the cycle at which the fluorescence signal reaches the detection threshold (smaller C_T values indicate more target DNA starting material). To convert C_T values to relative gene expression levels, the C_T for each target gene is first normalized to a reference gene measured in the same strain ($\Delta C_T = C_{T,\ \text{target gene}} - C_{T,\ \text{reference gene}}$). The relative gene expression level is then calculated from $2^{-\Delta\Delta C_T}$, where $\Delta\Delta C_T = (\Delta C_{T,\ \text{exp. strain}} - \Delta C_{T,\ \text{control strain}})$ (*see* **Note 12**).

Once guide RNA plasmids targeting individual genes have been validated, a single plasmid containing multiple guide RNAs can be

3.5 Assemble Multiguide RNA Expression Vector

assembled to target multiple genes simultaneously. The protocol below describes how to construct a single plasmid expressing three guide RNAs using an In-Fusion assembly strategy (Fig. 3) (*see* **Note 13**).

1. Select three guide RNA plasmids that you wish to combine into a single expression vector. Choose one plasmid to digest with PstI. The guide RNA contained in this plasmid corresponds to guide RNA 1 in Fig. 3.
2. Gel purify the ~5 kb linearized plasmid.
3. From the other two plasmids, PCR amplify the guide RNA cassettes (Fig. 3). Use *oligo9* and *oligo10* to generate guide RNA 2 and use *oligo11* and *oligo4* to generate guide RNA 3 (Subheading 2.2) (*see* **Note 14**).
4. Gel purify the ~0.5 kb PCR products (exact size depends on the template guide RNA).
5. Prepare an In-Fusion assembly reaction. Use a total volume of 5 μL with ~50 ng of plasmid backbone and ~100 ng of each PCR fragment.

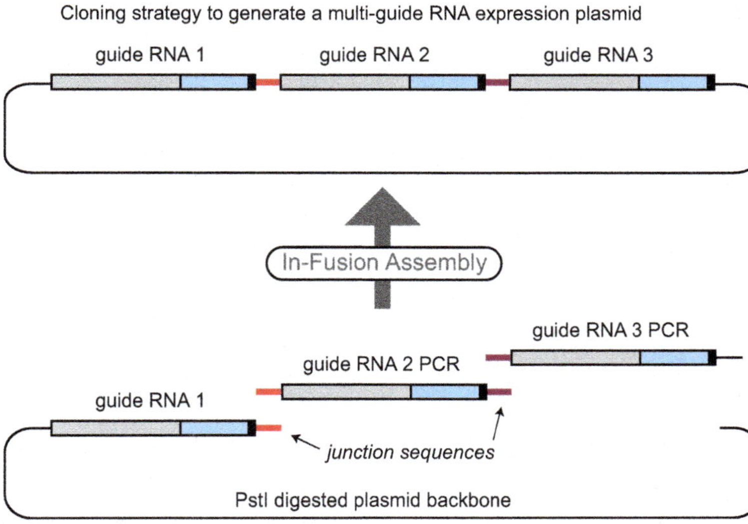

Fig. 3 Multiple guide RNA expression cassettes can be cloned into a single plasmid to enable multigene control with one vector. Each guide RNA expression cassette contains a promoter (*grey*), guide RNA sequence (*blue*), and terminator (*black*). Unique junction sequences indicated by the *red* and *purple lines* between guide RNA cassettes ensure proper assembly. The plasmid is assembled with a three-component In-Fusion reaction from a linearized guide RNA plasmid and two PCR fragments containing additional guide RNA cassettes

5× In-Fusion premix	1 μL
Digested plasmid backbone (100 ng/μL)	0.5 μL
PCR guide RNA fragment 1 (~100 ng/μL)	1 μL
PCR guide RNA fragment 2 (~100 ng/μL)	1 μL
Water	1.5 μL
Total	5 μL

6. Incubate the reaction at 50 °C for 15 min. Place the reaction on ice.
7. Using the entire volume of the In-Fusion reaction, transform to competent *E. coli* cells.
8. Plate cells on LB + carbenicillin (50 μg/mL carbenicillin).
9. Grow at 37 °C overnight. Expect ~100 colonies.
10. Pick colonies and inoculate in 5 mL LB + 50 μg/mL carbenicillin (*see* **Note 15**).
11. Grow at 37 °C overnight (for standard *E. coli* strains) or 6 h (for fast-growing strains like NEB Turbo *E. coli*).
12. Purify plasmids using any standard miniprep kit.
13. Verify DNA sequence using the M13 forward and reverse sequencing primers (*see* **Note 16**).
14. Use this multiguide RNA plasmid to transform yeast following Subheading 3.3.

4 Notes

1. In addition to MS2, the alternative RNA recruitment domains PP7 and com are available [7]. To use these effectors, obtain pJZC518 (Addgene #62277) for constitutive dCas9 expression with integration at *leu2*. pJZC504 (Addgene #62280) and pJZC506 (Addgene #62281) contain PCP-VP64 and Com-VP64, respectively, and both integrate at *his3*. The corresponding CEN/ARS plasmids with *ura3* selection for scRNA expression are pJZC548 (1× PP7, Addgene #62316), pJZC603 (2× PP7, Addgene #62317), and pJZC572 (1× Com, Addgene #62318). No 2× com designs are successful. pJZC593 (Addgene #62319) has both MS2 and PP7 hairpins for multivalent, heterologous recruitment.
2. Supplement media with 0.002–0.008% adenine for *ade2* strains.
3. Synthetic dropout mix media should be sterile filtered through a 0.2 μm membrane, not autoclaved. For plates, sterile filter a 2× media mix containing all components except the agar.

Autoclave a 2× agar solution, mix with the filtered media mix, and pour plates.

4. Genome-wide and library screens in human cells [5, 18, 28] have identified specificity rules for dCas9 binding that also likely apply in yeast. One general observation is that mismatches far from the PAM site are sometimes tolerated and allow off-target binding, while mismatches proximal to the PAM site are deleterious.
5. Check for other genes or features upstream of the gene of interest. Be aware that if there is a nearby gene (i.e., <500 bases) upstream and oriented in the opposite direction, it could also be subject to transcriptional activation.
6. For the first In-Fusion assembly reaction with a new plasmid backbone, always run a negative control with the digested plasmid alone. The total In-Fusion reaction volume can easily be scaled down to 2 μL to conserve reagents.
7. To identify properly assembled plasmids prior to DNA minipreps, PCR screen colonies with M13 forward and reverse primers. The expected PCR product size is ~0.5–0.6 kb depending on the exact guide RNA being constructed.
8. Aliquots of ssDNA can be boiled ~3 times before they should be discarded for fresh material.
9. Alternatively, the supernatant can be discarded by inverting the tube. To fully remove the PEG solution, repeat wash **steps 17–19** twice.
10. If yields are low, make sure the 50% PEG stock is fresh (<2 months old).
11. Extra care in primer design is necessary if the gene of interest has high homology to any other genes in the yeast genome.
12. Check to ensure that signal is low in the no-RT control reaction (high C_T value). Ideally, the no-RT control reaction should give a C_T value that is more than five cycles greater than the +RT reactions. If the no-RT C_T value is comparable to the +RT reactions, genomic DNA contamination may be significant and calculated gene expression levels may be unreliable.
13. These plasmids contain multiple repeats and can potentially recombine to eliminate guide RNA cassettes. In the absence of any selective pressure, recombination frequencies are relatively low (≤1%), but recombination can increase dramatically if the guide RNAs modify gene expression in a way that has any adverse effect on growth. In such cases, it may be necessary to separate guide RNA cassettes on plasmids with different selection markers (i.e., *his3* or *trp1*) [16].

14. If a plasmid with only two guide RNAs is desired, amplify one guide RNA cassette with *oligo9* and *oligo4* (Subheading 2.2) for assembly into a guide RNA plasmid digested with PstI.
15. Single colonies can be PCR screened with M13 forward and reverse primers. Expect ~1.4 kb for a complete three guide RNA construct. For two guide RNA cassettes, expect ~1.0 kb.
16. When comparing sequence data to the expected template file, be aware that some alignment programs will fail to correctly align sequences with multiple repeats. If the alignment appears incorrect, be sure to manually inspect the alignment before concluding that the assembly failed.

Acknowledgments

The author would like Brian Fan, Robin Kirkpatrick, and Jingwen Sun for validating and reproducing several of the methods described here, as well as members of the Lim, Qi, Dueber, and Weissman labs for technical support and discussions. This work was supported by a Career Award at the Scientific Interface from the Burroughs Wellcome Fund.

References

1. Mali P, Esvelt KM, Church GM (2013) Cas9 as a versatile tool for engineering biology. Nat Methods 10:957–963. doi:10.1038/nmeth.2649
2. Dominguez AA, Lim WA, Qi LS (2016) Beyond editing: repurposing CRISPR-Cas9 for precision genome regulation and interrogation. Nat Rev Mol Cell Biol 17:5–15. doi:10.1038/nrm.2015.2
3. Wright AV, Nuñez JK, Doudna JA (2016) Biology and applications of CRISPR systems: harnessing nature's toolbox for genome engineering. Cell 164:29–44. doi:10.1016/j.cell.2015.12.035
4. Qi LS, Larson MH, Gilbert LA, Doudna JA, Weissman JS, Arkin AP, Lim WA (2013) Repurposing CRISPR as an RNA-guided platform for sequence-specific control of gene expression. Cell 152:1173–1183. doi:10.1016/j.cell.2013.02.022
5. Mali P, Aach J, Stranges PB, Esvelt KM, Moosburner M, Kosuri S, Yang L, Church GM (2013) CAS9 transcriptional activators for target specificity screening and paired nickases for cooperative genome engineering. Nat Biotechnol 31:833–838. doi:10.1038/nbt.2675
6. Gilbert LA, Larson MH, Morsut L, Liu Z, Brar GA, Torres SE, Stern-Ginossar N, Brandman O, Whitehead EH, Doudna JA, Lim WA, Weissman JS, Qi LS (2013) CRISPR-mediated modular RNA-guided regulation of transcription in eukaryotes. Cell 154:442–451. doi:10.1016/j.cell.2013.06.044
7. Zalatan JG, Lee ME, Almeida R, Gilbert LA, Whitehead EH, La Russa M, Tsai JC, Weissman JS, Dueber JE, Qi LS, Lim WA (2015) Engineering complex synthetic transcriptional programs with CRISPR RNA scaffolds. Cell 160:339–350. doi:10.1016/j.cell.2014.11.052
8. Mumberg D, Müller R, Funk M (1994) Regulatable promoters of *Saccharomyces cerevisiae*: comparison of transcriptional activity and their use for heterologous expression. Nucleic Acids Res 22:5767–5768
9. Konermann S, Brigham MD, Trevino AE, Joung J, Abudayyeh OO, Barcena C, Hsu PD, Habib N, Gootenberg JS, Nishimasu H, Nureki O, Zhang F (2015) Genome-scale transcriptional activation by an engineered CRISPR-Cas9 complex. Nature 517:583–588. doi:10.1038/nature14136
10. Shechner DM, Hacisuleyman E, Younger ST, Rinn JL (2015) Multiplexable, locus-specific

targeting of long RNAs with CRISPR-Display. Nat Methods 12:664–670. doi:10.1038/nmeth.3433

11. Wang S, Su J-H, Zhang F, Zhuang X (2016) An RNA-aptamer-based two-color CRISPR labeling system. Sci Rep 6:26857. doi:10.1038/srep26857
12. Ma H, Tu L-C, Naseri A, Huisman M, Zhang S, Grunwald D, Pederson T (2016) Multiplexed labeling of genomic loci with dCas9 and engineered sgRNAs using CRISPRainbow. Nat Biotechnol 34:528–530. doi:10.1038/nbt.3526
13. Shao S, Zhang W, Hu H, Xue B, Qin J, Sun C, Sun Y, Wei W, Sun Y (2016) Long-term dual-color tracking of genomic loci by modified sgRNAs of the CRISPR/Cas9 system. Nucleic Acids Res 44:e86. doi:10.1093/nar/gkw066
14. Cheng AW, Jillette N, Lee P, Plaskon D, Fujiwara Y, Wang W, Taghbalout A, Wang H (2016) Casilio: a versatile CRISPR-Cas9-Pumilio hybrid for gene regulation and genomic labeling. Cell Res 26:254–257. doi:10.1038/cr.2016.3
15. DiCarlo JE, Norville JE, Mali P, Rios X, Aach J, Church GM (2013) Genome engineering in *Saccharomyces cerevisiae* using CRISPR-Cas systems. Nucleic Acids Res 41:4336–4343. doi:10.1093/nar/gkt135
16. Sikorski RS, Hieter P (1989) A system of shuttle vectors and yeast host strains designed for efficient manipulation of DNA in *Saccharomyces cerevisiae*. Genetics 122:19–27
17. Sherman F (2002) Getting started with yeast. Methods Enzymol 350:3–41
18. Gilbert LA, Horlbeck MA, Adamson B, Villalta JE, Chen Y, Whitehead EH, Guimaraes C, Panning B, Ploegh HL, Bassik MC, Qi LS, Kampmann M, Weissman JS (2014) Genome-scale CRISPR-mediated control of gene repression and activation. Cell 159:647–661. doi:10.1016/j.cell.2014.09.029
19. Liu H, Wei Z, Dominguez A, Li Y, Wang X, Qi LS (2015) CRISPR-ERA: a comprehensive design tool for CRISPR-mediated gene editing, repression and activation. Bioinformatics 31:3676–3678. doi:10.1093/bioinformatics/btv423
20. Smith JD, Suresh S, Schlecht U, Wu M, Wagih O, Peltz G, Davis RW, Steinmetz LM, Parts L, St Onge RP (2016) Quantitative CRISPR interference screens in yeast identify chemical-genetic interactions and new rules for guide RNA design. Genome Biol 17:45. doi:10.1186/s13059-016-0900-9
21. David L, Huber W, Granovskaia M, Toedling J, Palm CJ, Bofkin L, Jones T, Davis RW, Steinmetz LM (2006) A high-resolution map of transcription in the yeast genome. Proc Natl Acad Sci U S A 103:5320–5325. doi:10.1073/pnas.0601091103
22. Jinek M, Chylinski K, Fonfara I, Hauer M, Doudna JA, Charpentier E (2012) A programmable dual-RNA-guided DNA endonuclease in adaptive bacterial immunity. Science 337:816–821. doi:10.1126/science.1225829
23. Larson MH, Gilbert LA, Wang X, Lim WA, Weissman JS, Qi LS (2013) CRISPR interference (CRISPRi) for sequence-specific control of gene expression. Nat Protoc 8:2180–2196. doi:10.1038/nprot.2013.132
24. Hawkins JS, Wong S, Peters JM, Almeida R, Qi LS (2015) Targeted transcriptional repression in bacteria using CRISPR interference (CRISPRi). Methods Mol Biol 1311:349–362. doi:10.1007/978-1-4939-2687-9_23
25. Gietz RD, Schiestl RH (2007) High-efficiency yeast transformation using the LiAc/SS carrier DNA/PEG method. Nat Protoc 2:31–34. doi:10.1038/nprot.2007.13
26. Gietz RD, Schiestl RH (2007) Quick and easy yeast transformation using the LiAc/SS carrier DNA/PEG method. Nat Protoc 2:35–37. doi:10.1038/nprot.2007.14
27. Livak KJ, Schmittgen TD (2001) Analysis of relative gene expression data using real-time quantitative PCR and the $2^{(-\Delta\Delta C(T))}$ method. Methods 25:402–408. doi:10.1006/meth.2001.1262
28. Wu X, Scott DA, Kriz AJ, Chiu AC, Hsu PD, Dadon DB, Cheng AW, Trevino AE, Konermann S, Chen S, Jaenisch R, Zhang F, Sharp PA (2014) Genome-wide binding of the CRISPR endonuclease Cas9 in mammalian cells. Nat Biotechnol 32:670–676. doi:10.1038/nbt.2889

Chapter 23

Using Planar Phi29 pRNA Three-Way Junction to Control Size and Shape of RNA Nanoparticles for Biodistribution Profiling in Mice

Farzin Haque, Congcong Xu, Daniel L. Jasinski, Hui Li, and Peixuan Guo

Abstract

RNA is rapidly emerging as a versatile building block for nanoparticle assembly due to its simplicity in base pairing, while exhibiting diversity in function such as enzymatic activity similar to some proteins. Recent advances in RNA nanotechnology have generated significant interests in applying RNA nanoparticles for various applications in nanotechnology and nanomedicine. In particular, assessing the effect of size and shape on cell entry and intracellular trafficking as well as in vivo biodistribution of nanoparticles is challenging due to the lack of nanoparticles rich in structure while varying in size and shape. RNA nanotechnology exemplified by the packaging RNA (pRNA) of bacteriophage phi29 DNA packaging motor has provided a different prospect in nanoparticle designs. Of note, there is a robust three-way junction (3WJ) motif in pRNA which can serve as an adaptable scaffold to construct thermodynamically stable 2D planar and 3D globular RNA architectures with tunable shapes and sizes, and harboring various targeting, therapeutic, and imaging modules. This chapter focuses on the methods for constructing pRNA-3WJ based nanoparticles with controllable sizes and shapes, and assessment of their biodistribution profiles in cancer mouse models after systemic injection and ocular mouse models following subconjunctival injection.

Key words Nanobiotechnology, RNA nanoparticle, 3WJ, pRNA, Subcutaneous and orthotopic xenograft, Ocular delivery, Subconjunctival injection

1 Introduction

Nature has evolved smart building blocks for constructing sophisticated higher-ordered structures with unique functionalities. One of the burgeoning areas of nanobiotechnology is to exploit these biological building blocks for engineering nanosystems with de novo functionalities in order to probe and reprogram cells for therapeutic intervention. Lipid-based nanoparticles have been extensively

Authors "Farzin Haque" and "Congcong Xu" have contributed equally to this work.

Eckart Bindewald and Bruce A. Shapiro (eds.), *RNA Nanostructures: Methods and Protocols*, Methods in Molecular Biology, vol. 1632, DOI 10.1007/978-1-4939-7138-1_23, © Springer Science+Business Media LLC 2017

studied in the nanomedicine field due to their intrinsic structural simplicity and ease of encapsulating a variety of cargoes [1, 2]. However, they generally lack thermodynamic stability and often display off-target effects. Peptide and protein based scaffolds have propensity to aggregate, and lack scaffold addressability and structural fidelity, which limits their overall utility to generate programmable nanostructures [3]. Nucleic acids on the other hand are structurally highly programmable, and recent advances in nucleic acid chemistry have enabled the production of DNA and RNA nanoparticles with improved chemical and biological stability under physiological conditions [4, 5]. Rational design and assembly of DNA nanostructures was first conceived by Seeman in 1983 [6] and since then various applications of DNA nanostructures particularly in materials sciences have emerged [7, 8]. The principle is based on a simple set of construction rules relying on canonical Watson--Crick interactions (A–T and G–C) to generate programmable DNA nanostructures. As an alternative to DNA, RNA has emerged as a more powerful building block for fabricating nanoparticles due to its diversity in structure and function, and high thermodynamic stability [5].

RNA nanotechnology refers to the study of nanoscale RNA architectures with its major frame composed mainly of RNA [5]. Scaffold, targeting ligands, therapeutic modules and regulators can be all composed of RNA nucleotides. Different from classical studies on RNA structure and function focused on intramolecular interactions and 2D/3D structures, current exploration of RNA in nanotechnology field pays close attention to intermolecular interactions and quaternary structures. RNA nanostructure formation is much more complex relying on both canonical and noncanonical (such as G–U) base pairing, base stacking, and tertiary interactions [9–12]. Furthermore, RNA has special functionalities, such as ribozyme [13], riboswitch [14], siRNA [15], miRNA [16], and RNA aptamers [17], all of which can be seamlessly incorporated into ultrastable RNA nanoparticle scaffolds via sequence fusion for therapeutic intervention in vivo [5, 18–20].

The feasibility of RNA nanotechnology was first demonstrated in 1998 by Peixuan Guo using reengineered pRNA molecules [21] derived from the bacteriophage phi29 DNA packaging motor to form dimer, trimer, and hexamer complexes [22]. The pRNA in its native state forms a hexameric ring to gear one of the most powerful nanomotors known to date. The pRNA monomer has several structural features suitable for constructing multifunctional RNA nanoparticles with defined size, shape, and stoichiometry. In particular, there is an ultrastable three-way junction (3WJ) motif [23, 24] which can be assembled from three short fragments with very high efficiency in the absence of any metal ions. The pRNA-3WJ nanoparticles are homogeneous in size and structure; are thermodynamically [23, 25, 26] and chemically (after 2′-F modifications) stable [23, 27–29]; are resistant to denaturation in 8 M urea and remain intact at ultralow concentrations [23, 27, 28]; retain authentic folding and independent functionalities of all incorporated modules (RNA aptamer, siRNA, miRNA, or ribozyme) [23, 30]; are nontoxic and biocompatible [31, 32]; are highly soluble; and exhibit favorable biodistribution and pharmacological profiles [23, 27, 28, 31–37]. We have solved the crystal structure [24] of the pRNA-3WJ, thus facilitating the design of multifunctional RNA nanoparticles with tunable size, shape, and stoichiometry.

In construction of nanoparticles, the rigidity of scaffolds is crucial to keep the nanoassemblies well-structured with defined stoichiometry, while the flexibility of angle in each module makes it feasible to form complex structures with controllable sizes and shapes. This unique property of a versatile building block is exemplified by the pRNA-3WJ motif, which adopts an A-type geometry consisting of three helices (H1, H2, and H3) with a planar orientation (Fig. 1A, B). The helices H1 and H3 are coaxially stacked, and form a ~180° angle, while the H1 and H2 helices form a 60° angle

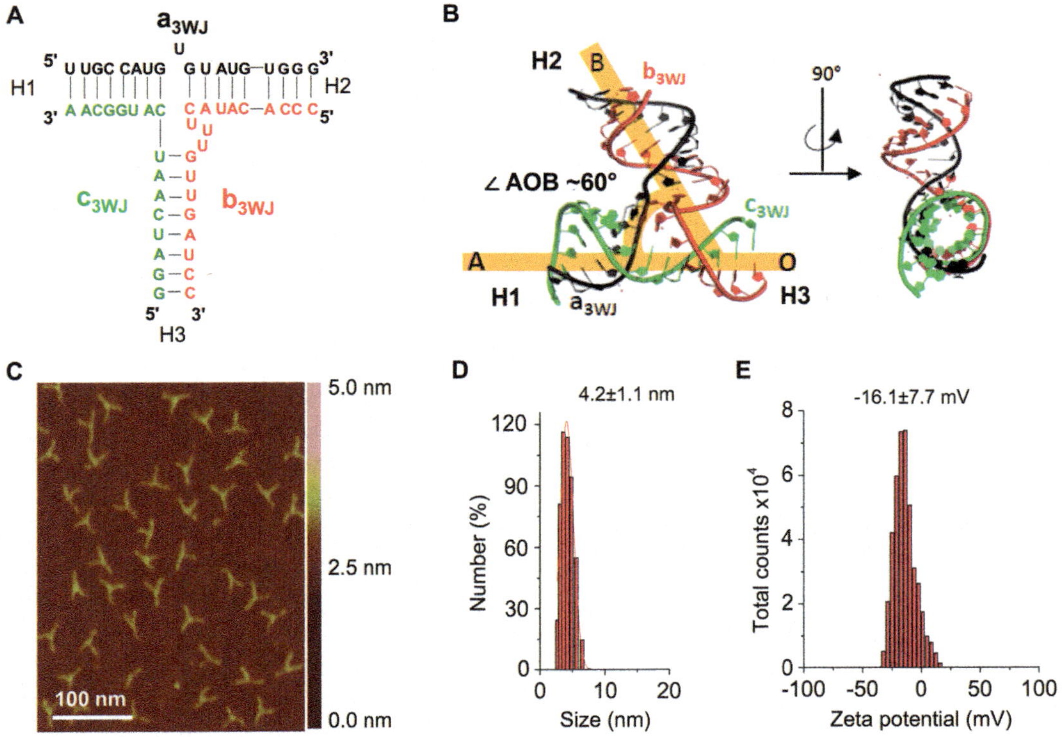

Fig. 1 (**A**) Sequence and (**B**) crystal structure of pRNA-3WJ motif. The 3WJ is composed of three RNA oligomers in *black* (a_{3WJ}), *red* (b_{3WJ}), and *green* (c_{3WJ}). Helical segments are represented as *H1*, *H2*, and *H3*. The internal ∠AOB angle used to construct polygons is shown. (**C**) AFM image of the extended pRNA-3WJ nanoparticles. Dynamic light scattering assay for characterizing the (**D**) size and (**E**) zeta potential of pRNA-3WJ core. Figure reproduced with permission from: (**A**) ref. 23 © Nature Publishing Group; (**B**) ref. 38 © Oxford University Press; (**C–E**) ref. 33 © American Chemical Society

[24]. The 3WJ scaffold can be extended to form branched architectures such as X-motif [27] or dendrimer-like 3D globular structures [39] by simple extension of the helical arms. The 3WJ can also serve as a building block to form 2D planar polygonal structures by simply stretching the intrahelical H1-H2 angle to form a triangle (∠60°), square (∠90°) and pentagon (∠108°) pRNA nanoparticles. Each vertex of the polygon contains a 3WJ, which are then linked by double stranded stem regions to form a closed polygon. More recently, we extended the 2D work to design and construct 3D RNA nanoparticles, such as a tetrahedron [37] and prism [40] architectures. Thus, the tunable features of 3WJ structure make it an attractive and unique building unit to construct defined RNA architectures with controllable shapes and sizes.

Physicochemical characteristics of nanoparticles such as size, shape, and surface charge have significant implications in their cellular uptake and in vivo fate after intravenous administration. Generally speaking, nanoparticles with a hydrodynamic diameter of <10 nm are rapidly filtered by the kidneys, while nanoparticles >100 nm are entrapped by the liver Kupffer cells and lung/liver/spleen macrophages. The optimal size range is therefore in the 10–100 nm range to ensure that nanoparticles can enter cells via receptor-mediated endocytosis while avoiding entrapment and rapid excretion. The effect of size and shape on cell entry and trafficking as well as in vivo behavior of nanoparticles is difficult to assess as finding suitable nanoparticles rich in structure while varying in shape and size is challenging. The pRNA-derived nanoparticles with different shape, size, stoichiometry, targeting ligand, and functionalities coupled with their flexibility, versatility, and programmability provide opportunities for these studies. This is exemplified by our recent studies showing that RNA nanoparticles of different size displayed different accumulation in cancer and healthy organs, and RNA nanoparticles with comparable size but different shapes enter different cell types: (1) In cancer xenograft and metastatic mouse models, after systemic injection, pRNA-3WJ nanoparticles accumulated in the tumor microenvironment via EPR (Enhanced Permeability and Retention) effects, strongly bound to tumors and entered the cancer cells via receptor-mediated endocytosis, and was cleared out of vital healthy organs within 8 hours post-administration; (2) Two fluorescent RNA nanoparticles were tested for penetration into ocular tissues through subconjunctival injection into mice [41]. Fluorescent imaging of eye tissue sections revealed that pRNA-3WJ RNA nanoparticles [23] entered the cornea, but not the retina, while the X-shaped pRNA nanoparticles [27] with comparable size as 3WJ, but with different shape entered both the cornea and the retina [41]; (3) In mouse models, triangular shaped RNA nanoparticles with CpG modules have different immunostimulatory effects compared to square and pentagon shaped RNA nanoparticles [38]. These three key findings indicated that the size and shape of RNA nanoparticles will have a profound impact on targeting of specific cells types. Moreover, the overall negative charge of RNA nanoparticles leads to extensive hydration, and minimizes nonspecific interactions with negatively charged cell membranes and also avoids interactions with serum proteins.

In this chapter, we focus on methods for constructing pRNA nanoparticles of various shapes and sizes using the 3WJ motif as a core scaffold. The physicochemical properties of the RNA nanoparticle can be controlled by the nanoparticle designer from particle inception to completion, due to the bottom-up assembly approach and the numerous toolkits and design strategies that have been developed for RNA nanoparticle construction. These principles are illustrated using the pRNA-3WJ core, pRNA-X motif, 2D square, and 3D tetrahedral RNA nanoparticles. The assembled nanoparticles are then evaluated in cancer and ocular mouse models to demonstrate how the size and shape of RNA nanoparticles influence their biodistribution profiles.

2 Materials

All the reagents must be analytical grade and RNase/DNase free to prevent any nucleic acid degradation. All the solutions and buffers should be prepared using Millipore water (resistivity 18.2 MΩ/cm at 25 °C) treated with diethylpyrocarbonate (DEPC) and then autoclaved. All tubes and glassware used in preparation of buffers and reagents should be autoclaved. Gloves and lab coats should be worn at all times. Prepare and store reagents at room temperature unless otherwise noted.

2.1 Reagents

1. Ammonium persulfate (AP).
2. Tetramethylethylenediamine (TEMED) (*see* **Note 1**).
3. Ethidium bromide (EtBr): 1% Solution, Molecular Grade (*see* **Note 2**).
4. Magnesium chloride hexahydrate ($MgCl_2 \cdot 6H_2O$) (*see* **Note 3**).
5. Urea, for molecular biology, DNase−/RNase−/ and protease-free.
6. Tris–HCl.
7. Sodium chloride (NaCl).
8. Potassium chloride (KCl).
9. Disodium hydrogen phosphate (Na_2HPO_4).
10. Monopotassium phosphate (KH_2PO_4).
11. Ethanol.

12. Boric acid.
13. Bis-acrylamide (*see* **Note 4**).
14. Acrylamide, 99%, electrophoresis grade (*see* **Note 5**).
15. Agarose.
16. Xylene cyanol.
17. Bromophenol blue.
18. Manganese chloride ($MnCl_2$).
19. Spermidine.
20. Dithiothreitol (DTT).
21. Sodium acetate.
22. RNA nucleotide triphosphates (rNTP).
23. Ethylenediaminetetraacetic acid (EDTA).
24. Male/female athymic nude$^{nu/nu}$ (6–8 weeks old) mice, Taconic.
25. Male/female (7–10 weeks old) C57BL/6 mice, Charles River Laboratories.
26. Isoflurane.
27. Buprenorphine.
28. Xylazine.
29. Ketamine.
30. 16% w/v paraformaldehyde solution, Polysciences.
31. Tissue-Tek O.C.T Compound, Andwin Scientific.
32. Hoechst 33342.
33. ProLong Gold Antifade Mountant with DAPI, Life Technologies, Inc.
34. Aqueous mounting medium, Dako.
35. Glass cover slides.
36. Microscope slides.
37. 10% povidone–iodine, Betadine.
38. Phosphoramidites, Glen Research.
39. DNA and RNA Oligo Synthesizer, Azco Biotech.
40. DNA primers, dissolved in water at 100 μM stock concentration, IDT.
41. PCR reaction kit with GoTaq FlexiDNA Polymerase, Promega.
42. Fluorine-modified UTP and CTP at 2′ hydroxyl ribose (2′F-UTP and 2′F-CTP), Trilink.
43. Fetal bovine serum (FBS).

44. Wild type or Y693F mutant T7 RNA polymerase: expressed in-house or acquired commercially, Epibio.
45. MasterCycler Gradient Thermal Cycler Eppendorf 5331, Eppendorf.
46. Typhoon FLA 7000 imaging system, GE Healthcare.
47. Temperature gradient gel electrophoresis (TGGE) system, Biometra.
48. Hand-held Mineralight lamp UVGL-25, UVP.
49. Eppendorf 5424 Centrifuge, Eppendorf.
50. Agarose gel electrophoresis system, mini-sub cell GT system, Bio-Rad.
51. Savant DNA120 SpeedVac concentrator, Thermo Electron Corporation.
52. ISOTEMP 115 Water Bath 15–460-15, Fisher Scientific.
53. NanoDrop 2000 spectrophotometer, Thermo Scientific.
54. Laser scanning confocal microscope, Olympus or Zeis.
55. IVIS spectrum imaging system, PerkinElmer.
56. Atomic force microscope (AFM), Veeco.
57. Dynamic light scattering (DLS), Malvern.
58. High Pressure Liquid Chromatography (HPLC), Agilent.
59. Confocal microscopy, Olympus or Zeis.
60. Alexa Fluor 488 phalloidin (Invitrogen).
61. TO-PRO-3 iodide (642/661) (Invitrogen).
62. Image processing software: Living Image 3.1 software (Caliper Life Sciences) or equivalent.

2.2 Reagent Preparation

1. RNA oligos: Chemically synthesize 2′-F modified RNA oligos (less than 60 nucleotides in length) with or without end-modifications, such as Folate or Alexa 647, using an Oligo synthesizer or custom order from Trilink; Prepare RNA oligos longer than 60 nucleotides by in vitro transcription.
2. DNA primers: Chemically synthesize DNA oligos using an Oligo synthesizer or custom order from *IDT*. Dissolve in DEPC H_2O at desired concentration.
3. DEPC aqueous solution, 0.05% (vol/vol): Add 0.05 ml of DEPC to 99.5 ml water and shake the solution vigorously. Incubate the solution overnight at 37 °C. Autoclave to remove DEPC. This reagent can be stored at room temperature (25 °C) for 1 year.
4. 1× Tris–borate–EDTA (TBE) buffer: 89 mM Tris base, pH 8.0, 86 mM boric acid, and 2 mM EDTA. This buffer can be stored at room temperature for 1 year.

5. 1× Tris–borate–magnesium (TBM) buffer: 89 mM Tris base, pH 8.0, 86 mM boric acid, and 5 mM $MgCl_2$. This buffer can be stored at room temperature for 1 year.
6. 1× Tris–acetate–EDTA (TAE) buffer: 40 mM Tris base, pH 8.0, 20 mM acetic acid, and 1 mM EDTA. This buffer can be stored at room temperature for 1 year.
7. 1× Tris–magnesium–saline (TMS) buffer: 50 mM Tris–HCl (pH 8.0), 100 mM NaCl, and 10 mM $MgCl_2$. Autoclave the solution before use. This buffer can be stored at room temperature for 1 year.
8. Sodium acetate (NaOAc, 3 M, pH 6.5): Autoclave the solution before use. This buffer can be stored at room temperature for 1 year.
9. Magnesium chloride ($MgCl_2{\cdot}6H_2O$, 2 M): This buffer can be stored at room temperature for 1 year.
10. APS: 10% (wt/vol). This reagent can be stored at 4 °C for 1 week.
11. Urea-denaturing PAGE gel, 10–15% (wt/vol): 10–15% (wt/vol; 37.5:1) acrylamide, 8 M urea, 10% (wt/vol) APS and TEMED. The gel needs to be freshly prepared.
12. Native PAGE gel (TBM or TBE): 10–15% (wt/vol): 10–15% (wt/vol; 37.5:1) acrylamide, 1× Tris–borate buffer (pH 7.8), 10 mM $MgCl_2$ (or 2 mM EDTA), 10% (wt/vol) APS and TEMED. The gel needs to be freshly prepared.
13. 1× RNA Elution buffer: 0.5 M ammonium acetate, 10 mM EDTA and 0.1% (wt/vol) SDS in 0.05% (vol/vol) DEPC-treated water. Autoclave the solution before use. This buffer can be stored at room temperature for 6 months.
14. Washing buffer (ethanol 70%): This buffer can be stored at 4 °C for 1 year.
15. 6× Native gel loading buffer: 40% (wt/vol) sucrose, 0.1% (wt/vol) xylene cyanol, and 0.1% (wt/vol) bromophenol blue. This buffer can be stored at −20 °C for 1 year.
16. 1× PBS buffer: 137 mM NaCl, 2.7 mM KCl, 10 mM Na_2HPO_4, and 2 mM KH_2PO_4 (pH 7.4). Autoclave the solution before use. This buffer can be stored at room temperature for 1 year.
17. Tissue fixation buffer: 4% Paraformaldehyde and 10% sucrose in 1× PBS buffer. Use immediately or store at 4 °C up to 1 week.
18. RNA nucleotide triphosphates (rNTP) 100 mM, adjust pH to 7.5 with NaOH. This reagent can be stored at −20 °C for 1 year.

19. 1× 2′-F transcription buffer: 40 mM Tris–acetate, pH 8.0, 0.5 mM DTT, 1 mM EDTA, 10 mM $Mg(OAc)_2$, 0.5 mM $MnCl_2$, 16 mM spermidine. This buffer can be stored at −20 °C for 1 year.
20. 1× T7 RNA polymerase transcription buffer: 40 mM HEPES–KOH, pH 7.5, 24 mM $MgCl_2$, 2 mM spermidine, 2 mM, 40 mM DTT. This buffer can be stored at −20 °C for 1 year.

3 Methods

General procedures such as PCR amplification, in vitro RNA transcription, solid-phase RNA synthesis, and RNA nanoparticle assembly have been discussed in detail previously [42–46].

3.1 Assembly of pRNA-3WJ Core

The 3WJ core is composed of three short strands a_{3wj}, b_{3wj}, and c_{3wj} (Fig. 1A, B). The three helical branches are denoted H1, H2, and H3, respectively. The 3WJ serves as the core scaffold for constructing RNA squares and RNA tetrahedrons outlined here.

1. Chemically synthesize three RNA oligo components of the pRNA-3WJ: a_{3wj}, b_{3wj}, and c_{3wj}.
2. Mix the three strands in equimolar ratio in DEPC-treated water or TBE/TMS buffer. No metal ions (such as NaCl or $MgCl_2$) or annealing procedure (heating to 95 °C and slowly cooling to room temperature) is necessary since the assembly occurs almost spontaneously in water.
3. Load the assembled complex(3WJ-a/b/c) and control strands (a_{3wj}, b_{3wj}, c_{3wj}, a_{3wj}/b_{3wj}, b_{3wj}/c_{3wj}, a_{3wj}/c_{3wj}) onto an 15% (wt/vol) native PAGE gel and run the gel in TBM or TBE buffer at 100 Vand 4 °C for around 2 h. Visualize the RNA by ethidium bromide staining using Typhoon FLA 7000. Since the 3WJ assembles with high efficiency, only one band should be observed and it will run slightly ahead of xylene cyanol.
4. Characterize the structure of 3WJ complex by atomic force microscope (*see* **Note 6**) (Fig. 1C); and dynamic light scattering (*see* **Note 7**) for measuring the hydrodynamic size and zeta potential (Fig. 1D, E).

3.2 Assembly of pRNA-3WJ with Functionalities and Extension to Form pRNA-X Scaffold

Folate or RNA aptamer can be used as targeting ligand; Alexa 647 or other fluorophores, especially near infrared dyes, can be used as imaging module.

1. Chemically synthesize three RNA oligo components of the pRNA-3WJ-Folate-Alexa647 nanoparticle with end-modifications: a_{3wj}-Folate, b_{3wj}, and c_{3wj}-Alexa 647.

2. Mix the three strands in equimolar ratio in DEPC-treated water or TBE/TMS buffer at room temperature.
3. Load the assembled complex and control strands onto an 15% (wt/vol) native PAGE gel and run the gel in TBM or TBE buffer at 100 V and 4 °C for around 2 h. Visualize the RNA by ethidium bromide staining or Alexa 647 fluorescence using Typhoon FLA 7000. If necessary, purify the complex from the gel and elute the RNA using the elution buffer for 4 h at 37 °C, followed by ethanol precipitation overnight.
4. Characterize the structure of the 3WJ complex by atomic force microscope (*see* **Note 6**); and dynamic light scattering (*see* **Note 7**) for measuring the hydrodynamic size and zeta potential.
5. The X-shaped motif builds on the 3WJ core by branch extension and consists of four strands, denoted a, b, c, and d (Fig. 2A). The four helical regions are denoted H1, H2, H3,

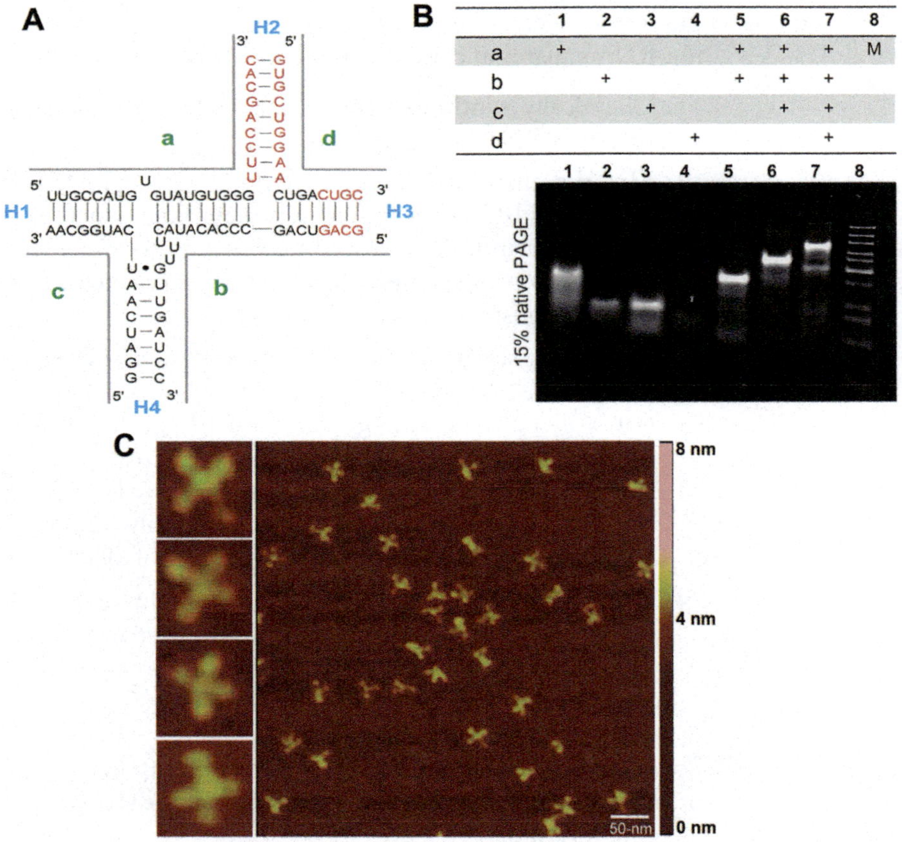

Fig. 2 (**A**) Sequence of pRNA-X motif. The nucleotides in *red* indicate new sequences that were added to the original pRNA sequences in *black* [27]. The core is composed of four RNA oligos (denoted *a*, *b*, *c*, and *d*). Helical segments are represented as *H1*, *H2*, *H3*, and *H4*. (**B**) 15% native PAGE showing the stepwise assembly of the pRNA-X core from component strands. In the table '+' indicates the presence of the strand in the complex. (**C**) AFM images of pRNA-X constructs harboring four pRNA monomer units at the branch ends. Figure reproduced with permission from ref. 27 © Elsevier

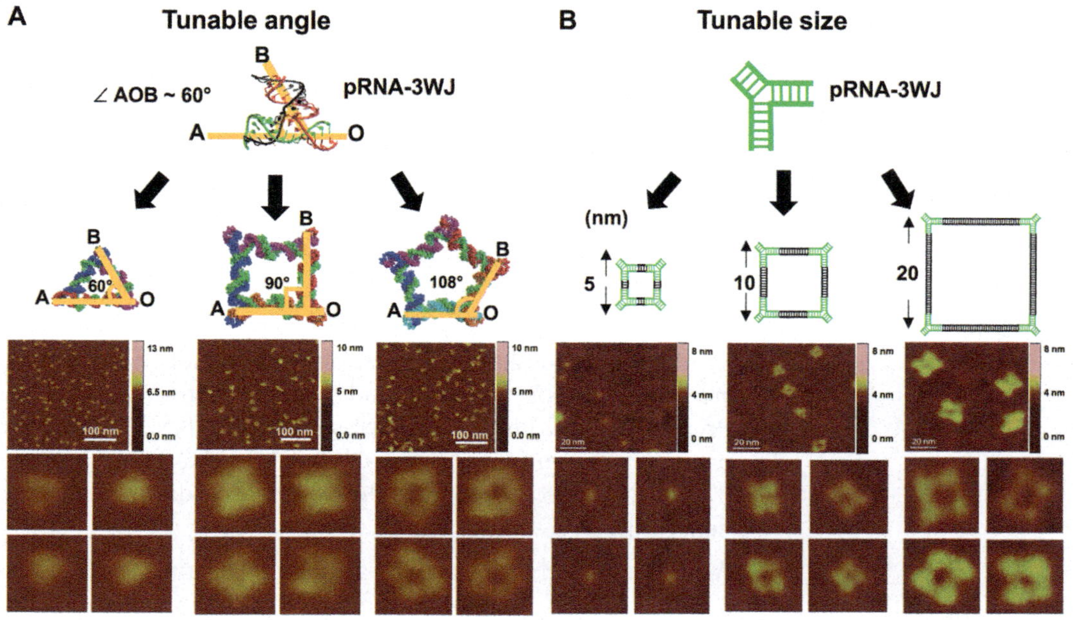

Fig. 3 Design and characterization of RNA nanoparticles with tunable shape and size. (**A**) AFM images of *triangle* (∠AOB 60°), *square* (∠AOB 90°) and *pentagon* (∠AOB 108°) nanoparticles derived from the pRNA-3WJ motif. (**B**) AFM images of small (5 nm), medium (10 nm), and large (20 nm) RNA squares derived from pRNA-3WJ. Figure reproduced with permission from: (**A**) ref. 38 © Oxford University Press; (**B**) ref. 48 © American Chemical Society

and H4. Synthesize the strands chemically and then proceed with the assembly (Fig. 2B), purification, and characterization (Fig. 2C) steps outlined in **steps 1–4** above.

3.3 Construction of 2D Planar RNA Squares with Same Shape but Variable Size Using 3WJ as Modules

Previously it has been shown that the intrahelical H1–H2 angle of 3WJ can be stretched to form 2D planar triangle (∠60°), square (∠90°) and pentagon (∠108°) pRNA nanoparticles [38, 47] (Fig. 3a). Herein, the RNA square [48] is used as an example for constructing RNA nanoparticles of various sizes (Fig. 3b).

1. Design RNA nanosquares using the Swiss PDB viewer. Open four copies of the pRNA-3WJ crystal structure (PDB ID: 4KZ2) and four copies of 8 bp RNA duplex, which will act as a connecting duplex between H1 and H2 of each 3WJ. Align each 3WJ in a planar fashion in square configuration and insert RNA duplex in-between H1 and H2 of each pRNA-3WJ.
2. Extend each corner of the square, H3 of the 3WJ, by five base pairs to increase thermodynamic stability and strand diversity (*see* **Note 8**). To construct the square, use a total of five strands: Four short strands (42 nt) that make up each edge of the square, and one long strand (88 nt) that makes up the middle core strand, connecting each of the strands. In this form, the edges are about 10 nm in length from arm to arm.

3. To construct different size of RNA squares, simply decrease or increase the number of base pairs that connect each 3WJ at the vertices. To make a small square decrease each edge by 11 bp in length. This will result in external strands of 31 nt in length and an internal strand with length of 44 nt. The resulting square will be about 5 nm in length from arm to arm on each edge. Conversely, to design a square of about 20 nm along each edge, increase each edge by 22 nt, two turns of RNA duplex. This results in external strands of 66 nt in length and an internal strand with a length of 176 nt (*see* **Note 9**).
4. Synthesize the designed RNA strands chemically if the strands are <60 nt. For longer strands, use in vitro transcription. Prepare the DNA templates including the T7 RNA polymerase promoter region (sequence: 5′-TAA TAC GAC TCA CTA TA-3′) by PCR (*see* **Note 10**). Purify the strands by HPLC or denaturing PAGE.
5. Assemble the RNA squares by annealing (heating at 95 °C for 5 min followed by slow cooling to 4 °C) the five strands in equimolar ratio in 1× TMS buffer.
6. Perform a stepwise assembly process (as in Fig. 2B) to determine the migration rate of the final square construct. Mix one additional strand for each sample and load onto an 8% native PAGE gel. Run the gel for 100 min at 100 V constant current at 4 °C in TBM buffer to preserve native conditions. Use total RNA staining with ethidium bromide to visualize RNA bands. Observe a decrease in migration rate going from monomer to dimer, to trimer, to tetramer, and finally to square nanoparticle. Characterize the RNA squares of different sizes using AFM imaging (Fig. 3) and DLS (as in Fig. 2D, E).

3.4 Construction of pRNA-3WJ Based 3D RNA Tetrahedrons of Different Sizes

1. Design the RNA tetrahedrons using UCSF Chimera and Swiss PDB viewer. Open the PDB file of the pRNA-3WJ motif (PDB ID: 4KZ2). Since the 3D conformation of the pRNA-3WJ motif is flexible, use UCSF Chimera to manually change the angles among the three-branched RNA double helices in order to fit the structure into the corners of the designed RNA tetrahedron. Then use Swiss PDB Viewer to align four pRNA-3WJ motifs and six RNA double helices to construct the computational model of the designed RNA tetrahedron. In the final form, the RNA tetrahedron will be composed of a total of four strands with a pRNA-3WJ motif positioned in the corners and RNA double helices positioned along the edges of the tetrahedron (Fig. 4A). Optimize the sequences using *mfold* [49] (*see* **Note 9**). For incorporating targeting ligands, such as RNA aptamer, simply incorporate the sequences at the 3′-end of the core strands, as illustrated with EGFR targeting RNA aptamers [37] (Fig. 4B).

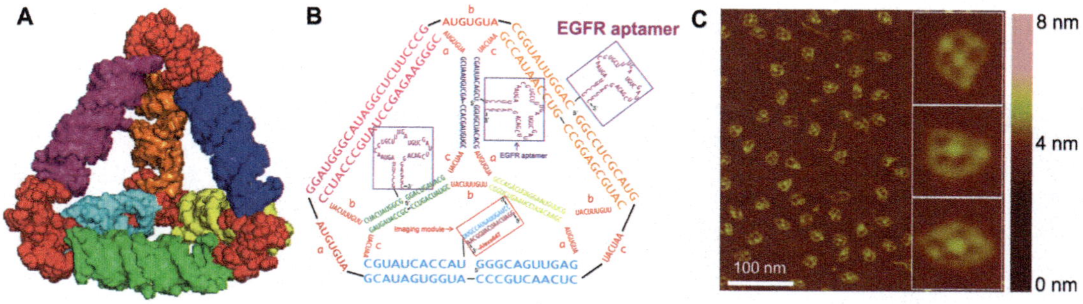

Fig. 4 (**A**) 3D computational model of RNA tetrahedrons. (**B**) Sequence of RNA tetrahedrons harboring imaging module and EGFR targeting RNA aptamers. (**C**) AFM imaging of RNA tetrahedrons. Figure reproduced with permission from ref. 37 © John Wiley & Sons, Inc.

2. For designing smaller or larger RNA tetrahedrons, simply decrease or increase every edge of the tetrahedron by 11 bp (one helical turn), respectively.
3. After finalizing the designs, synthesize the four RNA strands by in vitro transcription. Prepare the DNA templates including the T7 RNA polymerase promoter region by PCR. Purify the strands by 8 M urea, 8% denaturing PAGE.
4. For generating RNA tetrahedrons resistant to serum degradation, during transcription, use Y693F mutant T7 polymerase and 2′-F modified cytosine (C) and uracil (U) nucleotides in the sequence.
5. Assemble the RNA tetrahedrons by mixing the four strands in equimolar concentrations in 1× Tris buffer (100 mM NaCl, 50 mM Tris, pH 8.0), and heated to 95 °C for 5 min and slowly cooled to 4 °C over 45 min by using a thermal cycler.
6. Observe the stepwise assembly of the RNA tetrahedrons by running the RNA tetrahedrons and assemble intermediates in native PAGE gels run in 1× TBM buffer at 90 V, 4 °C. Visualize the RNA after ethidium bromide staining by Typhoon FLA 7000.
7. Characterize the RNA tetrahedrons of different sizes using AFM imaging (Fig. 4C) and DLS assay.

3.5 In Vitro Cell Binding and Internalization of pRNA-3WJ Based Nanoparticles Assayed by Confocal Microscopy

1. For assaying folate-labeled RNA nanoparticles, use folate receptor positive cells, such as KB cells. Maintain the cells in complete culture medium (RPMI-1640 with 10% fetal bovine serum).
2. One day before the binding study, trypsinize the cells using 0.25% Trypsin-EDTA and seed 1×10^5 cells in 24-well plate with clean glass cover slide in each well using RPMI-1640 complete culture medium containing no folate (*see* **Note 11**).

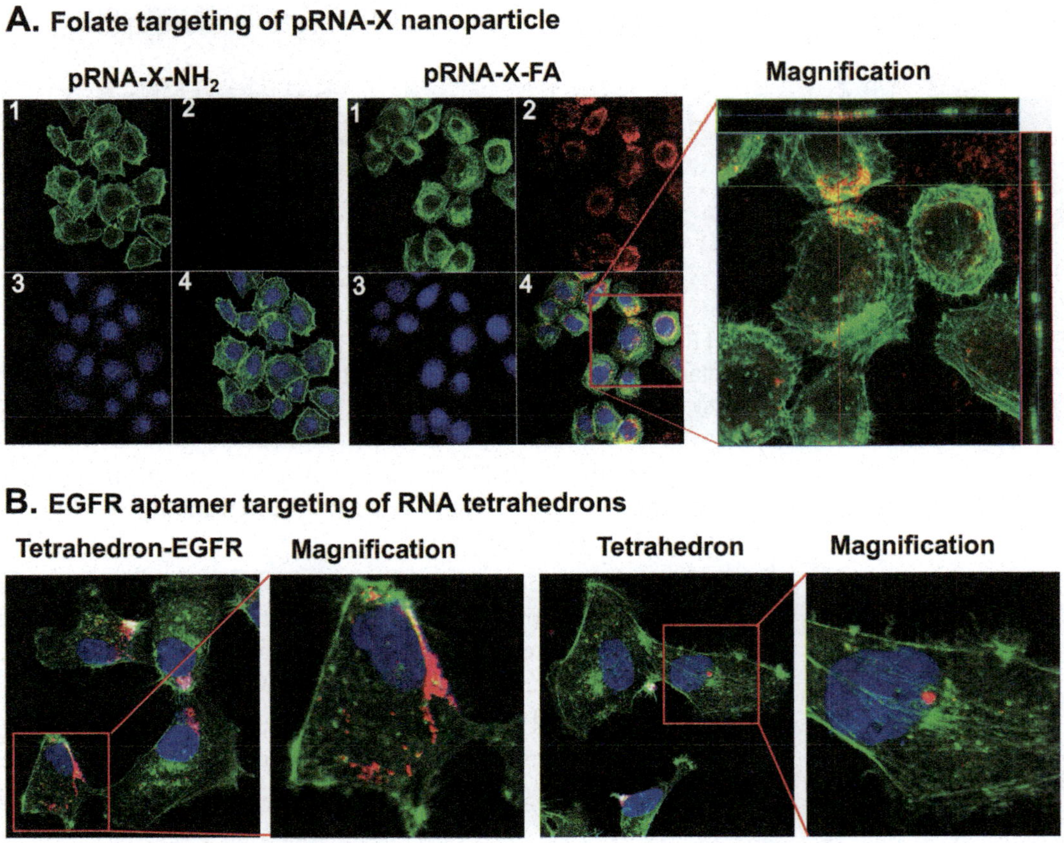

Fig. 5 Confocal microscopy images showing targeting of (**A**) pRNA-X nanoparticles harboring folate (or amine as control) to folate receptor positive KB cells and (**B**) RNA tetrahedrons (with and without EGFR aptamers) to MDA-MB-231 cells. In the images, nucleus (*blue*), cytoplasm (*green*), and RNA nanoparticles (*red*) are shown. In (**B**), only the overlapped images are shown. Figure reproduced with permission from: (**A**) ref. 27 © Elsevier; (**B**) ref. 37 © John Wiley & Sons, Inc.

3. Once the cell attaches to the glass slide, rinse the cells with blank RPMI-1640 medium containing no folate.
4. Incubate cells with 200–400 nM of pRNA-X-Folate-Alexa647 RNA nanoparticles (or with control RNA nanoparticles without folate) at 37 °C for 1–4 h. Similar assays can be done using pRNA-3WJ-Folate-Alexa647 nanoparticles.
5. After washing with PBS, fix the cells with 4% paraformaldehyde and stain by Alexa Fluor 488 phalloidin for the cytoskeleton and TO-PRO-3 iodide (642/661) for the nucleus. Assess cell binding and internalization with a laser scanning confocal microscope (Fig. 5A).
6. For assaying the EGFR aptamer mediated targeting in vitro, grow EGFR-positive cells, such as MDA-MB-231 cells on glass cover slides in 24-well plates in DME–F-12 (1:1) medium with 10% FBS at 37 °C in humidified air containing 5% CO_2

overnight. Dilute RNA tetrahedrons harboring EGFR aptamers (or controls) in Opti-MEM medium to 100 nM and incubate with the cells for 1 h at 37 °C. Once the cell attaches to the glass slide, rinse the cells with blank Opti-MEM medium and conduct step [5] (Fig. 5b).

3.6 Biodistribution Assessment of pRNA-3WJ, pRNA-X, and RNA Tetrahedral Nanoparticles in Cancer Xenograft Mouse Models After Systemic Injection

1. Obtain male/female athymic nude$^{nu/nu}$ (6–8 weeks old) mice. Keep the mice in clean, pathogen-free rooms in an environment with controlled temperature (27 °C), humidity, and a 12 h. light–dark cycle. Feed the mice with a folate-free diet (Harlan Laboratories; Indianapolis, IN) for at least 2 weeks before the experiment. Make sure that a proper protocol regarding animal experiments is in place at the institution to conduct the animal studies.
2. Inject the mouse subcutaneously with KB cells (3×10^6 cells in 100 μl PBS). Pinch the skin of the mouse with forceps and pull the skin away from the body of the mouse. Inject slowly and evenly into the pouch created by pulling the skin. Try to create a single bubble of cells beneath the skin (*see* **Note 12**). Gently remove the needle and hold the injection site with forceps for 10 s to prevent retrograde cell suspension flow.
3. Monitor the tumor growth for a few weeks using calipers until the tumors are about 500 mm^3.
4. Anesthetize the mice using isoflurane gas (2% in oxygen at 0.6 L/min flow rate).
5. Intravenously inject through the tail vein with a single dose of 3 μg 2′-F U/C modified pRNA-3WJ-Folate-Alexa647 (or pRNA-X-Folate-Alexa647) per g of mouse body weight in 300 μL of PBS.
6. Conduct whole-body imaging (excitation: 650 nm; emission: 668 nm) at 1, 4 and 8 h after RNA nanoparticle administration on an IVIS Spectrum station (Fig. 6a, b).
7. Obtain composite images comprising of black and white digital photos with an overlay of images reflecting fluorescent activity. Generate a density map (photons/second/cm^2/steradian; p/s/cm^2/sr) using the image processing software and represent it as a color gradient centered at the maximal spot.
8. Following CO_2 asphyxiation at 8 h time point, remove the tumor, liver, heart, lung and kidney of the mice for ex vivo imaging of the organs (Fig. 6A, B).
9. To generate orthotopic Triple Negative Breast Cancer xenograft tumors, inject 2×10^6 MDA-MB-231 cells suspended in PBS directly into the mammary fat pad of the mice. Typically, the xenograft tumors will be apparent within 2 weeks after the implantation. Follow **steps 4–8** to assay RNA tetrahedrons harboring EGFR targeting aptamers (Fig. 6C).

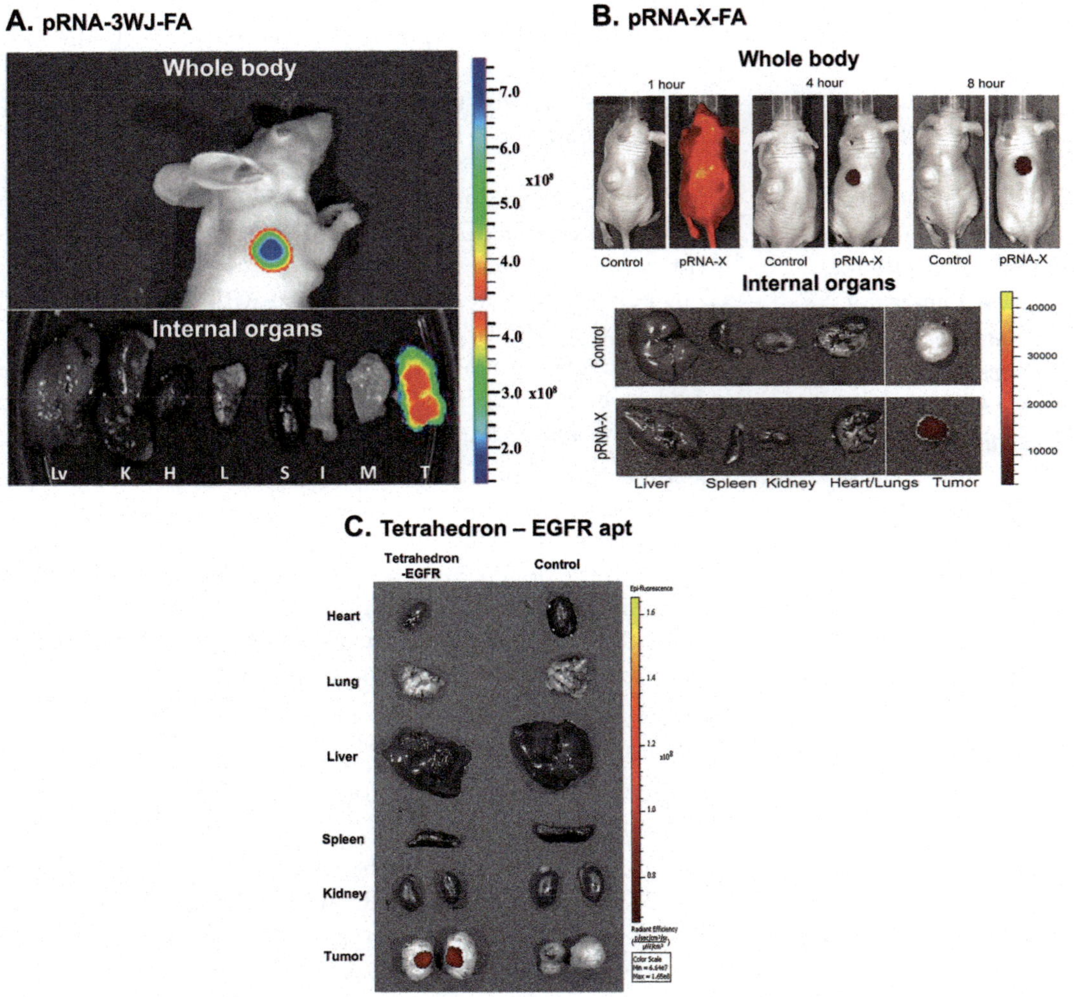

Fig. 6 Biodistribution assay showing targeting of (**A**) pRNA-3WJ-Folate-Alexa647 and (**B**) pRNA-X-Folate-Alexa647 nanoparticle to Folate receptor(+) subcutaneous tumor xenografts after systemic injection in nude mice. *Upper panel*: whole body; *lower panel*: organ imaging (*Lv* liver, *K* kidney, *H* heart, *L* lung, *S* spleen, *I* intestine, *M* muscle, *T* tumor). (**C**) Internal organ images showing the targeting of RNA tetrahedrons harboring EGFR aptamers to orthotopic MDA-MB-231 breast tumors after tail vein injection. Figure reproduced with permission from: (**A**) ref. 23 © Nature Publishing Group; (**B**) ref. 27 © Elsevier; (**C**) ref. 37 © John Wiley & Sons, Inc.

3.7 Assessment of Ocular Biodistribution of pRNA-3WJ Based Nanoparticles After Subconjunctival Injection in Mouse Models

1. Obtain 7–10 weeks old C57BL/6 male/female mice. Keep the mice in clean, pathogen-free rooms in an environment with controlled temperature (27 °C), humidity, and a 12 h light–dark cycle.
2. Anesthetize the mouse by intraperitoneal injection of 10 mg/kg xylazine and 80 mg/kg ketamine.
3. Gently pull the conjunctiva from the sclera with a pair of forceps. Inject 10 μL of pRNA-3WJ, pRNA-X, dsRNA, or

Alexa 647 dye into the superior subconjunctival region using a microsyringe with a 33 G needle.

4. Perform whole-body fluorescence imaging after subconjunctival injection (*see* **Note 13**) (Fig. 7A). Keep the mice anesthetized with 2% isoflurane during the imaging. Use the following imaging parameters: 630 nm (excitation) and 700 nm (emission), field of view (FOV) of 120 mm, f-stop of 2.51, 4 × 4 pixel binning, and exposure time of 1 s. For higher resolution, use the same parameters except with FOV of 22 mm, 1 × 1 pixel binning, and exposure time of 5 s.
5. For anatomical reference, perform X-ray (exposure time of 30 s) and surface optical imaging (exposure time of 0.175 s) with the same FOV as the fluorescence imaging.
6. To monitor ocular clearance of the pRNA-3WJ, pRNA-X nanoparticles, dsRNA (control), and Alexa 647 (dye only as control), perform fluorescence imaging at predetermined time points (e.g., 0.5, 1, 2, 4, 6, 7.5, and 9.5 h) after the injection. After each imaging scan, return the mice to their cages.
7. Overlay the fluorescence images with the corresponding anatomical reference images and manually select region of interest (ROI) to calculate the mean fluorescence intensities of the ROI at different time points.
8. Sacrifice the mice by cervical dislocation under anesthesia at predetermined time points (e.g., 6, 12, and 20 h after subconjunctival injection) (Fig. 7B–F). Collect the eyes including the conjunctiva and eyelids and wash them with PBS for 2 × 30 min at room temperature to remove blood and other contaminations. Fix the eyes on a piece of wax in 4% paraformaldehyde solution for 2 h at 4 °C to ensure that the conjunctivas are flattened.
9. Dissect the eyes and isolate tissues of conjunctiva, cornea, retina, and sclera. Stain with 0.01% DAPI solution overnight at room temperature. Rinse the tissues with PBS for 3 × 10 min and mount on a glass slide with Mowoil (PVA) for fluorescence imaging.
10. Image the tissues using wavelength of 461 nm for the cell nucleus stained by DAPI and 665 nm for the Alexa 647 marker. Analyze the images by counting the numbers of cells with the cell nucleus surrounded by or overlapped with Alexa647 signals (for assessing cell internalization) in 0.42 mm × 0.33 mm FOV at the same imaging plane in the tissues (Fig. 7B–F).

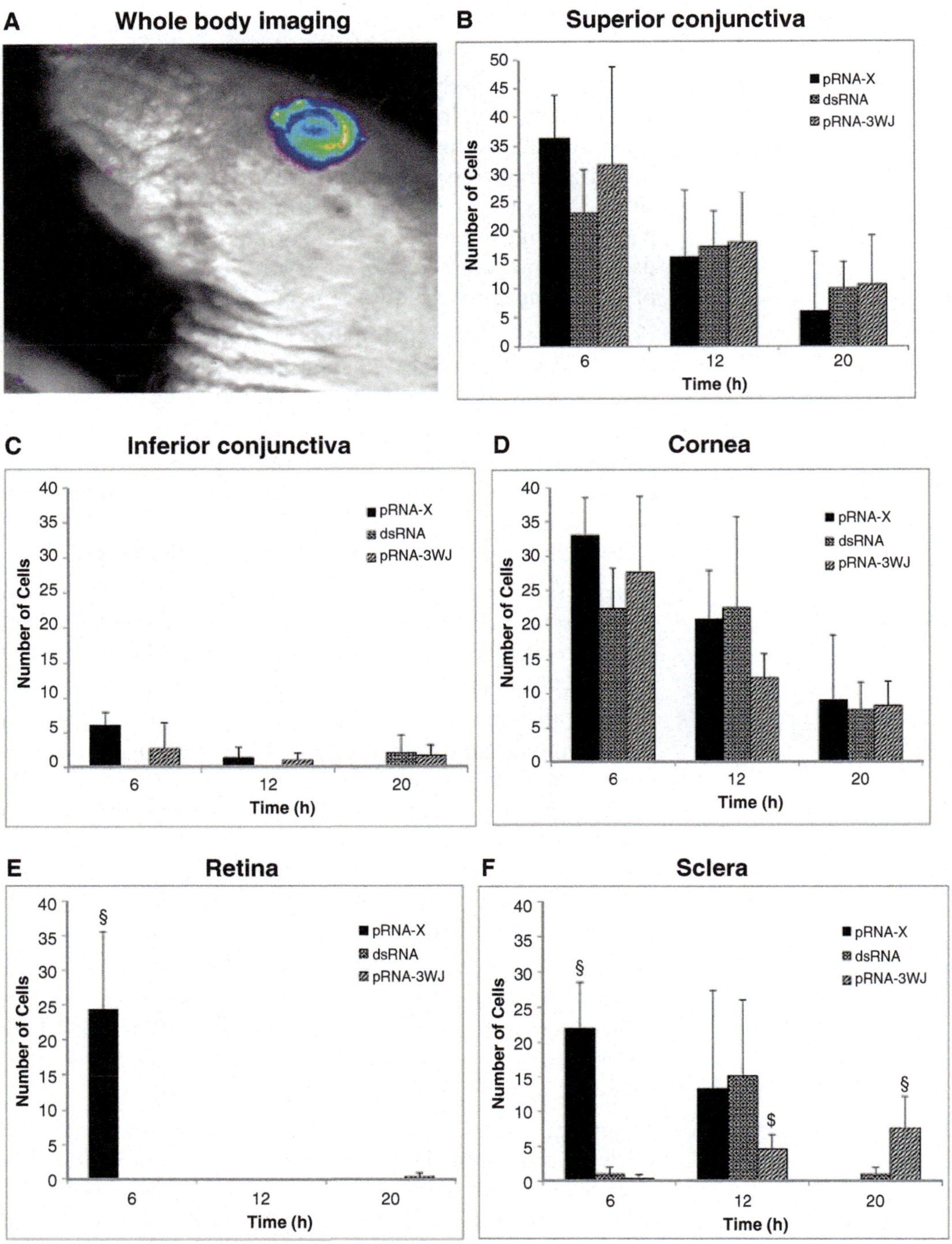

Fig. 7 Assessment of ocular delivery of pRNA nanoparticles in mouse models. (**A**) Whole body imaging of the eye after subconjunctival injection of pRNA nanoparticle (2 h time point). (**B–F**) Numbers of cells with pRNA-3WJ, pRNA-X, and dsRNA (control) internalization in different cells after subconjunctival injection: (**B**) superior conjunctiva (near the injection site); (**C**) inferior conjunctiva; (**D**) cornea; (**E**) retina; (**F**) sclera. Figure reproduced with permission from ref. 41 © Springer

4 Notes

1. TEMED is a skin and respiratory tract irritant.
2. Ethidium bromide is mutagenic and moderately toxic, and must be handled with care. It is a strong carcinogen and is a skin, eye, and respiratory tract irritant. Wear gloves while dispensing it in a hood.
3. $MgCl_2$ will absorb moisture from the air. Be sure to handle as quickly as possible and then recap the reagent bottle tightly to prevent moisture absorption.
4. Bis-acrylamide causes eye, skin, and respiratory tract irritation. It is harmful if inhaled, swallowed or absorbed through skin. It may cause CNS effects. Wear gloves, goggles, and a mask when you are handling it.
5. Acrylamide may cause cancer and heritable genetic damages. It is harmful on inhalation and on contact with skin. It is toxic if swallowed. Wear gloves and a mask when you are handling it.
6. Due to the resolution limit of AFM imaging affected by the probe size of 4–10 nm, pRNA-3WJ and pRNA-X nanoparticles are too small to be reliably imaged with detailed structure and shape. To evaluate the global structures of the RNA nanoparticles derived from the 3WJ core, dsRNA is appended to the three arms of the 3WJ core (Fig. 1C). Similarly, for pRNA-X core, four pRNA monomers are incorporated for AFM imaging (Fig. 2C).
7. It is important to note that RNA nanoparticles are not globular in shape, and deviations from DLS measurements are expected, since the hydrodynamic size from DLS corresponds to the average of the three dimensions of RNA nanoparticles due to rapid tumbling in solution.
8. When choosing the nucleotide sequence for the 8 base pair connecting helix and 5 base pair extension sequences, it is important to introduce sequence diversity while also avoiding nonspecific base interactions into the structure. Optimize the sequences using the *mfold* RNA folding software [49].
9. To maintain planar geometry, it is important to decrease or increase each edge in multiples of 11 bp, as one complete turn of RNA duplex is equal to 11 bp.
10. Adding two guanosine nucleotides to the DNA template sequence immediately following the T7 promoter typically enhances the RNA transcription efficiency.
11. To ensure sufficient binding, the cells need to be kept in a folate-free medium for at least 12 h before the binding assay.

12. Try to avoid too much spreading of the cells. In addition, prepare large amount of cells to avoid inevitable loss during the experimental handling process.
13. Keep the ocular surface and surrounding area clean with saline before the first imaging to ensure that there is no significant contamination from the residues from the dosing.

Acknowledgments

The research was supported by NIH grants R01EB019036, U01CA151648 and U01CA207946. P.G.'s Sylvan G. Frank Endowed Chair position in Pharmaceutics and Drug Delivery is funded by the CM Chen Foundation. P.G. is the founder of P&Z Biological Technology and a consultant of Oxford Nanopore and Nanobiodelivery Pharmaceutical Co. Ltd.

References

1. Puri A, Loomis K, Smith B et al (2009) Lipid-based nanoparticles as pharmaceutical drug carriers: from concepts to clinic. Crit Rev Ther Drug Carrier Syst 26:523–580
2. Weijun L, Francis CS (2007) Lipid-based nanoparticles for nucleic acid delivery. Pharm Res 24:438–449
3. Vo-Dinh T (2005) Protein nanotechnology: the new frontier in biosciences. Methods Mol Biol 300:1–13
4. Zhang F, Nangreave J, Liu Y et al (2014) Structural DNA nanotechnology: state of the art and future perspective. J Am Chem Soc 136:11198–11211
5. Guo P (2010) The emerging field of RNA nanotechnology. Nat Nanotechnol 5:833–842
6. Seeman NC (1982) Nucleic acid junctions and lattices. J Theor Biol 99:237–247
7. Seeman NC (2010) Nanomaterials based on DNA. Annu Rev Biochem 79:65–87
8. Lin C, Liu Y, Yan H (2009) Designer DNA nanoarchitectures. Biochemistry 48:1663–1674
9. Sugimoto N, Nakano S, Katoh M et al (1995) Thermodynamic parameters to predict stability of RNA/DNA hybrid duplexes. Biochemistry 34:11211–11216
10. Searle MS, Williams DH (1993) On the stability of nucleic acid structures in solution: enthalpy-entropy compensations, internal rotations and reversibility. Nucleic Acids Res 21:2051–2056
11. Lemieux S, Major F (2002) RNA canonical and non-canonical base pairing types: a recognition method and complete repertoire. Nucleic Acids Res 30:4250–4263
12. Leontis NB, Westhof E (2001) Geometric nomenclature and classification of RNA base pairs. RNA 7:499–512
13. Scott WG (2007) Ribozymes. Curr Opin Struct Biol 17:280–286
14. Breaker RR (2012) Riboswitches and the RNA world. Cold Spring Harb Perspect Biol 4: a003566
15. Ryther RC, Flynt AS, Phillips JA III et al (2005) siRNA therapeutics: big potential from small RNAs. Gene Ther 12:5–11
16. Ha M, Kim VN (2014) Regulation of microRNA biogenesis. Nat Rev Mol Cell Biol 15:509–524
17. Zhou J, Bobbin ML, Burnett JC et al (2012) Current progress of RNA aptamer-based therapeutics. Front Genet 3:234
18. Shu Y, Pi F, Sharma A et al (2014) Stable RNA nanoparticles as potential new generation drugs for cancer therapy. Adv Drug Deliv Rev 66C:74–89
19. Guo P, Haque F, Hallahan B et al (2012) Uniqueness, advantages, challenges, solutions, and perspectives in therapeutics applying RNA nanotechnology. Nucleic Acid Ther 22:226–245

20. Li H, Lee T, Dziubla T et al (2015) RNA as a stable polymer to build controllable and defined nanostructures for material and biomedical applications. Nano Today 10:631–655
21. Guo P, Erickson S, Anderson D (1987) A small viral RNA is required for *in vitro* packaging of bacteriophage phi29 DNA. Science 236: 690–694
22. Guo P, Zhang C, Chen C et al (1998) Inter-RNA interaction of phage phi29 pRNA to form a hexameric complex for viral DNA transportation. Mol Cell 2:149–155
23. Shu D, Shu Y, Haque F et al (2011) Thermodynamically stable RNA three-way junctions for constructing multifuntional nanoparticles for delivery of therapeutics. Nat Nanotechnol 6:658–667
24. Zhang H, Endrizzi JA, Shu Y et al (2013) Crystal structure of 3WJ core revealing divalent ion-promoted thermostability and assembly of the Phi29 hexameric motor pRNA. RNA 19:1226–1237
25. Binzel DW, Khisamutdinov EF, Guo P (2014) Entropy-driven one-step formation of Phi29 pRNA 3WJ from three RNA fragments. Biochemistry 53:2221–2231
26. Binzel D, Khisamutdinov EF, Vieweger M et al (2016) Mechanism of three-component collision to produce ultra-stable pRNA three-way junction of Phi29 DNA-packaging motor by kinetic assessment. RNA 22:1710–1718
27. Haque F, Shu D, Shu Y et al (2012) Ultrastable synergistic tetravalent RNA nanoparticles for targeting to cancers. Nano Today 7:245–257
28. Shu Y, Haque F, Shu D et al (2013) Fabrication of 14 different RNA nanoparticles for specific tumor targeting without accumulation in normal organs. RNA 19:766–777
29. Liu J, Guo S, Cinier M et al (2011) Fabrication of stable and RNase-resistant RNA nanoparticles active in gearing the nanomotors for viral DNA packaging. ACS Nano 5:237–246
30. Shu D, Khisamutdinov E, Zhang L et al (2013) Programmable folding of fusion RNA complex driven by the 3WJ motif of phi29 motor pRNA. Nucleic Acids Res 42:e10
31. Abdelmawla S, Guo S, Zhang L et al (2011) Pharmacological characterization of chemically synthesized monomeric pRNA nanoparticles for systemic delivery. Mol Ther 19:1312–1322
32. Cui D, Zhang C, Liu B et al (2015) Regression of gastric cancer by systemic injection of RNA nanoparticles carrying both ligand and siRNA. Sci Rep 5:10726
33. Shu D, Li H, Shu Y et al (2015) Systemic delivery of anti-miRNA for suppression of triple negative breast cancer utilizing RNA nanotechnology. ACS Nano 9:9731–9740
34. Binzel D, Shu Y, Li H et al (2016) Specific delivery of MiRNA for high efficient inhibition of prostate cancer by RNA nanotechnology. Mol Ther 24(7):1267–1277
35. Rychahou P, Haque F, Shu Y et al (2015) Delivery of RNA nanoparticles into colorectal cancer metastases following systemic administration. ACS Nano 9:1108–1116
36. Lee TJ, Haque F, Shu D et al (2015) RNA nanoparticles as a vector for targeted siRNA delivery into glioblastoma mouse model. Oncotarget 6:14766–14776
37. Li H, Zhang K, Pi F et al (2016) Controllable self-assembly of RNA tetrahedrons with precise shape and size for cancer targeting. Adv Mater 34:7501–7507
38. Khisamutdinov E, Li H, Jasinski D et al (2014) Enhancing immunomodulation on innate immunity by shape transition among RNA triangle, square, and pentagon nanovehicles. Nucleic Acids Res 42:9996–10004
39. Sharma A, Haque F, Pi F et al (2015) Controllable self-assembly of RNA dendrimers. Nanomed Nanotechnol Biol Med 12:835–844
40. Khisamutdinov EF, Jasinski DL, Li H et al (2016) Fabrication of RNA 3D nanoprism for loading and protection of small RNAs and model drugs. Adv Mater 28(45): 10079–10087
41. Feng L, Li SK, Liu H et al (2014) Ocular delivery of pRNA nanoparticles: distribution and clearance after subconjunctival injection. Pharm Res 31:1046–1058
42. Shu Y, Shu D, Haque F et al (2013) Fabrication of pRNA nanoparticles to deliver therapeutic RNAs and bioactive compounds into tumor cells. Nat Protoc 8:1635–1659
43. Rychahou P, Shu Y, Haque F et al (2015) Methods and assays for specific targeting and delivery of RNA nanoparticles to cancer metastases. Methods Mol Biol 1297:121–135
44. Lee TJ, Haque F, Vieweger M et al (2015) Functional assays for specific targeting and delivery of RNA nanoparticles to brain tumor. Methods Mol Biol 1297:137–152
45. Jasinski D, Schwartz C, Haque F et al (2015) Large scale purification of RNA nanoparticles by preparative ultracentrifugation. Methods Mol Biol 1297:67–82

46. Zhang H, Pi F, Shu D et al (2015) Using RNA nanoparticles with thermostable motifs and fluorogenic modules for real-time detection of RNA folding and turnover in prokaryotic and eukaryotic cells. Methods Mol Biol 1297: 95–111
47. Khisamutdinov EF, Bui MN, Jasinski D et al (2015) Simple method for constructing RNA triangle, square, Pentagon by tuning interior RNA 3WJ angle from 60 degrees to 90 degrees or 108 degrees. Methods Mol Biol 1316: 181–193
48. Jasinski D, Khisamutdinov EF, Lyubchenko YL et al (2014) Physicochemically tunable poly-functionalized RNA square architecture with fluorogenic and ribozymatic properties. ACS Nano 8:7620–7629
49. Zuker M (2003) Mfold web server for nucleic acid folding and hybridization prediction. Nucleic Acids Res 31:3406–3415

INDEX

Eckart Bindewald and Bruce A. Shapiro (eds.), *RNA Nanostructures: Methods and Protocols*, Methods in Molecular Biology, vol. 1632, DOI 10.1007/978-1-4939-7138-1, © Springer Science+Business Media LLC 2017

X

Printed by Printforce, the Netherlands